# Apocynaceae Plants

Ravindra H. Patil • Mohini P. Patil •
Vijay L. Maheshwari

# Apocynaceae Plants

## Ethnobotany, Phytochemistry, Bioactivity and Biotechnological Advances

Springer

Ravindra H. Patil
Department of Microbiology
and Biotechnology
R. C. Patel Arts, Commerce and Science
College, Shirpur
Dhule, Maharashtra, India

Mohini P. Patil
Department of Microbiology
and Biotechnology
R. C. Patel Arts, Commerce and Science
College, Shirpur
Dhule, Maharashtra, India

Vijay L. Maheshwari
Department of Biochemistry,
School of Life Sciences
KBC North Maharashtra University
Jalgaon, Maharashtra, India

ISBN 978-981-99-5405-6 ISBN 978-981-99-5406-3 (eBook)
https://doi.org/10.1007/978-981-99-5406-3



This Springer imprint is published by the registered company Springer Nature Singapore Pte Ltd.
The registered company address is: 152 Beach Road, #21-01/04 Gateway East, Singapore 189721, Singapore
Paper in this product is recyclable.

*To,*
*Our student, late Dr. Bhushan*

# Preface

Plant biodiversity on Earth is incredibly diverse, with an estimated 391,000 plant species known to science as of 2021. This includes a wide variety of plants, ranging from tiny algae to towering trees, and from mosses to flowering plants. Angiosperm Phylogeny Group (APG) recognizes approximately 416 plant families, while other taxonomic systems may recognize a different number of families. Plant biodiversity is not evenly distributed across the planet. The tropics, particularly rainforests, are the most diverse regions, containing the highest numbers of plant species per unit area. Other areas with high plant biodiversity include Mediterranean-type ecosystems, temperate forests, and grasslands. Plants play a vital role in the environment, contributing to its health and sustainability in several ways. Conservation efforts are therefore crucial for maintaining plant biodiversity. These efforts may include protecting natural habitats, implementing sustainable land-use practices, promoting the use of native plant species in landscaping, and establishing seed banks and botanical gardens to preserve threatened species. By preserving plant biodiversity, we can help ensure the continued health and productivity of our planet and its ecosystems. Despite their ecological and economic importance, many species in the Apocynaceae family are threatened by habitat destruction, climate change, and other human activities. Conservation efforts, such as the establishment of protected areas and the cultivation of threatened species, are essential for preserving the biodiversity and ecological functions of this important plant family. Apocynaceae is one of the largest families in the plant kingdom. Robert Brown separated Asclepiadoideae from Apocynaceae for a practical reason. According to this updated classification, the family Apocynaceae consists of 424 genera with more than 4600 species distributed in five subfamilies: Rauvolfioideae, Apocynoideae, Periplocoideae, Secamonoideae, and Asclepiadoideae. Most of the members of this family are poisonous. Plants in the Apocynaceae family are diverse in appearance and habitat. Many are woody shrubs or small trees, while others are climbers or epiphytes as well as herbaceous. One of the most well-known genera in the Apocynaceae family is the genus Vinca, which includes the common garden plant, *Vinca minor*. Other familiar genera include Apocynum, which is also known as dogbane, and Nerium, which includes the ornamental plant, oleander. Chapter 1 provides an outline of the morphology, ecology, taxonomy, diversity, habitat, and geographical distribution of the Apocynaceae family. Applications of medicinal plants and their products for health management are an imperative part of tribal culture. Indian tribal people have a tremendous passion and knowledge of medicinal plants, and they use them for a wide range of health-related applications. Traditionally, local tribal communities have explored the medicinal uses of hundreds of plants found locally in their ecosystem. Today, all over the world, there is a growing appreciation of the value of traditional knowledge, and therefore, there is renewed interest among plant scientists to explore the heritage of traditional knowledge. It is believed that it can make a significant contribution to sustainable development. Most indigenous and local communities are situated in areas where the vast majority of the world's plant genetic resources are found. Many of them have been cultivated and used in a sustainable way for thousands of years. Chapter 2 highlights the traditional knowledge, ethnomedicinal importance, and practical uses of Apocynaceae members. Secondary metabolites are produced by plants through subsidiary pathways that are not directly involved in basic cellular processes and therefore are not essential for the host. However, they are responsible for mediating environmental interactions and are crucial for survival and reproduction. Plants are known to produce more than 50,000 different secondary metabolites. Secondary plant metabolites are the basis for the therapeutic effects of many modern medications and medicinal plants. The discovery of novel secondary metabolites is a continuous process that intends to discover a novel molecule with better therapeutic value. Chapter 3 highlights the role of the plant secondary metabolites as a source of new drugs. Many species in the Apocynaceae family contain chemical compounds with medicinal properties, including cardiac glycosides, alkaloids, and terpenoids. Chapter 4 describes the principle, methodology of extraction, and isolation of secondary metabolites from Apocynaceae plants.

The phytochemical tests are usually performed to determine the presence or absence of major classes of secondary metabolites such as alkaloids, flavonoids, terpenoids, tannins, and phenolic compounds. These tests are often performed using simple colorimetric tests or specific reagents that react with the target compounds. Chapter 5 provides details about phytochemical screening and chemical characterization of the bioactive metabolites. A large-scale separation and purification of metabolites can be carried out using modern characterization techniques such as ultraviolet spectroscopy (UV), infrared spectroscopy (IR), nuclear magnetic resonance spectroscopy (NMR), and mass spectrometry (MS) that are commonly used for the complete identification of the compound. Chapter 6 highlights the identification of bioactive secondary metabolites of Apocynaceae members using these modern spectroscopic and chromatographic techniques. Apocynaceae plants are rich in cardiac glycosides, terpenoids, flavonoids, tannins, steroids, proanthocyanidins, pigments, and other bioactive metabolites, which are widely used for various medicinal applications. Chapter 7 presents a detailed outlook of phytochemistry aspects of Apocynaceae members. Ethnomedicinal uses and traditional knowledge of Apocynaceae plants have resulted in the extensive evaluation of the pharmacological properties of many members of this family. Crude extracts as well as isolated compounds are known to possess a wide range of bioactivities, such as antioxidant, antimicrobial, anti-inflammatory, hepatoprotective, and cytotoxic activities. Chapter 8 provides a thorough update on the biological and pharmacological properties of Apocynaceae members. Agricultural insects and pests cause damage to crops, livestock, or stored products. Biopesticides are a type of pesticide derived from natural sources such as plants, microbes, and other naturally occurring substances. Apocynaceae plants have been widely used as a source of insecticidal metabolites. Chapter 9 highlights the agricultural importance of Apocynaceae members with emphasis on their pesticidal and insecticidal properties. Plant tissue culture has been used as a major platform for obtaining medicinally and industrially important secondary metabolites. Apocynaceae plants have been widely explored for micropropagation, *in vitro* propagation, and hairy root culture using tools and techniques of plant tissue culture. Chapter 10 highlights the plant tissue culture-based approaches for the production of pharmaceutically important bioactive compounds from Apocynaceae members. Enzymes and enzyme inhibitors have been used in various industries for many centuries. Chapter 11 explains the enzymes and enzyme inhibitors from Apocynaceae plants with reference to their industrial and biomedical applications. Endophytes are the microbes that inhabit the internal parts of plants regardless of the outcome of the association. Apocynaceae plants are known to harbor diverse endophytic microorganisms. Chapter 12 highlights the endophytes of Apocynaceae plants with reference to their biodiversity, bioactivities, and biotechnological significance. Due to overexploitation, many species of the family Apocynaceae are on the verge of extinction, and conservation efforts are urgently needed. Micropropagation is considered the best solution for the conservation of elite genotypes of plants. Chapter 13 explains the micropropagation-based conservation strategies for Apocynaceae plants. Nanotechnology is a branch of applied science and technology whose basis includes the control and synthesis of matter on a molecular level in the range of 1–100 nm. It involves the synthesis, characterization, and application of nanomaterial in various fields. Plant-based synthesis of nanoparticles using Apocynaceae plants is a green synthesis approach that utilizes Apocynaceae plants and their extracts to produce nanoparticles. This method is considered eco-friendly because it uses natural, renewable, and biodegradable materials and avoids the use of toxic chemicals and solvents. Chapter 14 thoroughly describes the principle, methodology, and practical applications of Apocynaceae plants in the field of nanotechnology.

We hope that the book will be of great interest to plant biologists, medicinal chemists, and natural product chemists. Moreover, this book offers a vast panorama of structures and bioactivities of natural products from the plants of the Apocynaceae family. It is hoped that the present book will be received with the same enthusiasm and interest as our previous work on different topics.

I would like to express my sincere thanks to Mr. Fumiko Yamaguchi, Editor, Books, Biomedical & Life Sciences, Springer Nature Singapore Pvt. Ltd., Singapore, for coordinating the initial review process of this book and giving us the opportunity to work on this project. We are grateful to the assistance of

Ms. Michelle Harding, Project Coordinator for Springer Nature books. We hope that the contents of the book will serve as a useful resource for all those working in the area and will be received with enthusiasm and interest.

We are thankful to Hon. Amrishbhai Patel, Hon. Bhupeshbhai Patel, and Hon. Rajgopal Bhandari and Principal of the college for their continued support, encouragement and for providing the necessary facilities to undertake research and write the books.' The support of my family members especially Utkarsh and Shreya, friends, and well-wishers is lovingly acknowledged.

Shirpur, Maharashtra, India — Ravindra H. Patil
Shirpur, Maharashtra, India — Mohini P. Patil
Jalgaon, Maharashtra, India — Vijay L. Maheshwari
May 2023

# Contents


**1 Morphology, Ecology, Taxonomy, Diversity, Habitat and Geographical Distribution of the Apocynaceae Family** . . . . . . . . . . 1
Introduction . . . . . . . . . . 1
Geographical Distribution . . . . . . . . . . 1
Habitat . . . . . . . . . . 2
General Characteristics of the Apocynaceae Family . . . . . . . . . . 3
Vegetative Characteristics . . . . . . . . . . 3
Major Genera and Species . . . . . . . . . . 4
Economic Importance . . . . . . . . . . 9
Ethnomedical Considerations . . . . . . . . . . 9
References . . . . . . . . . . 10

**2 Traditional Knowledge, Ethnomedicinal Importance and Practical Uses of Apocynaceae Members** . . . . . . . . . . 13
Introduction . . . . . . . . . . 13
Plant Metabolites in Modern Medicine . . . . . . . . . . 14
Ethnomedical Uses of Some Members of the Family Apocynaceae . . . . 15
References . . . . . . . . . . 19

**3 Plant Secondary Metabolites as a Source of New Drugs** . . . . . . . . . . 23
Introduction . . . . . . . . . . 23
Global Outlook of Communicable and Noncommunicable Diseases . . . 23
Communicable Diseases . . . . . . . . . . 24
Noncommunicable Diseases . . . . . . . . . . 24
Need for New and Effective Drugs . . . . . . . . . . 25
Natural Products as a Potential Source for New Metabolites . . . . . . . . . 26
Secondary Metabolites from Natural Sources . . . . . . . . . . 26
Soil . . . . . . . . . . 29
Marine Habitats . . . . . . . . . . 29
Insects . . . . . . . . . . 29
Plants: A Potential Source of Microbial Communities . . . . . . . . . . 30

Secondary Metabolites of Commercial Importance . . . . . 30
References . . . . . 34

4 **Extraction and Isolation of Secondary Metabolites from Apocynaceae Plants** . . . . . 37
Introduction . . . . . 37
Collection of Plant Material . . . . . 38
Extraction of Plant Material . . . . . 38
Activity-Guided Fractionation for Isolation of Bioactive Secondary Metabolites . . . . . 43
Advantages of Activity-Guided Fractionation . . . . . 49
References . . . . . 49

5 **Phytochemical Screening and Chemical Characterization of the Bioactive Metabolites** . . . . . 51
Introduction . . . . . 51
Phytochemical Screening Methods . . . . . 52
Tests for Proteins . . . . . 52
Test for Carbohydrates . . . . . 53
Test for Alkaloids . . . . . 53
Test for Flavonoids . . . . . 54
Test for Glycosides . . . . . 54
Test for Steroids and Triterpenoids . . . . . 55
Test for Saponins . . . . . 55
Test for Tannins . . . . . 56
Total Phenolics and Flavonoids . . . . . 56
Preliminary Phytochemical Characters of Different Extracts of Some Members of the Family Apocynaceae (*C. carandas* and *T. divaricate, C. procerra*) . . . . . 57
Chromatographic Separation of Compounds from a Mixture or Extracts . . . . . 59
Application of TLC and HPTLC in the Separation of Plant Extracts . . . 59
Column Chromatography . . . . . 62
High-Performance Liquid Chromatography (HPLC) . . . . . 64
References . . . . . 65

6 **Identification of Bioactive Secondary Metabolites of Apocynaceae Members** . . . . . 67
Introduction . . . . . 67
UV–Visible Spectroscopy . . . . . 67
Fourier Transform Infrared (FT-IR) Spectroscopy . . . . . 69
Nuclear Magnetic Resonance (NMR) Spectroscopy . . . . . 70
Mass Spectrometry . . . . . 72
Hyphenated Techniques in Phytoconstituent Identification . . . . . 74
Gas Chromatography–Mass Spectrometry (GC–MS) . . . . . 75

Practical Approaches for GC-HRMS Characterization of Extracts of *C. Procera* . . . . . 76
Practical Approaches for GC-HRMS Characterization of Extracts of C. Carandas . . . . . 77
References . . . . . 79

**7 Phytochemistry of Apocynaceae Members** . . . . . 83
Introduction . . . . . 83
Phytochemistry . . . . . 84
Alkaloids . . . . . 84
Flavonoids . . . . . 96
Terpenes . . . . . 97
Glycosides . . . . . 97
Sterols . . . . . 100
References . . . . . 101

**8 Biological and Pharmacological Properties of Apocynaceae Members** . . . . . 105
Introduction . . . . . 105
Antioxidant Activity . . . . . 105
Anti-Inflammatory/Analgesic Activity . . . . . 111
Anticancer/Cytotoxic Activity . . . . . 112
Cardioprotective Activity . . . . . 113
Antidiabetic Activity . . . . . 113
Antimalarial Activity . . . . . 113
References . . . . . 114

**9 Agricultural Importance of Apocynaceae Members: Insecticidal Properties** . . . . . 119
Introduction . . . . . 119
Insects and Pests . . . . . 121
Insects and Agricultural Losses . . . . . 121
Traditional Pest Control Methods and their Drawbacks . . . . . 122
Botanicals as Pesticides: An Overview . . . . . 123
Plants of the Apocynaceae Family as Bioresources for Pesticidal Molecules . . . . . 124
General Procedure Isolation of Insecticidal Metabolites(S) from *Calotropis Procera* . . . . . 125
Insecticidal Metabolites from Apocynaceae Plants . . . . . 125
Botanical Pesticides: A Way Forward . . . . . 129
References . . . . . 130

**10 Plant Tissue Culture-Based Approaches for the Production of Pharmaceutically Important Bioactive Compounds from Apocynaceae Members** . . . . . 135
Introduction . . . . . 135
Large-Scale Production of Plant Secondary Metabolites Using Callus and Cell Suspension Culture . . . . . 136
Media for Callus and Cell Suspension Culture . . . . . 137
Practical Approaches for Callus Induction in Apocynaceae Plants (*T. divaricata* and Seeds of *C. carandas*) . . . . . 139
Sterilization of Explants . . . . . 139
Cultural Conditions for Callus Induction . . . . . 139
Inoculation and Maintenance of in Vitro Cultures . . . . . 140
Subculture . . . . . 140
Extraction of Phytochemicals from Callus Tissue . . . . . 141
Cell Suspension Culture . . . . . 141
Biological Activities of the Callus Extract . . . . . 145
Hairy Root Culture . . . . . 145
References . . . . . 147

**11 Enzymes and Enzyme Inhibitors from Apocynaceae Plants** . . . . . 151
Introduction . . . . . 151
Proteases . . . . . 152
Amylases . . . . . 152
Lipases . . . . . 153
Latex Bearing Apocynaceae Members as an Important Source of Biomedical Enzymes . . . . . 153
Enzyme/Protease Inhibitors from Apocynaceae Plants . . . . . 155
Therapeutic Applications of PIs . . . . . 155
α-Glucosidase Inhibitors (AGIs) for Diabetes Control . . . . . 157
α-Amylase Inhibitors for Pest Control (α-AIs) . . . . . 157
$H^+/K^+$-ATPase Inhibitors . . . . . 159
Inhibitors of Other Enzymes of Biomedical Importance . . . . . 159
References . . . . . 165

**12 Endophytes in Apocynaceae Plants** . . . . . 173
Introduction . . . . . 173
Plant Endophyte Interactions . . . . . 174
Strategies for Metabolite Production Using Endophytic Microorganisms . . . . . 177
Bioprocess Optimization for the Production of Active Metabolites . . . . . 178
Screening of Bioactivities of the Crude Extracts and Isolated Compounds . . . . . 179
Antimicrobial Compounds from Endophytic Microorganisms . . . . . 179
Endophytic Metabolites with Antioxidant Potential . . . . . 180
Anticancer Compounds . . . . . 180

Other Activities of Endophyte-Derived Metabolites . . . . . . . . . . . . . 181
Biotechnological Potential of Endophytes from Apocynaceae Plants . . . 182
References . . . . . . . . . . . . . . . . . . . . . . . . . . . . . . . . . . . . . . . . 187

**13 Micropropagation-Based Conservation Strategies for Apocynaceae Plants** . . . . . . . . . . . . . . . . . . . . . . . . . . . . . . . . . . 191
Introduction . . . . . . . . . . . . . . . . . . . . . . . . . . . . . . . . . . . . . . . 191
Micropropagation Based Conservation of Apocynaceae Plants . . . . . . . 192
Advantages of Micropropagation . . . . . . . . . . . . . . . . . . . . . . . . . 193
Applications and Merits of Micropropagation Over Conventional Plant Breeding . . . . . . . . . . . . . . . . . . . . . . . . . . . . . . . . . . . . . 195
*In Vitro* Propagation of Apocynaceae Members . . . . . . . . . . . . . . . . 196
Multiple Shoot Induction . . . . . . . . . . . . . . . . . . . . . . . . . . . . . . 196
*In Vitro* Rooting of Shoots . . . . . . . . . . . . . . . . . . . . . . . . . . . . 200
Stepwise Acclimatization (hardening) of *In Vitro* Raised Plantlets . . . . 201
References . . . . . . . . . . . . . . . . . . . . . . . . . . . . . . . . . . . . . . . . 202

**14 Practical Applications of Apocynaceae Plants in Nanotechnology** . . 205
Introduction . . . . . . . . . . . . . . . . . . . . . . . . . . . . . . . . . . . . . . . 205
Methods of Nanoparticle Synthesis . . . . . . . . . . . . . . . . . . . . . . . . 206
Physical Methods . . . . . . . . . . . . . . . . . . . . . . . . . . . . . . . . . . . 206
Chemical Methods . . . . . . . . . . . . . . . . . . . . . . . . . . . . . . . . . . 209
Biological Methods . . . . . . . . . . . . . . . . . . . . . . . . . . . . . . . . . . 210
Microorganism-Mediated Nanoparticle Synthesis . . . . . . . . . . . . . . 210
Plant-Mediated Synthesis Of Nanoparticles . . . . . . . . . . . . . . . . . . 211
Applications of Nanoparticles . . . . . . . . . . . . . . . . . . . . . . . . . . . 213
Medical Applications . . . . . . . . . . . . . . . . . . . . . . . . . . . . . . . . 213
Antioxidant Activity . . . . . . . . . . . . . . . . . . . . . . . . . . . . . . . . . 214
Anticancer Activity . . . . . . . . . . . . . . . . . . . . . . . . . . . . . . . . . 214
Antimicrobial Activity . . . . . . . . . . . . . . . . . . . . . . . . . . . . . . . . 214
Applications in Biosensing . . . . . . . . . . . . . . . . . . . . . . . . . . . . . 215
Applications in Catalysis . . . . . . . . . . . . . . . . . . . . . . . . . . . . . . 216
Characterization of Nanoparticles . . . . . . . . . . . . . . . . . . . . . . . . . 216
Silver Nanoparticles . . . . . . . . . . . . . . . . . . . . . . . . . . . . . . . . . 217
Copper Nanoparticles (CuNPs) . . . . . . . . . . . . . . . . . . . . . . . . . . 221
Ferrous Nanoparticles . . . . . . . . . . . . . . . . . . . . . . . . . . . . . . . . 228
Gold Nanoparticles (AuNPs) . . . . . . . . . . . . . . . . . . . . . . . . . . . 234
Zinc Nanoparticles (ZnNPs) . . . . . . . . . . . . . . . . . . . . . . . . . . . . 234
Titanium Nanoparticles (TiNPs) . . . . . . . . . . . . . . . . . . . . . . . . . 237
Practical Approaches for the Biosynthesis of Metal Nanoparticles Using Apocynaceae Plants . . . . . . . . . . . . . . . . . . . . . . . . . . . . 242
Characterization of Biosynthesized MNPs . . . . . . . . . . . . . . . . . . . 243
UV–Visible Spectrophotometer . . . . . . . . . . . . . . . . . . . . . . . . . . 243
Fourier Transform Infrared Spectroscopy (FTIR) . . . . . . . . . . . . . . 246
Scanning Electron Microscopy (SEM) . . . . . . . . . . . . . . . . . . . . . 247

Energy Dispersive X-ray Analysis (EDX) . . . . . . . . . . . . . . . . . . . . 247
X-ray Diffraction (XRD) . . . . . . . . . . . . . . . . . . . . . . . . . . . . . 247
Dynamic Light Scattering (DLS) . . . . . . . . . . . . . . . . . . . . . . . . . 248
Zeta Potential . . . . . . . . . . . . . . . . . . . . . . . . . . . . . . . . . . 248
Conclusion . . . . . . . . . . . . . . . . . . . . . . . . . . . . . . . . . . . . 248
References . . . . . . . . . . . . . . . . . . . . . . . . . . . . . . . . . . . . 248

# About the Authors

**Ravindra H. Patil** is working as Professor in the Department of Microbiology and Biotechnology at R. C. Patel Arts, Commerce and Science College, Shirpur, MS, India. His areas of research are in microbial and plant secondary metabolites and endophyte derived natural products. He has over eighteen years of teaching experience in microbiology and biotechnology for graduate students. Six students have been awarded with PhD degrees under his guidance and four are currently working. He has published over fifty research and review articles in national and international peer-reviewed journals. He has authored three textbooks for undergraduate students. He has also served as an editor for two books published by Springer Nature Singapore Pvt. Ltd., Singapore. Dr. Patil has received the Young Scientist award from the Govt. of India, Dept. of Science & Technology, in the form of a research grant. He has successfully handled several research projects funded by various federal funding agencies.

**Mohini P. Patil** is working as Asst. Professor in the Department of Microbiology and Biotechnology at R. C. Patel Arts, Commerce and Science College, Shirpur, MS, India. She has over ten years of teaching and research experience in microbiology and biotechnology. Dr. Patil has received Junior Research Fellowship from the Department of Science and Technology (DST, New Delhi), Government of India, in 2013. She has completed her doctoral research on endophytic microorganisms and their bioactive metabolites. Dr. Patil was also awarded a postdoctoral fellowship from the Indian Council of Medical Research (ICMR, New Delhi), Government of India, for her work on novel enzyme inhibitors of biomedical importance. Her areas of research are endophytic fungi and plant secondary metabolites of biological and pharmacological significance. She has guided over twenty students for their postgraduate dissertations. She has published over 15 research and review articles in national and international peer-reviewed journals. She has coauthored one textbook for undergraduate students.

**Vijay L. Maheshwari** is currently serving as the Vice-Chancellor at Kavayitri Bahinabai Chaudhari North Maharashtra University, Jalgaon, MS, India. He has teaching experience of over 30 years. Prof. Maheshwari has received the "Young Scientist" award by IUBMB, India, in 1994. He is also a recipient of the BOYSCAST fellowship of Govt. of India, Dept. of Science & Technology for postdoctoral research at the University of Virginia, USA. He was Norman E. Borlaug fellow of US Department of Agriculture, USA, and undertook advanced training in Risk Assessment in Genetically Modified Food. He has served as the Regional Editor for the journal *Physiology and Molecular Biology of Plants*. He is a honorary Fellow of the Society of Science and Environment (FSSE) and the Maharashtra Academy of Sciences. Prof. Maheshwari has received the Best Teacher Award (2009–2010) from the Government of Maharashtra for his outstanding contribution in teaching and research. He has developed and transferred one technology to a local industry on the isolation of a pesticidal principle from seeds of *Annona squamosa* and provided consultancy to industries. He has attended more than 20 national and international conferences. He has published around one hundred research papers and review articles in journals of national and international repute. He has also served as editor for two books published by Springer Nature Singapore Pvt. Ltd., Singapore, and authored three textbooks. He has successfully handled several research projects and supervised 22 students for doctoral degrees.

His areas of research/expertise include protein biochemistry, plant tissue culture, and secondary metabolites and their biological activities. (A detailed profile is available on www.nmu.ac.in.)

# Chapter 1
# Morphology, Ecology, Taxonomy, Diversity, Habitat and Geographical Distribution of the Apocynaceae Family

## Introduction

Apocynaceae is an important family in angiosperms, also known as the dogbane family of flowering plants. It is a large family comprising approximately 5000 species and 415 genera according to updated classification (Endress et al., 2014). The family is widely distributed across tropical and subtropical regions of the world, although some species can also be found in temperate regions. Robert Brown (1810) separated the family Asclepediaceae from Apocynaceae for easier understanding of these two large families. Apocynaceae consists of five subfamilies: Rauvolfioideae, Apocynoideae, Periplocoideae, Secamonoideae, and Asclepiadoideae. Most of the members of this family are poisonous. Apocynaceae plants are rich in cardiac glycosides and various alkaloids and therefore are widely used for various medicinal applications.

## Geographical Distribution

Plants of the family Apocynaceae are native throughout India, Pakistan, China, Bangladesh, and Sri Lanka (Mahmood et al., 2011). Several members of this family are abundantly found in the tropics and subtropics (Fig. 1.1). Members of the Apocynaceae family are trees and shrubs or herbaceous plants with milky latex. The majority of the plants in this family are abundant in alkaloids and have significant therapeutic value. Additionally, many species are planted for decorative purposes. People in rural regions utilise some Apocynaceae species as food and some of them as poisons (Aiyambo, 2010). Due to the widespread use of Apocynaceae plants for medicinal purposes, members of this family are found in many different traditional medical practices, including Indian, Chinese, and Thai medicine (Van Beek et al., 1984; Bhat et al., 2012). The Indian subcontinent has a



R. H. Patil et al., *Apocynaceae Plants*,
https://doi.org/10.1007/978-981-99-5406-3_1

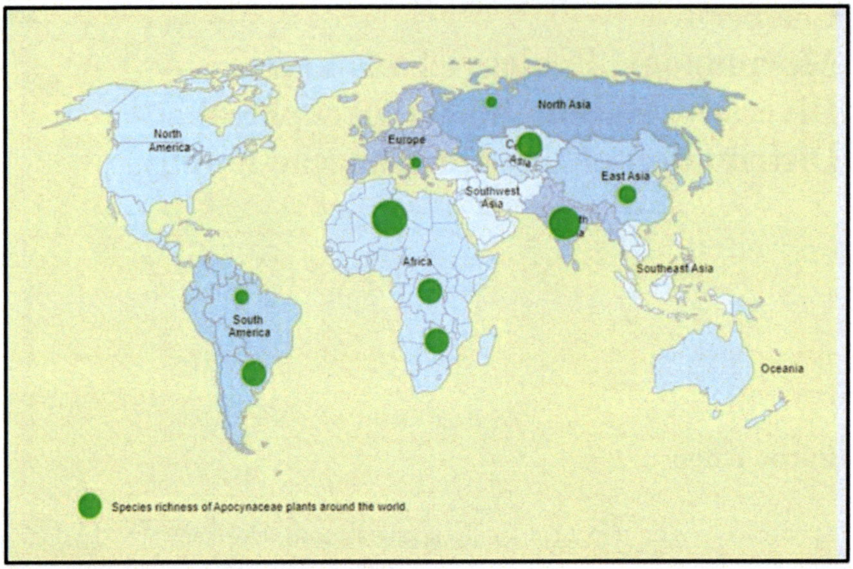


**Fig. 1.1** Species richness of Apocynaceae plants around the world

dominant presence of Apocynaceae plants. Approximately 84 species belonging to 30 genera are present in India.

## Habitat

Apocynaceae plants show universal distribution. Birds and anthropological activities were found to be responsible for their spread in many parts of the world. This family's members can be found in temperate, subtropical, and tropical regions. The majority of the vegetation in this family is trees and shrubs, while subshrubs and herbs are very infrequently encountered. According to Oze et al. (2007), *Alstonia scholaris* (R. Br.) is found all over the tropics and rainforests of West and Central Africa. According to Motwani et al. (2012), *Carissa carandas* (Linn.) is a natural and widespread plant in India, Sri Lanka, Java, Malaysia, Myanmar, and Pakistan. Originating from Madagascar Island, *Catharanthus roseus* (Linn.) has been widely farmed for hundreds of years and may now be found growing wild in most warm climates across the world (Aslam et al., 2010). *Holarrhena antidysentrica* (Wall) is another important member of this family distributed across India (Shah et al., 2010). *Nerium oleander* is a beautiful ornamental tree 10–12 ft. in height grown alongside roads and highways due to its asthetic appearance and beautiful flowers. It is native to North Africa and Southeast Asia (Dolgova & Ignateva, 1963). *Rauvolfia serpentina* (Benth. ex Kurtz) is native to Asia and found in Thailand and Burma. In India, it is abundantly present in Punjab, Haryana. The Himachal Pradesh, Sikkim and Himalayan regions include Bhutan (Vakil, 1955). According to Muzafar and Daulatabadi (1986), *Tabernaemontana divaricata* (Linn.) can be found in North India up to an elevation of 3000 feet in the Konkan, Western Ghats in Malabar, and other regions of India. In tropical and subtropical areas, *Thevetia peruviana*, sometimes known as yellow oleander, is found (Langford & Boor, 1996). *Wrightia tinctoria* (R. Br.) is particularly common in India. In southern India, it is widely referred to as Indrajav or a tree that treats jaundice (Ramchandra et al., 1993).

## General Characteristics of the Apocynaceae Family

The members of the family Apocynaceae are mostly herbs, and some are trees and shrubs bearing white and milky latex. Leaves are simple, alternate, opposite, or whorled, and the margin is entire. Flowers are actinomorphic to hermaphrodite and hypogynous. The calyx is either free or united. The corolla is gamopetalous, funnel-shaped, valvate, or twisted. The androecium is epipetalous and introrse, whereas the gynoecium is bicarpellary and syncarpous. The ovary is superior and united by the styles only. The fruits are of a follicle or berry type. The seeds usually have a crown of hairs.

### *Vegetative Characteristics*

**Root System:** Highly complex roots with an extensively branched tap root system.

**Stem:** Erect, herbaceous, erect, solid and woody, solid, branched to unbranched, succulent with latex.

**Leaves:** Simple, alternate, opposite, or whorled, the margin is entire.

#### Floral Characteristics

**Inflorescence:**

Cymose is the most common inflorescence among Apocynaceae plants, terminal or axillary, rarely cyme (Genus: Carissa) or umbellate cyme (Genus: Rauwolfia).

**Flower:**

Ebracteate or Bracteate, complete, pedicellate, hermaphrodite, tetra- or pentamerous actinomorphic.

**Calyx:**

Sepals 5, the calyx is of gamo- or polysepalous type, small, often granular, deeply lobed, imbricate, or valvate.

**Corolla:**

Gamopetalous, with five petals and a corona present (hairy scales or protrusion), usually twisted, occasionally imbricate, and infrequently valvate.

**Androecium:**

Stamens 4–5, alternipetalous, epipetalous, filament short, free, dithecous, connate to stigma, dehiscing longitudinally, introrse.

**Gynoecium:**

Ovaries are free, superior style is short, bicarpellary, syncarpous, and contained in a tube-shaped corolla. Distally, the stigma is thickened. With marginal placentation, the ovary is unilocular; however, when the ovaries are united, axile placentation takes place.

**Fruit:**

The fruits are of a follicle or berry type. A drupe (Thevetia), berry (Carissa), and follicle (Nerium, Amsonia) are also reported among the Apocynaceae members.

**Seed:**

The seeds usually have a crown of hairs. Endospermic, hairy or winged.

**Pollination:**

Entomophilous (through insects).

### General Outline of the Classification of the Family Apocynaceae

For practical considerations, Robert Brown (1810) separated the asclepiads from Jussieu's (1789) Apocynaceae and designated them as a new family, Asclepiadaceae. Traditionally, these two are recognised as separate but highly similar families for nearly 200 years. Based on morphological and molecular characteristics, Endress and Bruyns (2000) reunited the two families and accepted five subfamilies, namely, Periplocoideae, Rauvolfioideae, Asclepiadoideae, Secamonoideae, and Apocynoideae. Endress et al. (2014) presented an updated suprageneric classification for Apocynaceae based on recent morphological, molecular, and mainly phylogenetic-based results. According to this classification, Apocynaceae includes a total of 366 genera placed within five subfamilies, 25 tribes, and 49 subtribes (Fig. 1.2).

According to *The Plant List,* Apocynaceae contains 410 plant genera. The Plant List database includes scientific names of over 18,268 species for this family. Of these, 5556 are accepted species names. The Plant List includes a further 1996 scientific plant names of intraspecific rank for the family Apocynaceae. These are primarily included because species rank names are synonyms of accepted intraspecific names. Out of 18,268, 168 are recorded as invalid, 414 are illegitimate and 13 are found to be spelling variants.

(http://www.theplantlist.org/Apocynaceae/)

## *Major Genera and Species*

Over 300 genera of Apocynaceae are placed across the five subfamilies (Table 1.1). A complete list of Apocynaceae is provided in Table 1.1 (http://www.theplantlist.

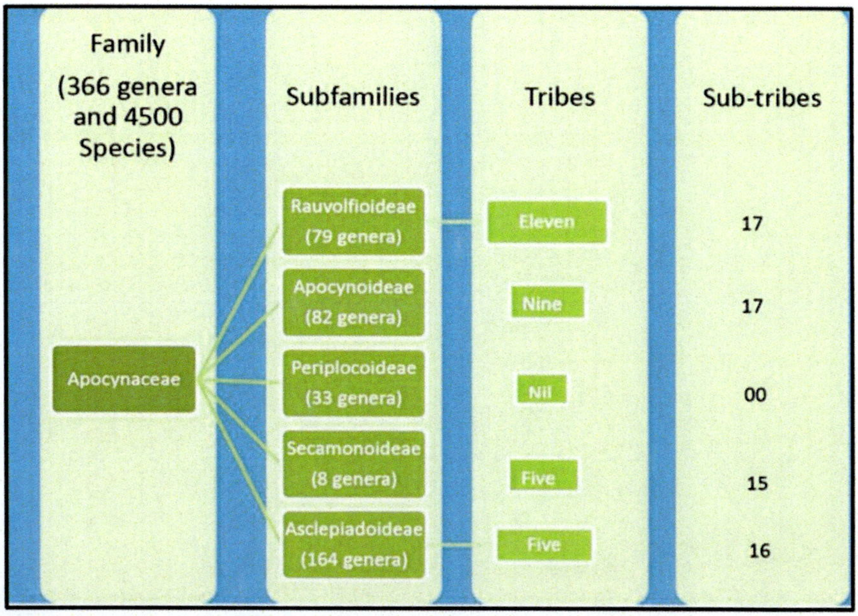


**Fig. 1.2** General outline of the classification of the Apocynaceae family

org/1.1/browse/A/Apocynaceae/). *Carissa carandus* (*H. Karaunda*) is a small spiny shrub. The fruits of *C. carandas* are eaten as wild raw fruits. They are also used as vegetables and pickles. *Catharanthus roseus* is the most popular ornamental plant worldwide. Another important member of this family is *Rauwolfia serpentina*. It is abundant in sub-Himalayan tracts. The roots of *R. serpentina* are hypnotic, sedative and used in managing blood pressure. *Thevetia peruviana* is an evergreen shrub with highly poisonous milky juice and bitter bark (Fig. 1.3). It is popular as an arrow poison. *Plumeria alba* is widely grown in India as an ornamental tree because of its pleasant smell. *Wrightia tomentosa*, a small deciduous tree, is popular for the antidote property of the bark of stem and roots. *Baissea multiflora* is a climber with beautifully scented flowers and ornamental value. *Nerium indicum* is widely cultivated as an ornamental shrub worldwide (Fig. 1.3). Many members of the *Trachelospermum* genus are lovely woody vines, including the star jasmine, *T. jasminoides*, *Mandevilla*, and *Allamanda*. *Hoodia*, *Huernia*, and carrion flower (*Stapelia*) produce unique smells that attract pollinating insects and flies. Arrow poisons are obtained from several members of this family. The poisonous alkaloids present in the genera *Strophanthus* and *Rauvolfia* are used in medicines due to their biological activities. Pollination in Apocynaceae generally takes place through insects.

**Table 1.1** Genera in *Apocynaceae*

| | | | | | | | | |
|---|---|---|---|---|---|---|---|---|
| *Absolmsia* | *Anechites* | *Baeolepis* | *Calathostelma* | *Dactylostelma* | *Ecdysanthera* | *Fanninia* | *Gonocrypta* | *Huerniopsis* |
| *Acerates* | *Angadenia* | *Baharuia* | *Callichilia* | *Daemia* | *Echidnopsis* | *Farquharia* | *Gonolobus* | *Hunteria* |
| *Acokanthera* | *Angolluma* | *Bahiella* | *Calocrater* | *Dalzielia* | *Echites* | *Fernaldia* | *Gonostemon* | *Hybanthera* |
| *Acrocoryne* | *Anisopus* | *Baissea* | *Calostigma* | *Decabelone* | *Ecliptostelma* | *Finlaysonia* | *Gothofreda* | *Hylaea* |
| *Acustelma* | *Anisotoma* | *Ballyanthus* | *Calotropis* | *Decaceras* | *Ectadiopsis* | *Fischeria* | *Goydera* | *Hypolobus* |
| *Adelostemma* | *Anodendron* | *Barjonia* | *Calyptranthera* | *Decalepis* | *Ectadium* | *Fockea* | *Graphistemma* | *Ibatia* |
| *Adenium* | *Anomotassa* | *Baroniella* | *Cameraria* | *Decanema* | *Ecua* | *Folotsia* | *Gunnessia* | *Ichnocarpus* |
| *Aganonerion* | *Anthanotis* | *Baseonema* | *Campestigma* | *Decanemopsis* | *Edithcolea* | *Fontellaea* | *Gymnanthera* | *Irmischia* |
| *Aganosma* | *Antiostelma* | *Basistelma* | *Camptocarpus* | *Decastelma* | *Elcomarhiza* | *Forsteronia* | *Gymnema* | *Ischnolepis* |
| *Ahouai* | *Antitoxicum* | *Batesanthus* | *Caralluma* | *Desmidorchis* | *Elytropus* | *Frerea* | *Gymnemopsis* | *Ischnostemma* |
| *Alafia* | *Antura* | *Baynesia* | *Carissa* | *Dewevrella* | *Emicocarpus* | *Funastrum* | *Gymnolaema* | *Isonema* |
| *Alexitoxicon* | *Aphanostelma* | *Beaumontia* | *Carpodinus* | *Dichaelia* | *Emplectanthus* | *Funtumia* | *Haemadictyon* | *Ixodonerium* |
| *Allamanda* | *Aphanostylis* | *Belostemma* | *Carruthersia* | *Dictyanthus* | *Ephippiocarpa* | *Fanninia* | *Haemax* | *Jasminanthes* |
| *Allomarkgrafia* | *Apocynum* | *Bidaria* | *Caruncularia* | *Dictyophleba* | *Epicion* | *Farquharia* | *Hancornia* | *Jobinia* |
| *Allowoodsonia* | *Araujia* | *Biondia* | *Carvalhoa* | *Dipladenia* | *Epigynum* | *Fernaldia* | *Hanghomia* | *Kamettia* |
| *Alstonia* | *Argelia* | *Blastotrophe* | *Cascabela* | *Diplocyatha* | *Epistemma* | *Finlaysonia* | *Haplophyton* | *Kanahia* |
| *Alyxia* | *Artia* | *Blepharanthera* | *Catharanthus* | *Diplolepis* | *Eriostemma* | *Fischeria* | *Haplostemma* | *Kasailo* |
| *Amalocalyx* | *Asclepias* | *Blepharodon* | *Centrostemma* | *Diplorhynchus* | *Ervatamia* | *Fockea* | *Hazunta* | *Kerbera* |
| *Ambelania* | *Asketanthera* | *Blyttia* | *Cerbera* | *Diplostigma* | *Eucorymbia* | *Folotsia* | *Heligme* | *Kibatalia* |
| *Amblyoglossum* | *Aspidoglossum* | *Bonafousia* | *Cerberiopsis* | *Dischidanthus* | *Eustegia* | *Fontellaea* | *Heliostemma* | *Kopsia* |
| *Amblyopetalum* | *Aspidonepsis* | *Boucerosia* | *Ceropegia* | *Dischidia* | *Excavatia* | *Forsteronia* | *Hemidesmus* | *Krebsia* |
| *Amblystigma* | *Aspidosperma* | *Bousigonia* | *Chamaeclitandra* | *Dischidiopsis* | *Ecdysanthera* | *Frerea* | *Hemipogon* | *Lacmellea* |
| *Ampelamus* | *Astelma* | *Brachystelma* | *Chilocarpus* | *Ditassa* | *Echidnopsis* | *Funastrum* | *Heterostemma* | *Lagarinthus* |
| *Amphidetes* | *Astephanus* | *Buckollia* | *Chlorocyathus* | *Dittoceras* | *Echites* | *Funtumia* | *Heynella* | *Landolphia* |
| *Amphineurion* | *Asterostemma* | *Baeolepis* | *Chonemorpha* | *Dolichopetalum* | *Ecliptostelma* | *Glossonema* | *Hiepia* | *Larryleachia* |
| *Amphistelma* | *Atherandra* | *Baharuia* | *Chthamalia* | *Drakebrockmania* | *Ectadiopsis* | *Glossostelma* | *Himatanthus* | *Lasiostelma* |
| *Amsonia* | *Atherolepis* | *Bahiella* | *Cibirhiza* | *Dregea* | *Ectadium* | *Glossostephanus* | *Holarrhena* | *Laubertia* |
| *Anabata* | *Australluma* | *Baissea* | *Cionura* | *Duvalia* | *Ecua* | *Gomphocarpus* | *Holostemma* | *Lavrania* |
| *Anantherix* | *Anechites* | *Ballyanthus* | *Cleghornia* | *Duvaliandra* | *Edithcolea* | *Gongronema* | *Hoodia* | *Laxoplumeria* |
| *Leachiella* | *Maclaudia* | *Microstemma* | *Oistonema* | *Pentanura* | *Plocostemma* | *Rhinolobium* | *Taberna* | *Urceola* |

| | | | | | | | | |
|---|---|---|---|---|---|---|---|---|
| *Leichardtia* | *Macoubea* | *Minaria* | *Omphalogonus* | *Pentarrhinum* | *Plumeria* | *Rhombonema* | *Tabernaemontana* | *Urechites* |
| *Lepinia* | *Macroditassa* | *Miraglossum* | *Oncinema* | *Pentasachme* | *Poacynum* | *Rhyncharrhena* | *Tabernanthe* | *Urostelma* |
| *Lepiniopsis* | *Macropelma* | *Mitolepis* | *Oncinotis* | *Pentasacme* | *Podanthes* | *Rhynchostigma* | *Tacazzea* | *Vahadenia* |
| *Leptadenia* | *Macropetalum* | *Mitostigma* | *Oncostemma* | *Pentastelma* | *Podostelma* | *Rhyssolobium* | *Tanulepis* | *Vallariopsis* |
| *Leuconotis* | *Macropharynx* | *Molongum* | *Ophionella* | *Pentatropis* | *Poicillopsis* | *Rhyssostelma* | *Tapeinostelma* | *Vallaris* |
| *Lhotzkyella* | *Macroscepis* | *Mondia* | *Orbea* | *Pentopetia* | *Polystemma* | *Rhytidocaulon* | *Tassadia* | *Vallesia* |
| *Lobostephanus* | *Madarosperma* | *Monolluma* | *Orbeanthus* | *Peplonia* | *Pootia* | *Riocreuxia* | *Tavaresia* | *Vietnamia* |
| *Lochnera* | *Mahawoa* | *Monostemma* | *Orbeopsis* | *Pergularia* | *Pottsia* | *Robbia* | *Telectadium* | *Vinca* |
| *Luckhoffia* | *Malouetia* | *Monothylaceum* | *Oreosparte* | *Periglossum* | *Prestonia* | *Rojasia* | *Telminostelma* | *Vincetoxicopsis* |
| *Lygisma* | *Mandevilla* | *Morrenia* | *Orthanthera* | *Periploca* | *Pseudibatia* | *Ruehssia* | *Telosma* | *Vincetoxicum* |
| *Lyonsia* | *Mangenotia* | *Mortoniella* | *Orthopichonia* | *Pervillaea* | *Pseudolithos* | *Saba* | *Temnadenia* | *Virchowia* |
| *Leachiella* | *Manothrix* | *Motandra* | *Orthosia* | *Pervinca* | *Pseudopectinaria* | *Sacleuxia* | *Tenaris* | *Voacanga* |
| *Leichardtia* | *Margaretta* | *Mucoa* | *Oxypetalum* | *Petalostelma* | *Pseudosarcolobus* | *Sarcolobus* | *Tetragonocarpus* | *Widgrenia* |
| *Lepinia* | *Marsdenia* | *Myriopteron* | *Oxystelma* | *Petchia* | *Pseusmagennetus* | *Sarcophagophilus* | *Tetraphysa* | *Willughbeia* |
| *Lepiniopsis* | *Mascarenhasia* | *Nanostelma* | *Pachycarpus* | *Petopentia* | *Pteralyxia* | *Sarcorrhiza* | *Thenardia* | *Winchia* |
| *Leptadenia* | *Matelea* | *Nautonia* | *Pachycymbium* | *Pezisicarpus* | *Ptycanthera* | *Sarcostemma* | *Thevetia* | *Woodia* |
| *Leuconotis* | *Meladerma* | *Neisosperma* | *Pachypodium* | *Phaeostemma* | *Pulvinaria* | *Sarmasikia* | *Thoreauea* | *Wrightia* |
| *Lhotzkyella* | *Melinia* | *Neobracea* | *Pacouria* | *Pherotrichis* | *Pycnobotrya* | *Sattadia* | *Thozetia* | *Xysmalobium* |
| *Lobostephanus* | *Mellichampia* | *Neocouma* | *Pala* | *Philibertella* | *Pycnobregma* | *Schistogyne* | *Thyrsanthella* | *Zschokkea* |
| *Lochnera* | *Melodinus* | *Neoschumannia* | *Papuechites* | *Philibertia* | *Pycnorhachis* | *Schistonema* | *Thyrsanthus* | *Zygostelma* |
| *Luckhoffia* | *Merrillanthus* | *Nephradenia* | *Parabarium* | *Phyllanthera* | *Pycnostelma* | *Schizoglossum* | *Tintinnabularia* | *Urceola* |
| *Lygisma* | *Mesechites* | *Nerium* | *Parahancornia* | *Physianthus* | *Quaqua* | *Schizonotus* | *Tonduzia* | *Urechites* |
| *Lyonsia* | *Metalepis* | *Notechidnopsis* | *Parameria* | *Physostelma* | *Quiotania* | *Schizostemma* | *Toxocarpus* | *Urostelma* |
| *Leachiella* | *Metaplexis* | *Notonerium* | *Parapodium* | *Piaranthus* | *Quisumbingia* | *Schizostephanus* | *Trachelospermum* | *Vahadenia* |
| *Leichardtia* | *Metastelma* | *Obesia* | *Parepigynum* | *Picralima* | *Raphionacme* | *Schizozygia* | *Trachomitum* | *Vallariopsis* |
| *Lepinia* | *Micholitzia* | *Ochrosia* | *Parsonsia* | *Pinochia* | *Raphistemma* | *Schlechterella* | *Traunia* | *Vallaris* |
| *Lepiniopsis* | *Micrechites* | *Odontadenia* | *Pectinaria* | *Plectaneia* | *Rauvolfia* | *Schubertia* | *Treutlera* | *Vallesia* |
| *Leptadenia* | *Microdactylon* | *Odontanthera* | *Peltastes* | *Pleiocarpa* | *Rhabdadenia* | *Secamone* | *Trichocaulon* | *Vietnamia* |
| *Leuconotis* | *Microloma* | *Odontostephana* | *Pentacyphus* | *Pleioceras* | *Rhazya* | *Secamonopsis* | *Trichosacme* | *Vinca* |
| *Lhotzkyella* | *Microplumeria* | *Oianthus* | *Pentalinon* | *Pleurostelma* | *Rhigospira* | *Secondatia* | *Trichosandra* | *Vincetoxicopsis* |

(Source: http://www.theplantlist.org/1.1/browse/A/Apocynaceae)

**Fig. 1.3** Common members of the Apocynaceae family found in India and Southeast Asia

## Economic Importance

Apocynaceae or the 'Dogbane' family has attracted attention because of many life-saving natural medicines and novel phytochemicals. Many plants of Apocynaceae have potential applications in the rubber and pesticide industry. The pulpy fruits of *Carissa carandus* are edible and used for making pickles. These wild fruits are widely used by tribal people in India. Apocynaceae members such as *Hancora, Dyera, and Urceola* are used as inexpensive and economic sources of rubber. Thevetine, a poisonous glycoside, is obtained from the seeds of *Thevetia.* Acokanthera latex is employed in the poisoning of arrows. Rat control is practiced by using a poison found in the bark and wood of *Nerium indicum*. The seeds of *Wrightia tinctoria* are used to obtain the dye. On the other hand, *nerium, Catharanthus* (syn. Vinca), *Thevetia, Wrightia,* and *Allamanda* are cultivated as ornamental plants in gardens.

### *Ethnomedical Considerations*

Plants from the Apocynaceae family are known all over the world for their vast ethnomedicinal uses. Bhadane et al. (2018) provided a detailed account of the traditional knowledge and ethnomedicinal applications of several members of this family. It highlighted the uses of *C. roseus*, *A. scholaris*, *R. serpentina*, *H. antidysentriaca*, *N. indicum* and *T. peruviana*. The ethnomedicnal properties of these members have been documented in classical Indian knowledge systems such as Ayurveda, Siddha, Unani and Homeopathy.

*C. carandas* produces nut-sized fruits that are used for their stomachic, antidiarrheal and antihelminthic properties. The stem of *C. carandas* is used to enhance the tendons; fruits are used for skin care, and leaves are used as antipyretic agents (Kirtikar & Basu, 1998). Traditionally, *T. divaricata* has been recommended for the management of fever, pain, and dysentery (Boonyaratanakornkit & Supawita, 2005). Its applications in the treatment of diarrhea, epilepsy, asthma, fever, fractures, headache, inflammation, leprosy, rheumatic pain, ulceration, and vomiting are well documented (Nadkarni, 1996). *A. boonei* stem bark is used in the treatment of febrile illness, renal infections, rheumatism and liver aliments (Asuzu & Anaga, 1991; Ojewole, 1984; Majekodunmi et al., 2008; Bello et al., 2009; Majekodunmi & Odeku, 2009). It is also used to control intestinal helminths (Olajide et al., 2000). *C. roseus* (Linn.) is used for the treatment of diabetes, fever, malaria, throat infections, and chest complaints. It is also used for the regulation of menstrual cycles (Santhi et al., 2016). The parts of *H. antidysentrica* have been used in the treatment of dysentery and diarrhea. In the Ayurvedic and Unani systems of medicine, it is used as an anthelmintic and antidiarrheal agent and is also used to treat skin diseases (Shah et al., 2010). According to traditional beliefs, *N. oleander* offers therapeutic properties for a number of ailments, including kustha (leprosy), valipalita (ageing), upadamsha (syphilis), indralupta (alopecia), and conjunctivitis (Chakara dutta) (Banerjee et al., 2011).

## References

Aiyambo, D. (2010). *Traditional uses of selected members of the Apocynaceae family in Namibia*. Ministry of Agriculture, Water and Forestry.

Aslam, J., Khan, S. H., Siddiqui, Z. H., Fatima, Z., Maqsood, M., Bhat, M. A., Nasim, S. A., Ilah, A., Ahmad, I. Z., Khan, S. A., & Mujib, A. (2010). *Catharanthus roseus* (L.) G. Don. An important drug: it's applications and production. *Pharmacie Globale (IJCP), 4*(12), 1–6.

Asuzu, I. U., & Anaga, A. O. (1991). Pharmacological screening of the aqueous extract of *Alstonia boonei* bark. *Fitoterapia, 62*, 411–417.

Banerjee, A. A., Vasu, K. K., Pancholi, H., Rajani, M., & Nivsarkar, M. A. (2011). Detoxification of *Nerium indicum* roots based on Indian system of medicine: Phytochemical and toxicity evaluations. *Acta Poloniae Pharmaceutica, 68*, 905–911.

Bello, I. S., Oduola, T., Adeosun, O. G., Omisore, N. O. A., Raheem, G. O., & Ademosun, A. A. (2009). Evaluation of antimalarial activity of various fractions of Morinda lucida leaf extract and *Alstonia boonei* stem bark. *Global Journal of Pharmacology, 3*(3), 163–165.

Bhadane, B. S., Patil, M. P., Maheshwari, V. L., & Patil, R. H. (2018). Ethnopharmacology, phytochemistry, and biotechnological advances of family Apocynaceae: A review. *Phytotherapy Research, 32*(7), 1181–1210.

Bhat, P., Hegde, G., & Hegde, G. R. (2012). Ethnomedicinal practices in different communities of Uttara Kannada district of Karnataka for treatment of wounds. *Journal of Ethnopharmacology, 143*(2), 501–514.

Boonyaratanakornkit, L., & Supawita, T. (2005). *Names of medicinal plants and their uses* (p. 69). Department of Pharmacognosy, Faculty of Pharmacy, Chulalongkorn University.

Brown, R. (1810). *On the Asclepiadeae, a natural order of plants separated from the Apocineae of Jussieu* (Preprint of: memoires of the Wernerian Natural History Society) (Vol. 1, pp. 12–78) (1811).

Dolgova, A. A., & Ignateva, N. S. (1963). Morphologo-anatomical features of oleander leaves. *Aptechnoe Delo, 12*, 36–41.

Endress, M. E., & Bruyns, P. V. (2000). A revised classification of the Apocynaceae sl. *The Botanical Review, 66*(1), 1–56.

Endress, M. E., Liede-Schumann, S., & Meve, U. (2014). An updated classification for Apocynaceae. *Phytotaxa, 159*(3), 175–194. http://www.theplantlist.org/Apocynaceae

Jussieu, A. D. (1789). *Genera plantarum secundum ordines naturales disposita, juxta methoclum in horto regio parisiensi exaratum*. Viduam Herissant and Theophilum Barrois. 498 pp.

Kirtikar, K. R., & Basu, B. D. (1998). *Data on medicinal plants and chemical constituents. Indian medicinal plants* (pp. 2532–2541). VIMSAT Publishers.

Langford, S. D., & Boor, P. J. (1996). Oleander toxicity: an examination of human and animal toxic exposures. *Toxicology, 109*(1), 1–3.

Mahmood, T., Meer, F., Munir, F., Nazar, N., & Naveed, I. (2011). Genetic diversity of selected Apocynaceae species based on chloroplast gene rps11. *Journal of Medicinal Plants Research, 5*, 4382–4387.

Majekodunmi, S. O., Adegoke, O. A., & Odeku, O. A. (2008). Formulation of the extract of the stem bark of *Alstonia boonei* as tablet dosage form. *Tropical Journal of Pharmaceutical Research, 7*(2), 987–994.

Majekodunmi, S. O., & Odeku, O. A. (2009). Effects of interacting variables on the formulation of *Alstonia boonei* De Wild (Apocynaceae) tablets. *Acta Pharmaceutical Science, 51*, 141–148.

Motwani, S. M., Pandey, E. P., Sonchhatra, N. P., Desai, T. R., Patel, V. L., & Pandya, D. J. (2012). Pharmacognostic and phytochemical study of aerial parts of *Carissa carandas*. *International Journal of Biological & Pharmaceutical Research, 3*, 75–81.

Muzaffar, A., & Daulatabadi, N. (1986). Stapfinine, an indole alkaloid from *Ervatamia coronaria*. *Phytochemistry., 25*(7), 1781–1782.

Nadkarni, K. M. (Ed.). (1996). *[Indian materia medica]; Dr. KM Nadkarni's Indian materia medica: With Ayurvedic, Unani-Tibbi, Siddha, allopathic, homeopathic, naturopathic & home remedies, appendices & indexes* (Vol. 1). Popular Prakashan.

Ojewole, J. A. O. (1984). Studies on the pharmacology of echitamine, an alkaloid from the stem bark of *Alstonia boonei* L. (Apocynaceae). *International Journal of Crude Drug Research, 22*(3), 121–143.

Olajide, O. A., Awe, S. O., Makinde, J. M., Ekhelar, A. I., Olusola, A., Morebise, O., & Okpako, D. T. (2000). Studies on the anti-inflammatory, antipyretic and analgesic properties of *Alstonia boonei* stem bark. *Journal of Ethnopharmacology, 71*(1), 179–186.

Oze, G. O., Nwanjo, H. U., & Onyeze, G. O. (2007). Nephrotoxicity caused by the extract of *Alstonia boonei* (de wild) stem bark in Guinea Pigs. *The Internet Journal of Nutrition and Wellness., 3*(2), 1–7.

Ramchandra, P., Basheermiya, M., Krupadanam, G. L., & Srimannarayana, G. (1993). Wrightial, a new terpene from *Wrightia tinctoria*. *Journal of Natural Products, 56*(10), 1811–1812.

Santhi, M., Muthulakshmi, S., Amutha, L., & Karpakaselvi, E. (2016). A comparative study on phytochemical analysis and Vincristine content of Catharanthus roseus (L.) G. Don. in three different localities. *Imperial Journal of Interdisciplinary Research, 2*(11).

Shah, S. A., Usmanghani, K., Akhtar, N., Asif, H. M., Akram, M., Ahmed, K., & Sumreen, L. (2010). Monograph of *Holarrhena antidysenterica* (linn.) Wall. *International Journal of Phytomedicine, 2*(4).

Vakil, R. J. (1955). *Rauwolfia serpentina* in the treatment of high blood pressure: A review of the literature. *Circulation, 12*(2), 220–229.

Van Beek, T. A., Verpoorte, R., Svendsen, A. B., Leeuwenberg, A. J., Bisset, N. G., & Tabernaemontana, L. (1984). (Apocynaceae): A review of its taxonomy, phytochemistry, ethnobotany and pharmacology. *Journal of Ethnopharmacology, 10*(1), 1–56.

# Chapter 2
# Traditional Knowledge, Ethnomedicinal Importance and Practical Uses of Apocynaceae Members

## Introduction

Indian traditional systems of medicine, i.e., Ayurveda, Siddha and Unani, are practiced all over world. Moreover, it is known for its vast reservoir of living traditions of plant-based systems of medicine. The rural population of India, who has limited access to modern health services, mostly relies on traditional knowledge, home remedies and formulae being practiced for generations. Traditionally, Indian people and ethnomedicine systems relied on plant-derived medicines over medicines of animal origin due to the presence of abundant natural flora and the practice of vegetarian diets. The idea of disease prevention and treatment is well documented in ancient scripts such as *Ayurveda.* The emphasis on the maintenance of positive health or *Swastha Vrutta* is a distinguishing feature of *Ayurveda* (Lele, 2001). *Ayurveda* believes that any patient can be cured with the help of herbs present in the environment. The main approaches to illness in *Ayurveda* are preventive and curative (Upadhyay, 1997). Ayurveda advocates the harmonious balance between the three humors of the body, viz. *'Vata', 'Pitta'* and *Kafa',* which enhance positive health. The imbalance of these factors may cause disease(s) (Lakshmi et al., 2004).

Applications of medicinal plants and their products for health management are an imperative part of tribal culture. Indian tribal people employ medicinal plants for a variety of health-related purposes and have great interest and knowledge for them. Traditionally, local tribal communities have explored the medicinal uses of hundreds of plants found locally in their ecosystem. They have their own management system for health-related problems that is inherited from one generation to another. Although the process of creating traditional knowledge is not well documented scientifically, it has been practiced for hundreds of years in tribal India. There is a rising understanding of the importance of traditional knowledge, and as a result, plant biologists are once again interested in learning more about the history of traditional knowledge. It is anticipated that it will significantly advance sustainable development. The vast majority of the world's plant genetic resources are found in

 
R. H. Patil et al., *Apocynaceae Plants*,
https://doi.org/10.1007/978-981-99-5406-3_2

regions where the bulk of indigenous and local communities are located. Since ancient times, many of them have been grown and used sustainably. However, indigenous and local populations go well beyond their responsibilities as managers of natural resources to contribute to the preservation and sustainable use of biological variety. Their knowledge and methods serve as a vital resource for the whole community and thus serve as an important model for setting biodiversity goals.

## Plant Metabolites in Modern Medicine

Recently, plants have been used as fine chemical factories to produce recombinant proteins of pharmaceutical interest. Plant-derived natural chemicals have also served as leads for the synthesis of novel drugs (Rupesh & Gautaml, 2013). Digitoxin isolated from *Digitalis purpura* (Plantaginaceae) is used as a cardiotonic agent. 'Taxol', the world's first multibillion dollar anticancer compound, has been extracted from *Taxus brevifolia* (Taxaceae). The antimalarial drugs quinine and artemisinin were isolated from *Cinchona officinalis* (Rubiaceae) and *Artemisia annua* (Asteraceae), respectively. Other plant-assisted drug discoveries with bioactivities for hypertension and amoebiasis include *Rauwolfia serpentina* and *Holarrhena antidysenterica* (Apocynaceae), respectively. Curcumin from *Curcuma longa* (Zingiberaceae) is an anti-inflammatory agent, baccocsides from *Bacopa monnierri* (Scrophulariaceae) for memory enhancement and many other steroidal lactones as immunomodulators have been widely explored (Patwardhan, 2000). Apart from their medicinal applications, there is also an increase in demand for medicinal plants for the purpose of food supplements and cosmetics preparations (Ramawat & Goyal, 2008). In the USA, the food drug and cosmetic act characterizes herbal plants on the basis of their intended use as food (dietary supplement), drug (biological drugs) and cosmetics. According to the Food Drug and Cosmetics Act, in sect. 201 (ff), dietary supplementation means that a product (other than tobacco) containing various dietary ingredients, including herbs or another botanical substance, an amino acid, a mineral, and vitamins, is used by people to increase their total dietary intake (Chang, 1998). Many pharma and biotech companies have shown their interest in herbal preparations that can be marketed as nutraceuticals or health foods (Kamboj, 2000). Pigments such as anthocyanins, betalains, and carotenoids present in herbal plants are used as coloring agents in food preparations, and they also have bioactivities (Aberoumand, 2011). Currently, food industries are turning their attention to natural food colours from plant origin because certain artificial colours in food have been demonstrated to cause adverse effects on consumption (Bridle & Timberlake, 1997).

## Ethnomedical Uses of Some Members of the Family Apocynaceae

There is widespread knowledge that members of the Apocynaceae family have a variety of ethnomedical applications. The ethnomedicinal significance of numerous species in the Apocynaceae family is compiled in Table 2.1. Ayurveda, Siddha, Unani, and homoeopathy offer descriptions of some significant members of this family, including *C. roseus, A. scholaris, R. serpentina, H. antidysentriaca, N. indicum,* and *T. peruviana.* Indigenous systems of medicine use plants and their constituent parts to treat a variety of illnesses and ailments.

In African countries such as Ghana, Uganda, Nigeria and Senegal, Apocynaceae members such as *T. crassa, V. africana* and *Rauvolfia vomitoria* are commonly used for similar ethnomedicinal uses. The leaf, tem, root and flower extracts of these plants have been widely used for anti-arthritic activity, in the treatment of rheumatism, parasitic skin infections including leprosy, used as diuretic, insecticidal as well as in the management of fever, wound dressing, ringworm and as a local anaesthetic in bones (Siddiqui et al., 1995; Agwu & Akah, 1990).

In the Indian subcontinent, the Himalayan ranges of the southernmost mountains at altitudes of 600–1200 m above sea level are richly colonized by Apocynaceae plants. Western states of India are geographically and culturally rich and diverse and are widely exploring plant diversity for various ethnomedicinal purposes.

The Western Ghats of India are a biodiversity hotspot and rich in medicinal plants that harbour over 3000 species of medicinal importance (Venkatachalapathi et al., 2018). The tribals of Western Ghats are using these plants for their primary health care. Due to the diverse ecosystems in this area, the Western Ghats and Nilgiri districts of Tamilnadu have a significant diversity of medicinal plants (Ramachandran, 2007).

Another member of the Apocynaceae family, *Caralluma europaea*, is indigenous to Morocco, Algeria, Tunisia, Libya, Egypt, Jordan, Spain, and Italy. External parts of these plants have been utilized for centuries to treat a variety of conditions, including diabetes mellitus, kidney stones, cysts, kidney infections, cancer, respiratory infections, digestive disorders, urogenital infections, metabolic disorders, and cardiovascular issues (Ouassou et al., 2021).

**Table 2.1** Ethnomedicinal uses of selected members of Apocynaceae (Reproduced with permission from: Bhadane et al., 2018)

| Ailment | Plant species | Plant parts used | Remedy | References |
|---|---|---|---|---|
| Stomach problems | *Wrightia tinctoria* | Seeds | Oral use of seed juice | Muthu et al. (2006) |
| | *Carrisa edulis* | Root and stem bark | Decoctions administered orally | Gakuya et al. (2013) |
| Diarrhoea/ loose motions | *Holarrhena antidysenterica* | Fruits, leaves | | Kaul and Atal (1983) |
| | *Alstonia scholaris* | Bark extract | Bark extracts used in chronic diarrhea | Wiart (2006) |
| Fever | *Wrightia tinctoria* | Fruits | Raw fruits are eaten | Muthu et al. (2006) |
| | *Thevetia neriifolia Voacanga Africana* | Leaves | Raw leaves are eaten to control the fever | Ayoola et al. (2008) |
| | *Tabernaemontana divaricata* | Leaves | Decoctions administered orally | Rahmatullah et al. (2009) |
| | *Rauvolfia serpentina* (L.) | Leaf extract | Decoctions administered orally | Roy et al. (2022) |
| Skin care and skin infectons | *Wrightia tinctoria* | Oil and latex | Liquid latex used in dandruff control | Muthu et al. (2006) |
| | *Rauwolfia tetraphylla* | Leaf and bark extract | Bark powder and castor oil is applied on skin | Kirtikar and Basu (2003) |
| | *Geissospermum sericeum* | Stem bark. | Water extract applied on skin | Steele et al. (2002) |
| | *Nerium indicum* | Leaves | Crushed leaves applied on skin | Rahmatullah et al. (2009) |
| | | Leaves and roots | Paste of leaves and root used against ringworm | Upadhyay et al. (2007) |
| | *Rauvolfia serpentina* (L.) | Leaf extract | Paste of leaves and root used against ringworm | Roy et al. (2022) |
| | *Voacanga africana* | Leaf, stem and root extract | Crushed leaves applied on skin | Agwu and Akah (1990) |
| | *Tabernaemontana crassa* | Leaf and root extract | Crushed leaves applied on skin | Siddiqui et al. (1995) |
| | *Rauvolfia vomitoria Azfel* | Leaf, stem and root extract | Crushed leaves applied on skin | Agwu and Akah (1990) |
| Ameboic dysentery | *Holarrhena antidysenterica* | Leaves | Leaves boiled in water and the juice is taken orally | Rahmatullah et al. (2009) |
| Tooth and oral problems | *Voacanga Africana* | Stem | Bark used in toothache | Ayoola et al. (2008) |
| | *Alstonia scholaris* | Latex | Used for relief from toothache | Wiart (2006) |

(continued)

**Table 2.1** (continued)

| Ailment | Plant species | Plant parts used | Remedy | References |
|---|---|---|---|---|
| Wound healing | *Acokanthera schimperi* | Fresh leaves | Crushed fresh leaves applied on wound | Teklehaymanot and Giday (2007) |
| | *Holarrhena antidysenterica* | Latex | The liquid is applied on wound | Kaul and Atal (1983) |
| | *Rauvolfia serpentina* (L.) | Leaf extract | Applied on skin | Roy et al. (2022) |
| | *Rauvolfia vomitoria Azfel* | Leaf, stem and root extract | Applied on skin | Agwu and Akaha (1990) |
| | *Tabernaemontana crassa* | Leaf and root extract | Applied on skin | Siddiqui et al. (1995) |
| Malaria | *Carrisa edulis* | Root and stem bark preparation | Applied on wound | Gakuya et al. (2013) |
| Rheumatism | *Holarrhena antidysenterica* | Stem bark | Bark is used as plaster | Kaul and Atal (1983) |
| | *Carissa carandas* | Root and root bark | Root and root bark extracts are used in rheumatism and joint pain. | Upadhyay et al. (2007) |
| | *Tabernaemontana crassa* | Leaf and root extract | Used in joint pain | Siddiqui et al. (1995) |
| Allergy/ asthma | *Alstonia scholaris* | Flowers used in respiratory trouble. | Decoctions administered orally | Guha Bakshi et al. (1999) |
| Contraceptive/ birth control | *Nerium indicum* | Roots | Roots extract is used for abortion | Upadhyay et al. (2007) |
| | *Tabernaemontana crassa* | Leaf and root extract | Roots extract is used for abortion | Siddiqui et al. (1995) |
| Women health | *Holarrhena pubescens* | Bark | Bark powder is used to control vaginal discharge. | Hossan et al. (2010) |
| | *Alstonia scholaris* | Leaves and stem bark. | Preparation is used in menstrual problems. | Kala (2005) |
| Lactation | *Alstonia scholaris* | Leaves | Decoctions administered orally | Wiart (2006) |
| | | Fresh bark | Decoctions in milk administered orally | Sharma and Sharma (2010) |
| Pediatric care | *Holarrhena antidysenterica* | Seeds | Seed preparation given to young children | Kaul and Atal (1983) |
| Snake bite | *Carissa carandas* | Leaf extract | Extract given orally. | Desai (1975) |
| | *Rauwolfia serpentina* | Roots | Squeezed roots tied on bitten area. | Prakasha et al. (2010) |

(continued)

**Table 2.1** (continued)

| Ailment | Plant species | Plant parts used | Remedy | References |
|---|---|---|---|---|
| | *Alstonia scholaris* | Stem bark | Oral chewing of the bark | Prakasha et al. (2010) |
| | *Tabernaemontana divaricata* | Roots | Root paste given orally in water of buttermilk | Prakasha et al. (2010) |
| Urinary tract infection | *Hemidesmus indicus* | Leaves | Cosumed orally. | Hossan et al. (2010) |
| Premature ejaculation | *Carissa spinarum* | Unripened | Infusion is used | Belayneh and Bussa (2014) |
| | *Tabernaemontana crassa* | Leaf and root extract | Cosumed orally | Siddiqui et al. (1995) |
| Mouth, throat and coughing problems | *Acokanthera schimperi* | Stem and leaves | Gargling is effective in tonsillitis | Belayneh and Bussa (2014) |
| | *Alstonia scholaris* | Leaves paste | Taken orally with honey | Rahmatullah et al. (2009) |
| Cardiac disorders | *Thevetia neriifolia* *Voacanga Africana* *Stropanthus hispidus* | Stem bark and root bark | Used as cardiotonic | Ayoola et al. (2008) |
| Measles | *Thevetia neriifolia* | Leaves | Applied on skin | Ayoola et al. (2008) |
| Diabetes | *Catharanthus roseus* | Whole plant | Whole plant powder taken orally | Muthu et al. (2006) |
| | | Leaves | Boiled leaves infusion taken orally | Erasto et al. (2005) |
| | *Plumeria obtuse* | Leaves | Boiled leaves infusion taken orally | Siddiqui et al. (2004) |
| Ear pain | *Nerium oleander* | Leaves and roots | Boiled leaves infusion taken orally | Muthu et al. (2006) |
| Sexually transmitted infections | *Catharanthus roseus* | Roots | Boiled roots administered anally in Gonorrhoea infection | Pereira et al. (2010) |
| | *Alstonia scholaris* | Stem bark powder | Dried stem bark powder taken orally to control Gonorrhoea infection | Sahu (2011) |
| | *Rauvolfia vomitoria Azfel* | Leaf, stem and root extract | Dried stem bark powder taken orally to control Gonorrhoea infection | Agwu and Akah (1990) |
| Appetizer | *Carissa carandas* | Ripened fruits | Ripened fruits used as an appetitizer | Rahmatullah et al. (2009) |
| Insecticidal | *Tabernaemontana crassa* | Leaf and root extract | Water extracts are used as larvecidal | Siddiqui et al. (1995) |
| | *Voacanga africana* | Leaf, stem and root extract | Water extracts are used as larvecidal | Agwu and Akah (1990) |

(continued)

**Table 2.1** (continued)

| Ailment | Plant species | Plant parts used | Remedy | References |
|---|---|---|---|---|
| Local anaesthetic | *Tabernaemontana crassa* | Leaf and root extract | Applied on external tissues | Siddiqui et al. (1995) |
| | *Rauvolfia vomitoria Azfel* | Leaf, stem and root extract | Applied on external tissues | Agwu and Akah (1990) |
| Eye care | *Rauvolfia serpentina* (L.) | Leaf extract | Used for better sight and in eye infections | Roy et al. (2022) |

# References


Aberoumand, A. (2011). A review article on edible pigments properties and sources as natural biocolorants in foodstuff and food industry. *World Journal of Dairy & Food Sciences, 6*(1), 71–78.

Agwu, I. E., & Akah, P. A. (1990). *Tabernaemontana crassa* as a traditional local anaesthetic agent. *Journal of Ethnopharmacology, 30*(1), 115–119.

Ayoola, G. A., Folawewo, A. D., Adesegun, S. A., Abioro, O. O., Adepoju-Bello, A. A., & Coker, H. A. B. (2008). Phytochemical and antioxidant screening of some plants of apocynaceae from south West Nigeria. *African Journal of Plant Science, 2*(10), 124–128.

Belayneh, A., & Bussa, N. F. (2014). Ethnomedicinal plants used to treat human ailments in the prehistoric place of Harla and Dengego valleys, eastern Ethiopia. *Journal of Ethnobiology and Ethnomedicine, 10*(1), 18.

Bhadane, B. S., Patil, M. P., Maheshwari, V. L., & Patil, R. H. (2018). Ethnopharmacology phytochemistry and biotechnological advances of family Apocynaceae: A review phytotherapy research, *32*(7), 1181–1210. https://doi.org/10.1002/ptr.v32.710.1002/ptr.6066

Bridle, P., & Timberlake, C. F. (1997). Anthocyanins as natural food colours—Selected aspects. *Food Chemistry, 58*(1), 103–109.

Chang, J. (1998). Scientific evaluation of traditional Chinese medicine under DSHEA: A conundrum. *The Journal of Alternative and Complementary Medicine, 5*(2), 181–189.

Desai, W. G. (1975). *Aushadhisangrah* (Vol. 1, p. 223). Shree Gajanan Book Depot Prakashan.

Erasto, P., Adebola, P. O., Grierson, D. S., & Afolayan, A. J. (2005). An ethnobotanical study of plants used for the treatment of diabetes in the eastern Cape Province, South Africa. *African Journal of Biotechnology, 4*(12).

Gakuya, D. W., Itonga, S. M., Mbaria, J. M., Muthee, J. K., & Musau, J. K. (2013). Ethnobotanical survey of biopesticides and other medicinal plants traditionally used in Meru central district of Kenya. *Journal of Ethnopharmacology, 145*(2), 547–553.

Guha, D. N., Sensarma, P., & Pal, D. C. (1999). A lexicon of medicinal plants in India. *Naya Prokash, Calcutta, 1st Publication, 1*, 389.

Hossan, S., Agarwala, B., Sarwar, S., Karim, M., Jahan, R., & Rahmatullah, M. (2010). Traditional use of medicinal plants in Bangladesh to treat urinary tract infections and sexually transmitted diseases. *Ethnobotany Research and Applications, 8*, 061–074.

Kala, C. P. (2005). Ethnomedicinal botany of the Apatani in the Eastern Himalayan region of India. *Journal of Ethnobiology and Ethnomedicine., 1*, 1–8.

Kamboj, V. P. (2000). Herbal medicine. *Current Science-Bangalore, 78*(1), 35–38.

Kaul, M. K., & Atal, C. K. (1983). Studies on *Holarrhena antidysenterica* Wall. 1. Botany, medico-ethnobotany and distribution. *Journal of Ethnopharmacology, 8*(3), 349–356.

Kirtikar, K. R., & Basu, B. D. (2003). *Data on medicinal plants and chemical constituents*. Indian Medicinal Plants.

Lakshmi, B., Tilak, J. C., Adhikari, S., Devasagayam, T. P., & Janardhanan, K. K. (2004). Evaluation of antioxidant activity of selected Indian mushrooms. *Pharmaceutical Biology, 42*(3), 179–185.

Lele, R. D. (2001). *Ayurveda and modern medicine*. Bharatiya Vidya Bhavan.

Muthu, C., Ayyanar, M., Raja, N., & Ignacimuthu, S. (2006). Medicinal plants used by traditional healers in Kancheepuram District of Tamil Nadu, India. *Journal of Ethnobiology and Ethnomedicine, 2*(1), 43.

Ouassou, H., Bouhrim, M., Kharchoufa, L., Imtara, H., elhouda Daoudi, N., Benoutman, A., Bencheikh, N., Ouahhoud, S., Elbouzidi, A., & Bnouham, M. (2021). Caralluma europaea (Guss) NE Br.: A review on ethnomedicinal uses, phytochemistry, pharmacological activities, and toxicology. *Journal of Ethnopharmacology., 273*, 113769.

Patwardhan, B. (2000). Ayurveda: The designer medicine. *Indian Drugs, 37*, 213–227.

Pereira, D. M., Ferreres, F., Oliveira, J. M., Gaspar, L., Faria, J., Valentão, P., & Andrade, P. B. (2010). Pharmacological effects of Catharanthus roseus root alkaloids in acetylcholinesterase inhibition and cholinergic neurotransmission. *Phytomedicine, 17*(8), 646–652.

Prakasha, H. M., Krishnappa, M., Krishnamurthy, Y. L., & Poornima, S. V. (2010). Folk medicine of NR Pura taluk in Chikmagalur district of Karnataka. *Indian Journal of Traditional Knowledge, 9*, 55–60.

Rahmatullah, M., Noman, A., Hossan, M. S., Rashid, M. H., Rahman, T., Chowdhury, M. H., & Jahan, R. (2009). A survey of medicinal plants in two areas of Dinajpur district, Bangladesh including plants which can be used as functional foods. *American Eurasian Journal of Sustainable Agriculture, 3*(4), 862–876.

Ramachandran, V.S. (2007). *Wild edible plants of the Anamalais*. Coimbatore district, Western Ghats, Tamil Nadu.

Ramawat, K. G., & Goyal, S. (2008). The Indian herbal drugs scenario in global perspectives. In *Bioactive molecules and medicinal plants* (pp. 325–347). Springer.

Roy, M., Sarkar, B. C., Shukla, G., Debnath, M. K., Nath, A. J., Bhat, J. A., & Chakravarty, S. (2022). Traditional homegardens and ethnomedicinal plants: Insights from the Indian Sub-Himalayan region. *Trees, Forests and People, 8*, 100236.

Rupesh, K., & Gautam, D. S. (2013). Medicinal plants having anti-arthritic potential: A review. *International Journal of Pharmaceutical Sciences Review and Research, 19*(1), 96–102.

Sahu, P. K. (2011). Plants used by Gond and Baiga women in ethnogynaecological disorders in Achanakmar wild life sanctuary, Bilaspur. *CG International Journal of Pharmacy and Life Sciences, 2*(2), 559–561.

Sharma, R., & Sharma, H. K. (2010). Ethnomedicines of Sonapur, Kamrup district, Assam. *Indian Journal of Traditional Knowledge, 9*, 163–165.

Siddiqui, et al. (1995). Two cytotoxic pentacyclic triterpenoids from Nerium oleander. *Phytochemistry, 39*(1), 171–174. https://doi.org/10.1016/0031-9422(94)00855-N

Siddiqui, B. S., Ilyas, F., Rasheed, M., & Begum, S. (2004). Chemical constituents of leaves and stem bark of *Plumeria obtusa*. *Phytochemistry, 65*(14), 2077–2084.

Steele, J. C., Veitch, N. C., Kite, G. C., Simmonds, M. S., & Warhurst, D. C. (2002). Indole and β-carboline alkaloids from *Geissospermum sericeum*. *Journal of Natural Products, 65*(1), 85–88.

Teklehaymanot, T., & Giday, M. (2007). Ethnobotanical study of medicinal plants used by people in Zegie Peninsula, Northwestern Ethiopia. *Journal of ethnobiology and Ethnomedicine, 3*(1), 12.

Upadhyay, B., Roy, S., & Kumar, A. (2007). Traditional uses of medicinal plants among the rural communities of Churu district in the Thar Desert, India. *Journal of Ethnopharmacology, 113*(3), 387–399.

Upadhyay, S. (1997). *Immunomodulation*. Narosa Publishing House.

Venkatachalapathi, A., Sangeeth, T., Ali, M. A., Tamilselvi, S. S., Paulsamy, S., & Al-Hemaidc, F. M. (2018). Ethnomedicinal assessment of Irula tribes of Walayar valley of Southern Western Ghats, India. *Saudi Journal of Biological Sciences, 25*(4), 760–775.

Wiart, C. (2006). *Medicinal plants of the Asia-Pacific: Drugs for the future?* (pp. 447–450). World Scientific Co. Pte. Ltd..

# Chapter 3
# Plant Secondary Metabolites as a Source of New Drugs

## Introduction

An increasing number of new pathogens, the emergence of multidrug drug-resistant (MDR) and extremely drug-resistant (XDR) bacterial and fungal pathogens, diseases associated with unhealthy diets and insufficient physical activity have been major problems concerning human health globally. Despite continuous progress in new drug discovery, the battle against many life-threatening diseases is still far from over. The recent emergence of MDR and XDR *Mycobacterium, Pseudomonas*, and *Candida spp.* led to thousands of new infections and deaths. Furthermore, infections by common human and animal opportunistic pathogens such as *Aspergillus*, *Cryptococcus*, and *Candida* in immune-compromised patients are increasing at an alarming rate. Therefore, new drugs from natural sources can certainly benefit society from infectious and noninfectious diseases.

## Global Outlook of Communicable and Noncommunicable Diseases

The development of antibiotic-resistant infections and the appearance of novel diseases appear to be significant risks in human history, causing an unfathomable number of illnesses and fatalities. Recently, the global burden of various diseases, such as tuberculosis, HIV, H1N1, SARS, malaria, cancer, and cardiovascular disease, has been increasing. The WHO predicts that between 2004 and 2030, communicable disease fatalities in low-income countries will decrease by approximately half, but deaths from noncommunicable diseases (NCDs) are projected to increase (World Health Organization, 2008). Over 56 million deaths were reported worldwide in 2012, of which 38 million were due to NCDs. According to the World



R. H. Patil et al., *Apocynaceae Plants*,
https://doi.org/10.1007/978-981-99-5406-3_3

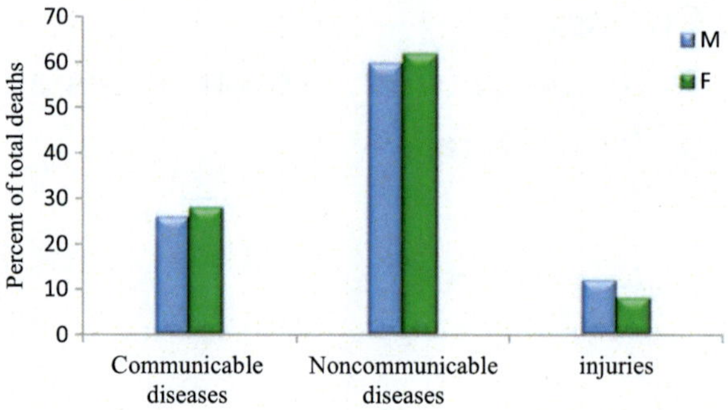


**Fig. 3.1** Global deaths according to cause and sex (2008) as per the World Health Organization (WHO)

Health Organization, approximately 60 million deaths occurred due to various communicable as well as noncommunicable diseases (Fig. 3.1).

## Communicable Diseases

Communicable diseases refer to diseases that can be transmitted through infective agents such as bacteria, fungi, viruses, parasites, and protozoa. The main communicable diseases in India with significant morbidity and mortality are diarrhea, HIV-AIDS, influenza, dengue and acute respiratory infections. Nearly one-fifth of all fatalities worldwide occur in children under the age of five, and diarrheal diseases are the second-greatest cause of infectious disease, accounting for almost 1000 million episodes and over two million deaths per year (Brewer, 2000). Similarly, TB bacilli are present in approximately one-third of the world's population. Over 33 million people have been identified to have HIV, and over two million have died in the past 10 years (Brewer, 2000). Malaria, a vector-borne disease, flourishes in poor tropical nations, leading to over one million deaths each year, and the majority of the deaths involve young children. *Nevertheless, a remarkable decline has been noticed in the morbidity and mortality rate of infectious scourges because of wider access to effective antimicrobial treatments and the dramatic reduction in their prices* (Concepcion et al., 2001).

## Noncommunicable Diseases

According to the World Health Organization, NCDs prematurely take 16 million lives annually. Cardiovascular diseases are the most important cause of mortality worldwide, with over 17.5 million deaths per year, which is 46% of total deaths caused by NCDs. Other diseases, such as cancers (8.2 million deaths), respiratory diseases (4.0 million deaths) and diabetes (1.5 million deaths), are among the main killers. Thus, these four major NCDs were responsible for 82% of total deaths due to

NCDs. According to the world economic forum, the economic losses due to NCDs in low- and mid-income countries during 2011–2030 are estimated at $47 trillion (Bloom et al., 2011).

The mortality rate due to diabetes and heart disease is gradually increasing worldwide. The global prevalence of diabetes in 2014 increased by 7%. According to projections from the WHO. The number of deaths due to cancer will rise from 8.2 million to 12.6 million annually, while the number of deaths due to cardiovascular diseases is expected to increase from 17.5 million in 2012 to 22.2 million in 2030 (World Health Organization, 2014).

## Need for New and Effective Drugs

Despite the ever-increasing economic burden, the treatment and prevention of communicable and noncommunicable diseases has not attracted global attention. The global load of chronic noncommunicable diseases cannot be reduced without development and appropriate access to effective medicines. Recent data on over 40 countries show that generic medicines for the treatment of acute diseases were available in 56.15% and 65.6% of the public and private sectors, respectively. For NCDs, medicine availability was only 36% and 54.7%, respectively (Gelders, 2006; Kotwani, 2009). For over 280 million people suffering from diabetes mellitus worldwide, the cost of basic oral treatment is unaffordable, especially in low- and middle-income countries. Access to insulin remains poor in many regions of the world due to high prices. Similarly, approximately 5.5 million cancer patients in developing countries are suffering without adequate treatment (Abegunde, 2007).

The development of effective medicines has played a pivotal role in the treatment of cardiovascular diseases, diabetes, chronic respiratory diseases, and different types of cancers, mental and neurological disorders. Nonetheless, due to the prevalence and global burden of these disorders, a large number of people with such problems do not receive proper treatment and care (Kohn, 2004; Meyer, 2010). Thus, the development of new, safe and affordable medicines is essential for the primary and secondary prevention of communicable and noncommunicable diseases, especially in the developing world.

According to WHO estimates, over 80% of people in underdeveloped countries rely on traditional medicines, the majority of which are made from plants, for their most basic medical requirements. The United States Food and Drug Administration (USFDA) has registered more than 49% of novel medications produced from natural materials or their derivatives. From 2000 to 2011, more than 63% of authorized anticancer medications were from natural sources, up from more than 60% of approved drug candidates during the years 1989–1995. The WHO has listed over 21,000 medicinal plants, most of which are temperate species. Over 2500 plant species that are utilized in traditional medicines can be found in abundance throughout the Indian subcontinent. Tribal culture is fundamentally based on the control of health through the use of medicinal plants. These people obviously possess extensive traditional knowledge of medicinal herbs and employ them for a variety of purposes relating to health. Numerous flora and plant-based substances have been used medicinally by indigenous tribes in the area according to long-standing customs. The people of tribal villages have systems for health-related problems and management that are generally transmitted from one generation to another.

## Natural Products as a Potential Source for New Metabolites

Ever since the first antibiotic was discovered in 1929, there has been a continuous increase in the demand for new and safe substances for the management of human illnesses. Bioactive compounds to treat various diseases can be obtained by chemical synthesis as well as from natural sources such as plants, animals, and microorganisms. A report by Bérdy (2012) described over ten million molecules with pharmaceutical potential that are mainly obtained by chemical synthesis (Table 3.1).

The approximate number of medicinal metabolites derived from natural sources is only approximately one-tenth that of compounds derived from synthetic sources (Table 3.1), and natural compounds, particularly those derived from microbes, typically have significantly higher bioactivity than synthetic compounds. Moreover, the structural and functional diversity of natural products is immense, and their use is safe and environmentally benign. Hence, after the discovery of penicillin, a metabolite from *Penicillium notatum*, several novel molecules were isolated and studied from new natural sources together with semisynthetic metabolites by the pharmaceutical industry, and hundreds of new antibiotics were discovered (Table 3.1). However, the trend of discovering new antibiotics has turned downwards in the last decade. However, the proportion of novel compounds to resistant infections has become unbalanced due to the increase in the number of resistant bacteria. Considering the pace of new molecule discovery, it is imperative to refocus our efforts on searching for new metabolites from natural sources.

## Secondary Metabolites from Natural Sources

Secondary metabolites are naturally derived substances and/or byproducts from plants, animals and microorganisms (Baker et al., 2000).

**Table 3.1** Approximate numbers of known synthetic compounds, natural products, and microbial products (Berdy, 2012)

| | Synthetic chemical compounds | Natural product | Microbial products |
|---|---|---|---|
| Numbers | 8–10 million | ~500,000 | ~70,000 |
| Drugs | 2000–2500 | 1200–1300 | 450–500 |
| Percentage | 0.005 | 0.6 | 1.6 |

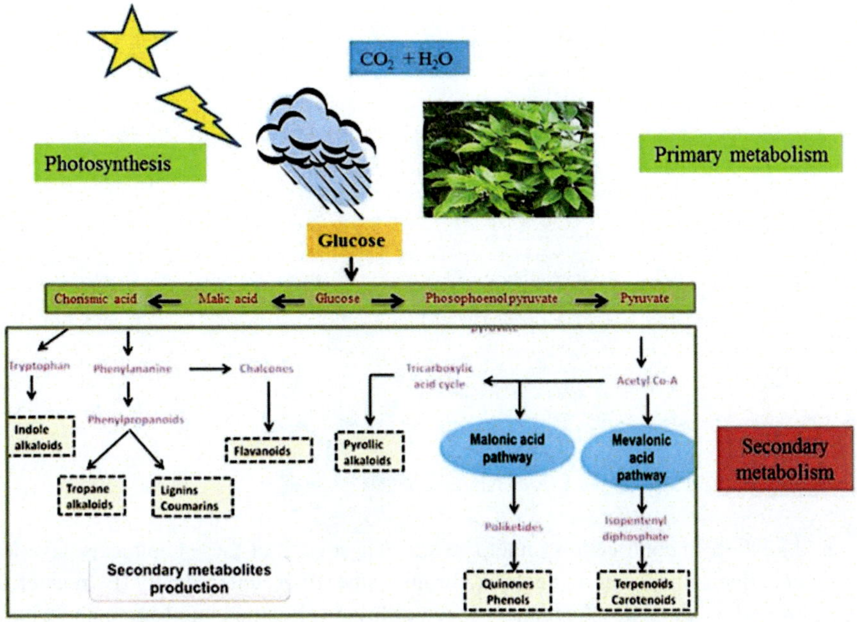


**Fig. 3.2** Primary and secondary metabolism in plants

Secondary metabolites are produced by plants through subsidiary pathways that are not directly involved in basic cellular processes and therefore are not essential for the host. However, they are responsible for mediating environmental interactions and are crucial for survival and reproduction. Plants are known to produce more than 50,000 different secondary metabolites. Plant secondary metabolites are considered the basis for the biological effects of several modern molecules and medicinal plants. The discovery of novel secondary metabolites is a continuous process that intends to discover a novel molecule with better therapeutic value. During evolution, plants have developed effective countermeasures to withstand herbivorous pests/diseases. These metabolites are species-specific and exert a wide range of effects on the plant itself and other living organisms. Living cells share primary metabolites such as carbohydrates, lipids, nucleotides, and peptides, which are essential to life activities. On the other hand, secondary metabolites, which are produced from primary metabolites (Fig. 3.2), serve to protect plants from external and environmental stresses such as disease, pests, and herbivores.

A number of plants produce secondary metabolites, such as alkaloids, phenolics and oils, which have been used for insect and pest control for a long time. There are several types of secondary plant metabolites with insecticidal/pesticidal properties, viz., repellents, antifeedants, phagostimulants and toxins (Krishnayya & Rao, 1996; Nivsarkar et al., 2001). Plant proteins and other secondary metabolites have been shown to have insecticidal and pesticidal properties and hence were proposed as possible candidates to develop insecticidal molecules (Simmonds et al., 1989; Kim

| Alkaloids | Terpenoids and steroids | Phenolic compounds | Sulfur-containing compounds |
|---|---|---|---|
| • Cyanogenic glucosides<br>• Non protein amino acids (Ex. Taxol) | • Hemiterpenes (single isoprene unit)<br>• Monoterpenes (two isoprene units)<br>• Sesquiterpenes (three isoprene units)<br>• Diterpenes (four isoprene units)<br>• Sesterterpenes (five isoprene units)<br>• Triterpenes (six isoprene units)<br>• Carotenoids (eight isoprene units) | • Lignin<br>• Flavonoids<br>• Isoflavonoids<br>• Tanins | • Glutathione<br>• Defensins<br>• Alenin |

**Fig. 3.3** Major classes of plant secondary metabolites

et al., 2003). In contrast to synthetic pesticides, several of these chemicals have less severe effects on insects and exhibit chronic rather than acute toxicity (Estruch et al., 1997; Park et al., 2003). Secondary metabolites are classified into four major groups: alkaloids, terpenoids, phenolic compounds and sulfur-containing compounds (Fig. 3.3).

After the 'wonders medication' penicillin was discovered, along with Pasteur's work on fermentation for the culture of microbial cells, the scientific community began to seriously look into microbes as a potential source of secondary metabolites. The early age (1940–1974) of microbial metabolite production was mainly focused on *Streptomyces,* while bacterial and fungal sources were less explored. The role of fungi as secondary metabolite producers was recognized by the early 1980s, and today, fungal metabolites account for the same percentage (45%) of metabolites as those from *Streptomyces* and *actinobacteria*.

Over the years, the discovery of bioactive secondary metabolites was centered on microorganisms that were largely obtained from soil. Selecting the ecological habitat for isolating prospective microbes is an important step in drug discovery. Thus, it is imperative to carefully select the biological source before proceeding instead of a random approach in the selection of biological sources. It is known that organisms are subjected to constant metabolic and abiotic stress that could produce novel secondary metabolites (Schutz, 2001). Several natural sources, such as soil, animals, plants and marine environments, are found to harbor microbial communities with 'novel chemistry'.

## *Soil*

The unanticipated discovery of penicillin from the contaminant fungus *Penicillium notatum* isolated from soil has prompted intensive investigation of soil as well as other natural habitats in the search for microorganisms with novel properties. Other novel compounds, such as cephalosporin (from *Cephalosporium cryptosporium*), tetracycline and other polyketides (from *Actinomycetes*), originate in soil-dwelling fungi. With the advent of new sophisticated isolation techniques and high-throughput screening methods, the isolation of microorganisms from soil has now become rapid, economical and more productive. However, it is estimated that less than 1% of soil microbes have only been cultured and characterized.

It appears that pharmaceutical companies have screened almost the entire world's soil in search of beneficial microbial communities. Therefore, it is necessary and interesting to explore and exploit alternate sources for microbes that could produce new compounds of therapeutic potential. Additionally, novel natural products and their producers offer better opportunities in the discovery of new drugs and agrochemicals. Thus, challenges to finding novel microbes remain.

## *Marine Habitats*

Due to extreme temperature, pressure and salinity, ocean water has been the source of microbiota that could produce natural compounds of therapeutic potential. However, it is difficult to grow marine bacteria in the laboratory and coax them to produce their full repertoire of therapeutic molecules. Methicillin-resistant *Staphylococcus aureus* (MRSA), a gram-positive bacterium, is an emerging challenge, particularly for public health, due to the rapid development of resistance towards all major antibiotics. Macrolactin, Pseudonocardians A, and Kandenol A, compounds reported for antibacterial activities against MRSA, were isolated from *Pseudomonas* species, a marine bacterium (Sung-Hwan et al., 2013). Halichondrine B, an anticancer drug active against lung cancer, was isolated from *Halichondria okadai*, a marine-derived sponge. Some of the most potent cytotoxic substances ever identified are produced by aquatic organisms, but their yields are often low, making it unlikely that natural sources will supply enough material for drug development investigations.

## *Insects*

Insects are the most diverse form of life present on Earth. Their extraordinary diversity in terms of numbers and morphological forms has led to the isolation of microorganisms that could produce natural compounds of therapeutic potential. The insect gut is the reservoir of complex microbial communities that can produce a variety of bioactive substances. Grasshopper gut-associated bacteria were screened for antagonistic activity against common human pathogens (Chellaram et al., 2012). A number of bioactive substances have been produced by insect pathogenic fungi (Masahiko et al., 2005).

## *Plants: A Potential Source of Microbial Communities*

Plant-based medicines have sparked renewed interest because they are relatively safe and rarely cause negative effects. Humans have been using plants and medications derived from plants to cure a wide range of diseases throughout history. However, widespread use of the world's plant diversity has led to the extinction of numerous species of medicinally important plants. Discovering other resources and methods to produce metabolites that are similar to those derived from plants is therefore essential. There are 3,00,000 nearby plant species on Earth, and each species, in turn, is hosted by a unique microbial community. The huge biodiversity of plants can be explored for isolating novel microbes for chemical and functional diversity.

## Secondary Metabolites of Commercial Importance

Secondary metabolites from plants have played a vital role in the treatment of many diseases and serve as building blocks for the creation of more potent medications. The oldest of the substances often employed in the therapy of cardiac disorders is digitaloxigenin, a steroid glycoside derived from *Digitalis purpura*. In addition to being used to treat congestive heart failure, it is also useful in the treatment of arrhythmia. An acceptable antiarrhythmic drug is quinidine, a stereoisomer of the antimalarial quinine found in cinchona bark. Atropine, a metabolite of *Atropa belladonna*, is used to treat slow heartbeat (bradycardia). Paclitaxel (Taxol), the first anticancer medication to reach the $1 billion mark, is a natural substance obtained from the yew plant *Taxus brevifolia* (Wani et al., 1971).

Artemisinine is an antimalarial sesquiterpene from the Chinese medicinal herb *Artemisia annua* that has been used in herbal remedies since ancient times (Arvigo & Balick, 1993; Farnsworth et al. 1985). Other bioactive constituents, such as aspirin and quinine, have also been isolated from medicinal plants.

Table 3.2 summarizes a few important secondary metabolites used as drugs, their sources and medical/pharmaceutical applications.

**Table 3.2** Important secondary metabolites used as drugs, their source and medical/pharmaceutical applications

| Sr. no. | Secondary metabolites (drugs) | Plant source | Medical/pharmaceutical application |
|---|---|---|---|
| 1. | Ajmalicine, solasodine, α-solanine, serpentine and reserpine, β-carboline, alstonine, reserpiline | *Rauwolfia* species | Insomnia, epilepsy, schizophrenia, antidote against snake bite |
| 2. | Oleanane triterpenoids, cinnamic caid, coumarins, flavonoids | *Baccharis megapotamica* | Invigorating blood circulation, antihelmintic, anti-inflammatory, anti-neoplastic, antioxidant, antiviral |
| 3. | Berberine, coptisine, jatrorrhizine | *Coptis* species | Anti-inflammatory, neuroprotective, antioxidant |
| 4. | Betalains, apigenin flavonoids like vitexin, vitexin-2-O-rhamnoside, | *Beta vulgaris* | Antihypertensive, hypoglycaemic, antioxidant, anticancer agent. |
| 5. | Borneol, α-terpineol, carvacrol, thymol, geraniol, carotenoids, rosmarinic acid | *Thymus* species | Anti-inflammatory, hypoglycemic, antihyperlipidemic activity, antioxidative capacity, flavouring agent |
| 6. | Camptothecin, 10-hydroxycamptothecin, 10-methoxycamptothecin | *Camptotheca acuminata* | Anti-cancer, cytotoxic |
| 7. | Cucurbitacins | *Trichosanthes*, *Cucurbita*, *Cucumis* and *Citrullus* | Anti-inflammatory, anti-diabetic activity, antitumour |
| 8. | Diosgenin | *Dioscorea doryphora* | Cancer therapy, hypercholesterolemia, |
| 9. | Diosgenin,, allantoin, phytic acid, coumaric acid, myricetin | *Dioscorea* species | Antitumour, anorexiant, antifungal, antimutagenic, hypoglycemic, immunomodulatory, antidotes. |
| 10. | Ephedrine alkaloids (1-ephedrine, *d*-pseudoephedrine, vicenin-1 and vicenin-2), proanthocyanidines | *Ephedra* | Appetite suppression, psychostimulant and hypotension caused by spinal anaesthesia and urinary incontinence |
| 11. | Ginsenosides (dammarane and oleanane) | *Panax ginseng* | Diabetes, anti-stress agents, cardioprotective, antioxidant, anxiolytic, anticancer, aphrodisiac |
| 12. | Hypericinn, hyperforin | *Hypericum perforatum* | Antidepressant, anti-inflammatory, antiviral, anticancer, antioxidant |
| 13. | Indole alkaloids, catharanthine | *Catharanthus roseus* | Cancer therapy |
| 14. | Kaempferol, isorhamnetin, quercetin, bilobalide, ginkgolide A, B, C, ginsenoside | *Ginkgo biloba* | Anti-inflammatory, cardiovascular diseases, cancer therapy |

(continued)

**Table 3.2** (continued)

| Sr. no. | Secondary metabolites (drugs) | Plant source | Medical/pharmaceutical application |
|---|---|---|---|
| 15. | Odorobioside G, verodoxin, glucodogifucoside, strospeside, digitoxin, digoxin, ouabain, oleandrin, proscillaridin | *Digitalis lanata* | Treatment of congestive heart failure, cytotoxic, antidiabetic, neuro, cardioprotective properties |
| 16. | Paclitaxel, essential oils, flavonoids, Taxol, baccatin III, lignans | *Taxus* species | Antiepileptic, anti-inflammatory, anticancer, antipyretic, analgesic, immunomodulatory, antimicrobial activities. |
| 17. | Pelletierine, ellagic acid, phenol, | *Punica granatum* | Anticancer, treat dental problems immune modulatory activity |
| 18. | Phenolics steroids, polysaccharides, quinochalcone C-glycosides, quinone-containing chalcones | *Carthamus tinctorius* | Cardiovascular diseases, cerebrovascular diseases anticoagulant, gastrointestinal, antioxidant, |
| 19. | Phloroglucinols like aspidin, deaspidin, filixic acid | *Dryopteris filix-mas* | Anti-inflammatory, treating rheumatoid arthritis, wounds, ulcers |
| 20. | Prenylated isoflavones (scandenone and auriculasin) | *Maclura pomifera* | Astringent, antioxidant and fungicidal agents, cardiovascular activity, insect repellant, natural dye |
| 21. | Quinolone alkaloids, furanocoumarins, acridone, flavonoids | *Ruta* species | Contraceptive preparations, anti-inflammatory, antipyretic, antioxidant, analgesic, antiviral, |
| 22. | Quinolozidine, alkaloids, nesodine, vertine, demethylsubine I & II | *Heimia salicifolia* | Antipyretic, laxative and diuretic activity, agents to reduce psychodysleptic activity |
| 23. | β-Sitosterol, forskolin, apigenin, chavicol, cavacrol, eugenol, quercetin | *Coleus blumei* | Cardiotonic, nervous system depressant, vasodialator, immunosuppressant, antioxidant |
| 24. | Scopalamine, hyoscyamine | *Scopolia parviflora* | Anticholinergic drug, antispasmodic drug hallucinogenic agent |
| 25. | Ajmalicine, vinblastine, vincristine, tabersonine, vindoline, vinceine | *Catharanthus roseus* | Hodgkin's disease, antineoplastic drug |
| 26. | Phytosterols, triterpenoids, lignans, phenolics, gallic acid, syringin, chlorogenic acid, ethyl gallat, rutin, isoquercitrin | *Saussurea medusa* | Spasmolytic, antimicrobial, CNS depressant, antioxidant |
| 27. | Shikonin | *Arnebia euchroma* | Antimicrobial, anticancer, antipyretic, anti-inflammatory |
| 28. | Stevioside, steviol rebaudioside, essential oils | *Stevia rebaudiana* | Cancer, renal disease, obesity, treating inflammatory bowel disease, dental caries |
| 29. | Kaempferol, syringetin, isorhamnetin, epicatechin, flavonols, vineferin | *Vitis vinifera* | Laxative, purgative, diuretic, antioxidative, anti-inflammatory, antithrombotic, neuroprotective agents, antianxietic, anticancer agents |

(continued)

**Table 3.2** (continued)

| Sr. no. | Secondary metabolites (drugs) | Plant source | Medical/pharmaceutical application |
|---|---|---|---|
| 30. | Tropane alkaloids (atropine, hyoscyamine and scopolamine) | *Atropa belladona* | Motion sickness, inflammation, antidote for snake bites |
| 31. | Atropine and scopolamine) | *Datura stramonium* | Treat epilepsy and depression, powerful mind altering drug, anti-inflammatory, antimicrobial |
| 32. | Tropane alkaloids(anisodamine, anisodine and hyoscyamine), piperidine, pyridine alkaloid | *Anisodus luridus* | Anticholinergic, treatment of septic shock |
| 33. | Phytosterols, saponins, tannins, phenolic compounds, anthraquinone glycosides, triterpenoids | *Rubia tinctorum* | Treatment of skin disease, spleen disorders, astringent, febrifuge, antithrombotic, antioxidants |
| 34. | Visnagin, khellin, γ-pyrones, pyranocoumarins, apigenin, luteolin, chrysoeriol, essential oils | *Ammi visnaga* | Treatment of renal colic, abdominal cramps, mild anginal symptoms |
| 35. | Syringetin, isorhamnetin | *Wrightia tinctoria* | Stomach aches |
| 36. | Syringetin, epigenin | *Carrisa edulis* | Stomach aches |
| 37. | Sterols, alkaloids | *Holarrhena antidysenterica* | Diarrhea, stomach aches |
| 38. | Rutin, quercetin | *Alstonia scholaris* | Diarrhea, diuretic |
| 39. | Periconicin A | *Taxus cuspidata.* | Skin infection, antibacterial, antifungal |
| 40. | Periconicin B | *Taxus cuspidata.* | Skin infection, antibacterial, antifungal |
| 41. | Methyl triacetic lactone | *Polygonum senegalense* | Skin infection, antibacterial, antifungal |
| 42. | Altersolanol A | *Urospermum picroides* | Motion sickness, inflammation, antidote for snake bites |
| 43. | Pestacin | *Ephedra fasciculata* | Spasmolytic, antimicrobial, CNS depressant, antioxidant |
| 44. | Isopestacin | *Ephedra fasciculata* | Spasmolytic, antimicrobial, CNS depressant, antioxidant |
| 45. | Pyrrocidine A and B | *Zea mays* | Antioxidant, antimicrobial |
| 46. | Phomol | *Erythrina crista-galli* | Antioxidant, antimicrobial, urinary tract infections |
| 47. | Guanacastepene A to L | *Daphnopsis Americana* | Antioxidant, antimicrobial, sour throat |
| 48. | Xylariaquinone A | *Sandoricum Koetjape,* | Cytotoxic, anticancer, antifungal |
| 49. | Penicillenols | *Aegiceras corniculatum* | Cytotoxic, anticancer, antifungal |
| 50. | Nigerasperone | *Colpomenia sinuosa* | Cytotoxic, anticancer, antifungal |
| 51. | Phomopsin A | *E. agallocha* | Cytotoxic, anticancer, antifungal |

(continued)

**Table 3.2** (continued)

| Sr. no. | Secondary metabolites (drugs) | Plant source | Medical/pharmaceutical application |
|---|---|---|---|
| 52. | Brefeldin A and B | *Taxus mairei* | Cytotoxic, anticancer, antifungal |
| 53. | Vincristine | *Catharanthus roseus* | Anticancer, antifungal |
| 54. | Vinblastine | *Catharanthus roseus* | Anticancer, antifungal |
| 55. | Paclitaxel (Taxol) | *T. brevifolia* | Anticancer, antifungal |
| 56. | Torreyanic acid, | *T. taxifolia* | Anticancer, antifungal |

# References

Abegunde, D. O. (2007). The burden and costs of chronic diseases in low-income and middle income countries. *Lancet, 370*(9603), 1929–1938.

Arvigo, R., & Balick, M. (1993). *Rainforest remedies: One hundred healing herbs of Belize*. Lotus Press.

Baker, D., Mocek, U., & Garr, C. (2000). Natural products vs. combinatorial: A case study. In S. K. Wrigley, M. A. Hayes, R. Thomas, E. J. T. Chrystal, & N. Nicholson (Eds.), *Biodiversity: New leads for pharmaceutical and agrochemical industries* (pp. 66–72). The Royal Society of Chemistry.

Bérdy, J. (2012). Thoughts and facts about antibiotics: Where we are now and where we are heading. *The Journal of Antibiotics, 65*, 385–395.

Bloom, D. E., Cafiero, E. T., & Jané-Llopis, E. (2011). *The global economic burden of noncommunicable diseases*. World Economic Forum. http://www.weforum.org/reports/global-economicburden-noncommunicable-diseases

Brewer, S. (2000). The relationship between natural products and synthetic chemistry in the discovery process. In S. K. Wrigley, M. A. Hayes, R. Thomas, E. J. T. Chrystal, & N. Nicholson (Eds.), *Biodiversity: New leads for pharmaceutical and agrochemical industries* (pp. 59–65). The Royal Society of Chemistry.

Chellaram, C., Venkatesh, S., Prem Anand, T., Kuberan, G., Alex John, A., & Priya, G. (2012). Antimicrobial properties of insect gut associated bacteria. *World Journal of Medical Sciences., 7*(4), 260–263.

Concepcion, G. P., Lazaro, J. E., & Hyde, K. D. (2001). Screening for bioactive novel compounds. In S. B. Pointing & K. D. Hyde (Eds.), *Bioexploitation of filamentous fungi* (pp. 93–130). Fungal Diversity Press.

Estruch, J. J., Carozzi, N. B., Desai, N., Duck, N. B., Warren, G. W., & Koziel, M. G. (1997). Transgenic plants: An emerging approach to pest control. *Nature Biotechnology, 15*, 137–141.

Farnsworth, N. R., Akerele, O., Bingel, A. S., Soejarto, D. D., & Guo, Z. (1985). Medicinal plants in therapy. *Bulletin of World Health Organ, 63*, 965–981.

Gelders, S. (2006). Price, availability and affordability. In *An international comparison of chronic disease medicines*. World Health Organization Regional Office for the Eastern Mediterranean.

Kim, S. I., Park, C., Ohh, M. H., Cho, H. C., & Ahn, Y. (2003). Contact and fumigant activities of aromatic plant extracts and essential oils against *Lasioderma serricorne* (Coleoptera: Anobiidae). *Journal of Stored Products Research, 39*, 11–19.

Kohn, R. (2004). The treatment gap in mental health care. *Bulletin of the World Health Organization, 82*(11), 858–866.

Kotwani, A. (2009). Availability, price and affordability of asthma medicines in five Indian states. *International Journal of Tuberculosis and Lung Diseases, 13*(5), 574–579.

Krishnayya, P. V., & Rao, P. J. (1996). Plumbagin: A natural bioactive principle against insects. *The Andhra Agriculture Journal, 43*, 1–5.

Masahiko, I., Prasat, K., Kanyawim, K., Hywel-Jones, N. L., & Yodhathai, T. (2005). Bioactive substances from insect pathogenic fungi. *Accounts of Chemical Research, 38*(10), 813–823.

Meyer, A. C. (2010). Global disparities in the epilepsy treatment gap: A systematic review. *Bulletin of World Health Organ, 88*(4), 260–266.

Nivsarkar, M., Cherian, B., & Padh, H. (2001). Alpha terthienyl: A plant derived new generation insecticide. *Current Science, 81*(6), 667–672.

Park, I., Lee, S., Cho, D., Park, J., & Ahn, Y. (2003). Insecticidal activities of constituents identified from the leaves of *Chamaecyparis obtuse* against *Callosobruchus chinensis* (L.) and *Sitophilus oryzae* (L.). *Journal of Stored Products Research, 39*, 375–384.

Schutz, B. (2001). Endophytic fungi: A source of novel biologically active secondary metabolites. In *British mycological society, international symposium proceedings: Bioactive fungal metabolites—Impact and exploitation* (p. 20). University of Wales.

Simmonds, M. S. J., Blaney, W. M., Ley, S. V., Savona, G., Bruno, M., & Rodriguez, B. (1989). The antifeedant activity of clerodane diterpenoids from Teucrium. *Phytochemistry, 28*, 1069–1070.

Sung-Hwan, E., Young-Mog, K., & Se-Kwon, K. (2013). Marine bacteria: Potential sources for compounds to overcome antibiotic resistance. *Applied Microbiology Biotechnology, 97*, 4763–4773.

Wani, M. C., Taylor, H. L., Wall, M. E., Goggon, I. P., & McPhail, A. T. (1971). Plant antitumour agents. VI. The isolation and structure of Taxol, a novel antileukemic and antitumour agent from *Taxus brevifolia*. *Journal of American Chemistry Society, 93*, 2325–2327.

World Health Organization. (2008). Global disparities in the epilepsy treatment gap: A systematic review. *Bulletin of World Health Organization, 88*(4), 260–266.

World Health Organization. (2014). *Bulletin of World Health Organization, 34*(4), 260–266.

# Chapter 4
# Extraction and Isolation of Secondary Metabolites from Apocynaceae Plants

## Introduction

Secondary metabolites are compounds produced by plants that are not directly involved in growth and development but instead are involved in various ecological functions, such as defense against predators, attraction of pollinators, and communication with other organisms. Plants produce a wide range of secondary metabolites, including alkaloids, terpenoids, flavonoids, phenolic compounds, and glucosinolates. These compounds can have a variety of biological activities, such as antimicrobial, antifungal, antiviral, antioxidant, anti-inflammatory, and cytotoxic effects. Common examples of plant secondary metabolites include alkaloids (caffeine in coffee, nicotine in tobacco, morphine in poppies), terpenoids (menthol in peppermint, carotenoids in carrots, cannabinoids in cannabis), flavonoids (quercetin in apples, anthocyanins in blueberries, catechins in green tea), polyphenols (resveratrol in grapes, curcumin in turmeric, ellagic acid in berries, glucosinolates (sulforaphane in broccoli, isothiocyanates in mustard, indole-3-carbinol in cruciferous vegetables). The isolation of secondary metabolites from plants is a long and laborious process that involves a series of steps that can vary depending on the type of metabolite being isolated and the plant source. However, some general steps in the isolation process include the following:

Extraction: This involves the removal of secondary metabolites from the plant material using a suitable solvent. Different solvents can be used depending on the polarity of the compounds of interest. For example, polar solvents such as methanol or ethanol are often used for flavonoid extraction, while nonpolar solvents such as hexane or chloroform are used for terpenoid extraction.

Filtration: After extraction, the solvent is filtered to remove any solid particles or impurities. Concentration: The extracted solution is then concentrated to reduce the volume of solvent and increase the concentration of secondary metabolites.

Purification: Depending on the complexity of the mixture, further purification steps may be necessary to isolate specific metabolites. This can include techniques

 
R. H. Patil et al., *Apocynaceae Plants*,
https://doi.org/10.1007/978-981-99-5406-3_4

such as column chromatography, thin-layer chromatography, or high-performance liquid chromatography (HPLC).

## Collection of Plant Material

The collection of plant material for secondary metabolite isolation should be done carefully to ensure the purity and quality of the compounds obtained. Plant collections are important because they serve as voucher specimens in taxonomic studies. Additionally, they aid in classifying the family, genus, and species. A herbarium is essentially a collection of botanical specimens that have been organized according to a recognized classification system and are available for use in scientific research and other applications. Plants including herbaceous (nonwoody), roots, bulbs, rhizomes, etc., should be gently dug up, and dirt should be removed with caution. Cut one or two small branches when gathering shrubs and trees. For each specimen, enough material should be collected for identification as well as for preparing the herbarium sheet. Plants that are too large to fit on a single sheet can be separated and then pressed as multiple sheets. Having duplicates is a good idea while collecting. This will enable the researcher to create multiple herbarium specimens, and it will be helpful in the event that one specimen is destroyed or lost. It should be kept in mind that the specimens should be representative of the population and should include the range of variation of the plants. The underground parts, such as roots, bulbs, and rhizomes, should be carefully removed. The specimen must include fruits, flowers, etc., for identification purposes. Standard procedures need to be followed for collecting plant material from healthy plants. Leaves and stem bark of healthy plants should be collected in a sterile polythene bag and brought safely to the laboratory. The plant parts should be adequately numbered or labelled in envelopes or packets with the main specimen. The bushy twigs should be pruned to have a flatter specimen so that it can be preserved properly. Thorny plants may first be placed under a board before pressing to prevent tearing the paper. After collection, the surface of plant materials is cleaned by washing under running tap water and then shade dried before extraction. Moreover, local regulations and laws regarding the collection of plant material should be followed to ensure ethical and sustainable collection practices. Additionally, proper documentation and record-keeping should be maintained to ensure the traceability of the plant material.

## Extraction of Plant Material

Extraction and preparation of plant extracts are the first and most important steps in medicinal plant research. It is important to note that extraction is a critical step in the isolation of secondary metabolites, and optimization of the extraction process can significantly impact the yield and purity of the final product. Plant material collected as per scientific methods is authenticated by an expert taxonomist and then subjected to extraction, fractionation, and isolation of bioactive constituents. Extraction of plant material involves the separation of bioactive secondary metabolites such as flavonoids, alkaloids, terpenoids, saponins, steroids, and glycosides using an appropriate solvent and standard extraction principles.

Several techniques are available for the extraction and fractionation of plant material. It includes cold and hot maceration, infusion, decoction, percolation, digestion and Soxhlet extraction, superficial extraction, ultrasound-assisted extraction, and microwave-assisted extraction. The choice of extraction method depends on the nature and type of the plant material, extraction solvent to be utilized, pH and polarity of the solvent, extraction temperature and solvent-to-sample ratio. The extracted secondary metabolites are further purified using partition or column chromatographic techniques such as thin-layer chromatography (TLC), high-performance liquid chromatography (HPLC), paper chromatography (PC), and gas chromatography (GC).

In brief, the plant material to be extracted, such as the leaves, roots and stem bark, etc., is dried under a shade and ground to obtain a fine powder. Successive extraction of plant material is carried out using Soxhlet extraction (Redfern et al., 2014). The fine powder of leaves and stem bark is extracted by the Soxhlet apparatus (Fig. 4.1) using organic solvents with increasing polarity, such as hexane, ethyl acetate, and methanol.

The selection of the extraction solvent depends on the type of plant material to be extracted, the nature of the phytoconstituents in the plant, and the availability of solvent. Generally, polar solvents such as water, ethanol, ethyl acetate, and methanol are preferred for the extraction of polar constituents, whereas hexane and dichloromethane are used in the extraction of nonpolar compounds (Pandey & Tripathi, 2014). Liquid–liquid extraction involves two miscible solvents: water–dichloromethane, water–ether, and water–hexane. Water is used at varying proportions due to its high polarity and miscibility with other organic solvents. The phytoconstituent of interest is extracted using liquid–liquid extraction and should be soluble in organic solvents but not in water for easier separation. To ease separation (Majekodunmi, 2015). The solvents used for the extraction of phytochemicals are selected based on their polarity as per the polarity and miscibility chart (Table 4.1). For example, *n*-hexane, which is the least polar compared to water. Water is the most polar or universal solvent which is commonly used for the extraction of phytoconstituents. The general strategy for successive extraction is depicted in Fig. 4.1. Initially, plant material is extracted using hexane for 48–72 h. The phytochemicals extracted in hexane are (hexane extract) collected in a separate container, and the residue remaining is further subjected to extraction using ethyl acetate and later on with methanol for 48–72 h. After each extraction step, the solvent of the collected extract is evaporated and concentrated under reduced pressure using a rotary vacuum evaporator (Fig. 4.1) to afford the solid/semisolid residue. The test extracts obtained after evaporation from both plants are properly labelled with different codes, usually with the name of the solvents, and the extracts are stored in a refrigerator at 4 °C until use. Overall, solvent extraction is a popular and

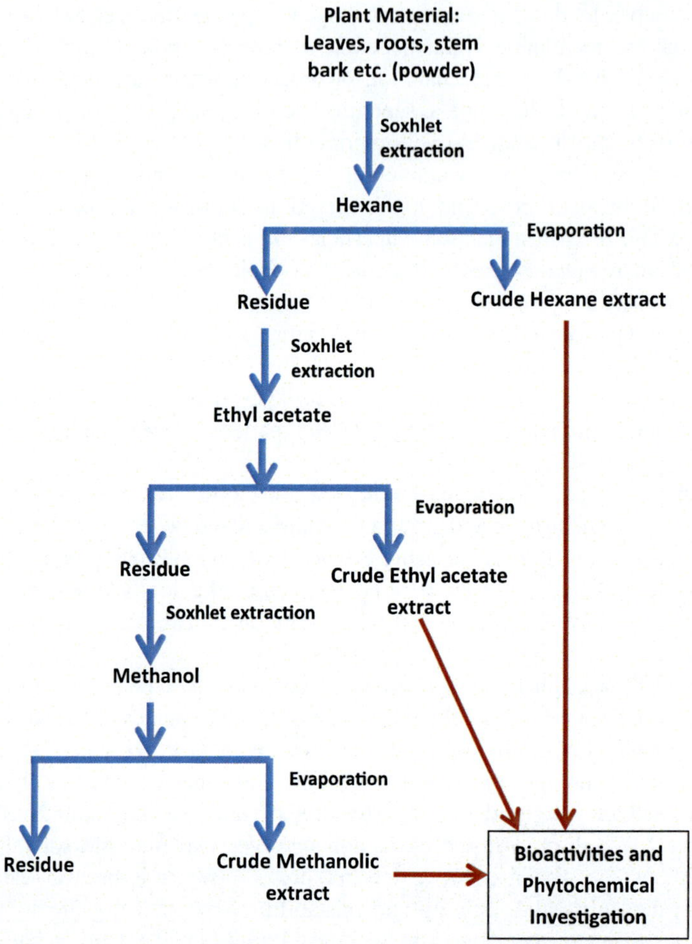


**Fig. 4.1** General strategy for the extraction of phytoconstituents from plant material

versatile method for the extraction of secondary metabolites from plants. However, it is important to consider the potential drawbacks, including safety concerns, environmental impacts, and the need for appropriate quality control measures. The extraction process depends on factors such as the choice of solvent, extraction time and temperature, and the type of plant material used. Optimization of these factors can significantly impact the yield and purity of the extracted compounds. The applications, advantages, and drawbacks of some commonly used solvents for the extraction of phytochemicals are summarized in Table 4.2.

Several procedures are commonly used for the extraction of medicinal materials (Ingle et al., 2017). Many methods are used with modifications as per the requirements of the researchers. Nonetheless, the choice of an appropriate extraction method is very important. This section highlights the principles and applications

**Table 4.1** Solvent polarity chart showing formula, boiling point, melting point and solubility in water

| Solvent | Formula | Boiling point (° C) | Melting point (° C) | Density (g/mL) | Solubility in $H_2O$[1] (g/100 g) | Relative polarity |
|---|---|---|---|---|---|---|
| Cyclohexane | $C_6H_{12}$ | 80.7 | −6.6 | 0.779 | 0.005 | 0.006 |
| Pentane | $C_5H_{12}$ | 36.1 | −129.7 | 0.626 | 0.0039 | 0.009 |
| Hexane | $C_6H_{14}$ | 69 | −95 | 0.655 | 0.0014 | 0.009 |
| Heptane | $C_7H_{16}$ | 98 | −90.6 | 0.684 | 0.0003 | 0.012 |
| Carbon tetrachloride | $CCl_4$ | 76.7 | −22.4 | 1.594 | 0.08 | 0.052 |
| Carbon disulfide | $CS_2$ | 46.3 | −111.6 | 1.263 | 0.2 | 0.065 |
| *p*-xylene | $C_8H_{10}$ | 138.3 | 13.3 | 0.861 | 0.02 | 0.074 |
| Toluene | $C_7H_8$ | 110.6 | −93 | 0.867 | 0.05 | 0.099 |
| Benzene | $C_6H_6$ | 80.1 | 5.5 | 0.879 | 0.18 | 0.111 |
| Ether | $C_4H_{10}O$ | 34.6 | −116.3 | 0.713 | 7.5 | 0.117 |
| Methyl *t*-butyl ether (MTBE) | $C_5H_{12}O$ | 55.2 | −109 | 0.741 | 4.8 | 0.124 |
| Diethylamine | $C_4H_{11}N$ | 56.3 | −48 | 0.706 | M | 0.145 |
| Dioxane | $C_4H_8O_2$ | 101.1 | 11.8 | 1.033 | M | 0.164 |
| N,N-dimethylaniline | $C_8H_{11}N$ | 194.2 | 2.4 | 0.956 | 0.14 | 0.179 |
| Chlorobenzene | $C_6H_5Cl$ | 132 | −45.6 | 1.106 | 0.05 | 0.188 |
| Anisole | $C_7H_8O$ | 153.7 | −37.5 | 0.996 | 0.10 | 0.198 |
| Tetrahydrofuran (THF) | $C_4H_8O$ | 66 | −108.4 | 0.886 | 30 | 0.207 |
| Ethyl acetate | $C_4H_8O_2$ | 77 | −83.6 | 0.894 | 8.7 | 0.228 |
| Ethyl benzoate | $C_9H_{10}O_2$ | 213 | −34.6 | 1.047 | 0.07 | 0.228 |
| Dimethoxyethane (glyme) | $C_4H_{10}O_2$ | 85 | −58 | 0.868 | M | 0.231 |
| Diglyme | $C_6H_{14}O_3$ | 162 | −64 | 0.945 | M | 0.244 |
| Methyl acetate | $C_3H_6O_2$ | 56.9 | −98.1 | 0.933 | 24.4 | 0.253 |
| Chloroform | $CHCl_3$ | 61.2 | −63.5 | 1.498 | 0.8 | 0.259 |
| 3-pentanone | $C_5H_{12}O$ | 101.7 | −39.8 | 0.814 | 3.4 | 0.265 |
| 1,1-Dichloroethane | $C_2H_4Cl_2$ | 57.3 | −97.0 | 1.176 | 0.5 | 0.269 |
| di-n-butyl phthalate | $C_{16}H_{22}O_4$ | 340 | −35 | 1.049 | 0.0011 | 0.272 |
| Cyclohexanone | $C_6H_{10}O$ | 155.6 | −16.4 | 0.948 | 2.3 | 0.281 |
| Pyridine | $C_5H_5N$ | 115.5 | −42 | 0.982 | M | 0.302 |
| Dimethylphthalate | $C_{10}H_{10}O_4$ | 283.8 | 1 | 1.190 | 0.43 | 0.309 |
| Methylene chloride | $CH_2Cl_2$ | 39.8 | −96.7 | 1.326 | 1.32 | 0.309 |
| 1-octanol | $C_8H_{18}O$ | 194.4 | −15 | 0.827 | 0.096 | 0.537 |
| Pentane | $C_5H_{12}$ | 36.1 | −129.7 | 0.626 | 0.004 | 0.009 |
| 1-pentanol | $C_5H_{12}O$ | 138.0 | −78.2 | 0.814 | 2.2 | 0.568 |

(continued)

**Table 4.1** (continued)

| Solvent | Formula | Boiling point (° C) | Melting point (° C) | Density (g/mL) | Solubility in $H_2O^1$ (g/100 g) | Relative polarity |
|---|---|---|---|---|---|---|
| 2-pentanol | $C_5H_{12}O$ | 119.0 | −50 | 0.810 | 4.5 | 0.488 |
| 3-pentanol | $C_5H_{12}O$ | 115.3 | −8 | 0.821 | 5.1 | 0.463 |
| 2-pentanone | $C_5H_{10}O$ | 102.3 | −76.9 | 0.809 | 4.3 | 0.321 |
| 3-pentanone | $C_5H_{12}O$ | 101.7 | −39.8 | 0.814 | 3.4 | 0.265 |
| 1-propanol | $C_3H_8O$ | 97 | −126 | 0.803 | M | 0.617 |
| Pyridine | $C_5H_5N$ | 115.5 | −42 | 0.982 | M | 0.302 |
| Tetrahydrofuran (THF) | $C_4H_8O$ | 66 | −108.4 | 0.886 | 30 | 0.207 |
| Toluene | $C_7H_8$ | 110.6 | −93 | 0.867 | 0.05 | 0.099 |
| *p*-xylene | $C_8H_{10}$ | 138.3 | 13.3 | 0.861 | 0.02 | 0.074 |
| Water | $H_2O$ | 100.00 | 0.00 | 0.998 | M | 1.000 |

*M* miscible. Source: Christian Reichardt, *Solvents and Solvent Effects in Organic Chemistry*, Wiley-VCH Publishers, 3rd ed., 2003

**Table 4.2** Applications, advantages, and drawbacks of some commonly used solvents for the extraction of phytochemicals (Reichardt, 2003)

| Sr. no. | Solvent | Extraction applications | Advantages | Drawbacks |
|---|---|---|---|---|
| 1. | Water | One of the most polar solvents. Used for extraction of a wide range of polar compounds | Highly polar, suitable for a wide range of phytoconstituents, cheap, nontoxic, and nonflammable | Causes hydrolysis and very high heat energy is required to concentrate the extract |
| 2. | Alcohol | Polar in nature, miscible with water, and could extract polar secondary metabolites | Nontoxic at low concentrations, less heat energy is required to concentrate the extract | Cannot be used for fats, gums, and wax; it is flammable and volatile. |
| 3. | Chloroform | A nonpolar used for the extraction of terpenoids, flavonoids, fats, and oils | Colorless, has a sweet smell, absorbed by living tissues | Known for its sedative and carcinogenic properties |
| 4. | Petroleum ether | A nonpolar used for the extraction of alkaloids, terpenoids, coumarins, and fatty acids | Testless, miscible with water, has a low boiling point | Highly volatile and flammable |
| 5. | *Ionic liquid* | Highly polar and extremely heat stable | Highly polar, suitable for microwave-assisted extraction, nonflammable, useful for liquid–liquid extraction | Not ideal for the preparation of tinctures. |

of commonly used extraction methods. A number of factors need to be considered before selecting the extraction method. For example, the heat sensitivity of the plant material, biochemical properties of the solvent, intended use of the drug/metabolite, cost of the product, duration of extraction, etc., must be considered when selecting the extraction method. Commonly used extraction methods, their principles and applications are summarized in Table 4.3.

## Activity-Guided Fractionation for Isolation of Bioactive Secondary Metabolites

The use of medicinal plants in the form of crude extracts involves several challenges. The concentration of bioactive phytoconstituent(s) in a given plant may vary depending on the geographical region, season and time of harvesting, age of plant, etc. Many plant bioactive compounds can be potent poisons if consumed in large amounts. On the other hand, the suboptimal dosage may not show the desired bioactivity if the plant extract has a lower concentration of bioactive compound(s). The medicinal potential of the plant also deteriorates during prolonged and inadequate storage conditions. Therefore, it is necessary to separate and isolate the bioactive compounds in plant extracts. It is possible to increase the efficacy and reduce the adverse effects of crude extracts through structural alteration of extracted and discovered bioactive chemicals from plant extracts. Research on phytochemical aspects in the pharmaceutical industry is concentrated on the development of new medications, novel/indigenous methods for existing medications, and the development of medications derived from plants (Patwardhan et al., 2004). The use of modern technologies has significantly increased the success rate of discovering new leads using plant-derived molecules. Several molecules of plant origin, such as quinine from Cinchona bark, morphine and codeine from the opium poppy, digoxin from *Digitalis purpura* and podophyllotoxin from the rhizome of *Podophyllum,* are in clinical use, and several others are used as lead compounds for the synthesis of new bioactive products (Bhadane et al., 2018). Activity-guided fractionation of plant extracts in combination with chromatographic separation techniques results in better and more effective extraction of biologically active compounds. The development of techniques for the fractionation of plant extracts based on biological activity rather than a specific class of the component is a result of the recent interest in photochemistry-based drug discovery. The majority of naturally occurring bioactive substances are secondary metabolites or species-specific chemical molecules that fall under a variety of categories. The isolated fraction/compound is simultaneously subjected to biological and/or chemical analysis (Phillipson, 2001). The recently discovered natural products such as paclitaxel, camptothecin, etoposide, mevastatin, and artemisinin as well as older ones such as quinine and morphine are the outcome of activity-guided fractionation techniques.

Bioassay-guided fractionation involves the sequential separation of extracted components based on differences in their physicochemical properties and evaluation of their biological activity. It is a widely used strategy to isolate a pure chemical

**Table 4.3** Commonly used extraction methods with their working principle, advantages and drawbacks

| Sr. No. | Extraction method | Brief description of the method | Application/ advantages | Drawbacks | Reference |
|---|---|---|---|---|---|
| 1. | Soxhlet extraction (continuous hot extraction) | Completely dried and finely powdered plant material is placed inside a porous cloth bag or Whatman filter paper. The solvent is poured into the bottom flask which is kept in the heating mantle. As the heating starts, the solvent evaporates and passes through the condenser where it condenses and flows down to the extraction chamber and extracts the drug from the plant material. When the level of solvent in the extraction chamber reaches the top of the siphon, the solvent and the extracted plant material flow back to the flask. The process is repeated for 48–72 h or until the drug is completely extracted. Impurities | Suitable for high-volume extraction. Needs a lesser amount of solvent. The solvent can be reused. No filtration is needed, and a high amount of heat could be applied | Lack of regular shaking and hence extraction takes a longer time. The method is not suitable for thermolabile materials | Ingle et al. (2017), Azwanida (2015), Pandey and Tripathi (2014), Majekodunmi (2015) |
| 2. | Maceration | The coarsely powdered plant material is placed inside a container; the menstruum is poured on top and then closed and kept for at least 3 days. The mixture is stirred periodically. At the | Maceration is suitable for thermo labile plant material. Can be carried out at room temperature. Compounds can be extracted depending on | Consume a large amount of solvent | Doughari (2012) |

(continued)

**Table 4.3** (continued)

| Sr. No. | Extraction method | Brief description of the method | Application/ advantages | Drawbacks | Reference |
|---|---|---|---|---|---|
| | | end of extraction, the micelle is separated by filtration or decantation, and the micelle is then separated from the menstruum by evaporation | the polarity of the solvent | | |
| 3. | Digestion | Digestion uses moderate heat for. The extraction solvent is poured into a clean container and then the powdered drug material is added. The mixture is placed over a water bath/ oven at approximately 50 °C. the heat is applied throughout the extraction process decreases the viscosity of the extraction solvent and separates the secondary metabolites | This method is suitable for soluble plant materials | | Ingale et al. (2017), Pandey and Tripathi (2014), Majekodunmi (2015) |
| 4. | Percolation | It is carried out using a percolator. The coarsely powdered plant material is moistened with the extraction solvent and kept for 4–6 h. the extraction solvent is poured to saturate the plant material. The liquid was allowed to drip slowly from the lower part of the percolator. At the same, the solvent is added continuously as the extraction | Carried out at room temperature, occasionally under heat | Consume a large amount of solvent | Pandey and Tripathi (2014), Majekodunmi (2015) |

(continued)

**Table 4.3** (continued)

| Sr. No. | Extraction method | Brief description of the method | Application/ advantages | Drawbacks | Reference |
|---|---|---|---|---|---|
| | | continues. The extract is separated by filtration or centrifugation | | | |
| 5. | Microwave-assisted extraction | Microwave-assisted extraction works on the principle of dipole rotation and ionic transfer by displacement of charged ions in plant material and solvent. The electromagnetic radiation in frequencies between 300 MHz and 300 GHz and wavelength between 1 cm and 1 m are bombarded onto the extraction material that produces heat which is transferred in the plant material leading to the extraction of phytoconstituents. Used | Uses little quantity of solvent and has less time of extraction as well as the increase in the outcome | It is suitable only for phenolics and flavonoids. Other compounds such as tannins and anthocyanins may be degraded because of high heating | Pandey and Tripathi (2014), Majekodunmi (2015), Altemimi (2017) |
| 6. | Ultrasound-assisted extraction | This method uses sound energy at a very high frequency (greater than 20 KHz) that disrupts the cells and membranes of plant cells increasing the drug surface area for solvent penetration which leads to the release of secondary metabolites. The coarsely powdered plant material is moistened with the extraction | It applies to small samples. Extraction time is less. Uses lesser amounts of solvents and maximizes the yield | Lesser reproducibility. The high amount of energy applied may degrade the phytochemicals | Pandey and Tripathi (2014), Majekodunmi (2015), Azwanida (2015) |

(continued)

**Table 4.3** (continued)

| Sr. No. | Extraction method | Brief description of the method | Application/ advantages | Drawbacks | Reference |
|---|---|---|---|---|---|
| | | solvent in an ultra-sonic extractor. The high-frequency sound waves enhance the extraction process by reducing the heat | | | |
| 7. | Supercritical fluid extraction | It uses supercritical fluids that can dissolve a wide variety of natural products. Supercritical fluids change their solvating properties due to changes in pressure and temperature. Supercritical carbon dioxide (S-$CO_2$) is widely used in supercritical fluid extraction because of its low critical temperature (31 °C), inertness, low cost; nontoxicity as well as it can extract thermolabile secondary metabolites | The low polarity of S-$CO_2$ makes it ideal for the extraction of nonpolar compounds such as lipid and volatile oils | | Zhang et al. (2018) |
| 8. | Enzyme assisted extraction | Enzyme-assisted extraction makes use of enzymes such bas Cellulases, α-amylase, proteases and pectinase. The hydrolytic action of these enzymes on plant cell components such as the cell wall and membranes, natural products inside the cell are released | Higher extraction yield. No use of toxic solvents | Enzymes are costlier as compared to solvents | Zhang et al. (2018) |

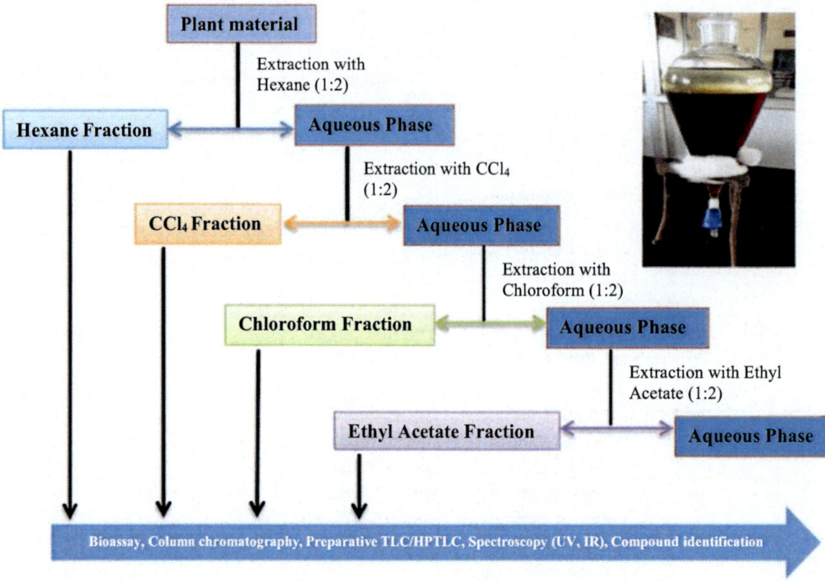


**Fig. 4.2** Liquid–liquid extraction scheme for the isolation of bioactive principles from plant extracts

agent from complex natural sources. The strategy for bioassay-guided fractionation is presented in Fig. 4.2.

The partial purification and separation of phytochemicals are carried out according to the polarity of crude plant material (Eloff, 1998). The crude plant extract is suspended in distilled water in a separating funnel and successively partitioned with hexane, chloroform, ethyl acetate, and water in this order (in increasing polarity). These solvents with increasing polarities will selectively extract different phytoconstituents that are soluble in a particular solvent. It will yield four to five fractions of active plant extract. The resulting fractions are further evaluated for intended biological activities (*in vitro* or *in vivo*). The fractions showing potent activities are used for the isolation of active principles by column chromatography or preparative HPTLC/TLC methods. The isolation of the bioactive phytochemicals from selected fractions of active medicinal plants can be carried out by column chromatography followed by preparative HPTLC/TLC. These separation techniques provide valuable information about phytoconstituents present in the crude extracts and are also invariably handsome techniques for the isolation of desired compounds from the crude plant material based on differences in polarity. Thus, bioassay-guided fractionation and purification techniques are essential for the successful discovery of bioactive natural products from medicinal plants.

## Advantages of Activity-Guided Fractionation

Activity-guided fractionation enables the identification of the specific bioactive compounds responsible for the observed biological activity. This allows for the isolation and purification of these compounds, which can be used for further studies or potential drug development. Activity-guided fractionation allows for the efficient use of resources, including time, effort, and materials. By targeting the most active fractions, the isolation and purification process can be streamlined, reducing the need for extensive separation and purification steps. Plant extracts can contain hundreds or thousands of different compounds, making it difficult to identify the specific compounds responsible for the observed activity. Activity-guided fractionation reduces the complexity of the extract by focusing on the most active fractions, making it easier to identify and isolate the target compounds. It enables the isolation and purification of the bioactive compounds, resulting in higher purity and quality of the final product. Activity-guided fractionation can lead to the discovery of new bioactive compounds with potential therapeutic applications. By identifying the specific compounds responsible for the observed biological activity, new drug candidates can be developed and tested.

In summary, activity-guided fractionation is an effective method for the isolation and purification of bioactive secondary metabolites from plant extracts. It allows for the identification of the specific compounds responsible for the observed activity, leading to higher purity and quality of the final product and the potential for drug discovery.

## References

Altemimi, A., Lakhssassi, N., Baharlouei, A., Watson, D. G., & Lightfoot, D. A. (2017). Phytochemicals: Extraction, isolation, and identification of bioactive compounds from plant extracts. *Plants, 6*, 42.

Azwanida, N. N. (2015). A review on the extraction methods use in medicinal plants, principle, strength, and limitation. *Medicinal & Aromatic Plants, 4*, 196.

Bhadane, B. S., Patil, M. P., Maheshwari, V. L., & Patil, R. H. (2018). Ethnopharmacology, phytochemistry, and biotechnological advances of family Apocynaceae: A review. *Phytotherapy Research, 32*(7), 1181–1210.

Doughari, J. H. (2012). Phytochemicals: Extraction methods, basic structures, and mode of action as potential chemotherapeutic agents, phytochemicals—A global perspective of their role in nutrition and health. In R. Venketeshwer (Ed.), *A global perspective of their role in nutrition and health*. InTech.

Eloff, J. N. (1998). Which extractant should be used for the screening and isolation of antimicrobial components from plants? *Journal of Ethnopharmacology, 60*(1), 1–8.

Ingle, K. P., Deshmukh, A. G., Padole, D. A., Dudhare, M. S., Moharil, M. P., & Khelurkar, V. C. (2017). Phytochemicals: Extraction methods, identification, and detection of bioactive compounds from plant extracts. *Journal of Pharmacognosy and Phytochemistry, 6*, 32–36.

Majekodunmi, S. O. (2015). Review of extraction of medicinal plants for pharmaceutical research. *Merit Research Journal of Medicine, 3*, 521–527.

Pandey, A., & Tripathi, S. (2014). Concept of standardization, extraction, and prephytochemical screening strategies for the herbal drug. *Journal of Pharmacognosy and Phytochemistry, 2*, 115–119.

Patwardhan, B., Vaidya, A. D., & Chorghade, M. (2004). Ayurveda and natural products drug discovery. *Current Science*, 789–799.

Phillipson, J. D. (2001). Phytochemistry and medicinal plants. *Phytochemistry, 56*(3), 237–243.

Redfern, J., Kinninmonth, M., Burdass, D., & Verran, J. (2014). Using soxhlet ethanol extraction to produce and test plant material (essential oils) for their antimicrobial properties. *Journal of Microbiology & Biology Education, 15*(1), 45–46.

Reichardt, C. (2003). *Solvents and solvent effects in organic chemistry* (3rd ed.). Wiley-VCH Publishers.

Zhang, Q. W., Lin, L. G., & Ye, W. C. (2018). Techniques for extraction and isolation of natural products: A comprehensive review. *Chinese Medicine, 13*, 20. https://doi.org/10.1186/s13020-018-0177-x

# Chapter 5
# Phytochemical Screening and Chemical Characterization of the Bioactive Metabolites

## Introduction

Preliminary phytochemical analysis of the crude extract is carried out for the detection of primary as well as secondary metabolites present in the crude extracts. The phytochemicals such as alkaloids, flavonoids glycoside, steroids and triterpenoids, saponins, tannins etc. are detected by various chemical methods. This proximate chemical analysis of the crude extract gives an idea about the presence or absence of certain metabolites in the extract which helps to decide the purification strategy for the compound of interest. In addition to revealing the components of plant extracts and which one predominates over the others, phytochemical screening is effective for finding bioactive compounds that can be used in the production of therapeutic pharmaceuticals.

In general phytochemical screening involves preliminary and specific tests. Preliminary tests are usually performed to determine the presence or absence of major classes of secondary metabolites such as alkaloids, flavonoids, terpenoids, tannins, and phenolic compounds. These tests are often performed using simple colorimetric tests or specific reagents that react with the target compounds. Once the presence of a particular class of compound is confirmed using a preliminary test, specific tests can be performed to identify the individual compounds within that class. These tests can include thin-layer chromatography (TLC), high-performance liquid chromatography (HPLC), gas chromatography (GC), and mass spectrometry (MS). After identifying the compounds present in the extract, quantitative analysis can be performed to determine the amount of each compound. This can be done using various analytical techniques such as UV-visible spectroscopy, HPLC, or GC. In the end, the results of the phytochemical screening can be analyzed to determine the potential medicinal or industrial uses of the plant extract. The compounds identified may have antibacterial, anti-inflammatory, antioxidant, or other biological activities.



R. H. Patil et al., *Apocynaceae Plants*,
https://doi.org/10.1007/978-981-99-5406-3_5

It is important to note that phytochemical screening is just one step in the process of evaluating the biological activity of plant extracts. Further studies, including in vitro and in vivo assays, are often needed to fully understand the potential medicinal or industrial applications of the compounds identified.

## Phytochemical Screening Methods

### *Tests for Proteins*

#### Biuret Test

Three millilitre of aqueous tuber extract is added with a few drops of NaOH (4%) and $CuSO_4$ solution. The appearance of violet or pink color indicates the presence of proteins.

#### Million's Test

Three millilitre of aqueous tuber extract is mixed with 5 mL of Million's reagent. The white precipitate is obtained. After warming the obtained precipitate, the development of a brick-red color indicates the presence of proteins.

#### Ninhydrin Test

In the 3 mL of aqueous tuber extract, three drops of ninhydrin solution (5%) are added. The mixed solution is heated in a boiling water bath for 10 min formation of the purple or bluish color indicates the presence of proteins.

#### Xanthoproteic Test

One millilitre of the extract was taken and placed into a test tube. Then few drops of nitric acid were added and shaken. The emergence of yellow color indicates the presence of protein.

## Test for Carbohydrates

### Molish Test

To 2–3 mL of aqueous tuber extract, a few drops of the alcoholic alpha-naphthol solution is added. The reaction mixture is shaken and the solution is allowed to stabilize. Afterward, the concentrated sulphuric acid from the sides of the test tube is added. The violet ring formed at the junction of two liquids confirms the presence of carbohydrates.

### Fehling Test

One millilitre of Fehling's solution A and 1 mL of Fehling's solution B are mixed and boiled for 1 min. An equal volume of aqueous tuber extract is added to it and heated in a boiling water bath for 5–10 min. The formation of a yellow, then brick-red precipitate indicates the presence of carbohydrates.

### Benedict's Test

One millilitre of aqueous tuber extract is mixed with 1 mL of benedict's reagent. The mixed solution is heated in a boiling water bath for 5 min. The solution appeared green, yellow, or red depending on the amount of reducing sugar present in the test solution indicating the presence of carbohydrates.

### The Tannic Acid Test for Starch

Formation of precipitate after adding a few drops of tannic acid solution (20%) to tuber extract indicated the presence of starch.

## Test for Alkaloids

### Mayer's Test

One millilitre of test extract is treated with 1–2 drops of Mayer's reagent. White or creamy precipitate in the reaction mixture indicates the presence of alkaloids (Kokate, 1994; Khandelwal, 2006).

#### Wagner's Test

One millilitre of test extract is treated with 1–2 drops of Wagner's reagent. Reddish brown precipitate in the reaction mixture indicates the presence of alkaloids.

#### Dragendorff's Test

One millilitre of test extract is treated with 1–2 drops of Dragendroff's reagent. Reddish brown precipitate in the reaction mixture indicates the presence of alkaloids.

#### Hager's Test

One millilitre of test extract is treated with 1–2 drops of Hager's reagent. Yellow precipitate in the reaction mixture indicates the presence of alkaloids.

### *Test for Flavonoids*

#### Shinoda Test

To the crude extract (2 mL), pieces of magnesium ribbons were added and simultaneously concentrated $H_2SO_4$ is added by the side wall of the test tube. Pink scarlet or crimson red color in the reaction mixture indicates the presence of flavonoids.

#### Zinc Hydrochloride Test

To the test extract (2 mL), a pinch of zinc dust is added, and then concentrated $H_2SO_4$ is added by the side wall of the test tube. Formation of red color after few min. Indicates the presence of flavonoids.

### *Test for Glycosides*

#### Borntrager's Test

To the crude extract (2 mL), 3 mL chloroform is added and shaken for a few min. A few drops of 10% ammonia solution were added to the reaction mixture. The pink color in the reaction mixture indicates the presence of glycosides.

### Keller-Killiani Test

Test extract (2 mL) is mixed with an equal volume of glacial acetic acid. One drop of 5% $FeCl_3$ is added to the mixture. A few drops of concentrated $H_2SO_4$ are added to the side wall of the test tube. The bluish-green colour in the reaction mixture indicates the presence of glycosides.

### Legals Test

One millilitre of an extract is taken, and then an equal volume of sodium nitroprusside is added followed by few quantity of sodium hydroxide solution and shaken. The formation of pink-to-blood-red precipitate signifies the existence of cardiac glycoside.

## *Test for Steroids and Triterpenoids*

### Salkowoski Test

Two millilitre extract is mixed with an equal volume of chloroform and treated with a few drops of concentrated $H_2SO_4$ and the mixture is vigorously shaken. The red color ring at the bottom of the test tube indicates the presence of steroids and the yellow colored lower layer indicates the presence of triterpenoids.

### Libermann Burchard's Test

This method is utilized for an alcoholic extract. The extract needs to dry out first through evaporation, then be extracted again with chloroform. Add a few drops of acetic anhydrite followed by sulfuric acid from the side of the test tube. The formation of violet-to-blue-colored rings at the junction of the two liquids indicates the presence of steroids.

## *Test for Saponins*

### Froth Formation Test

The crude extract is mixed with an equal volume of water and shaken vigorously for a few min. and left it to stand. The stable froth (foam) formation indicates the presence of saponins.

## *Test for Tannins*

### Ferric Chloride Test

A few drops of 5% $FeCl_3$ solution are added to the 2 mL test extract. The formation of blue color indicates the presence of hydrolyzable tannins and the formation of green color indicates the presence of condensed tannins.

### Skin Test for Gold Beater

Ox skin is used to make a Gold Beater's Skin. After being submerged in 2% hydrochloric acid, the skin of the Gold Beater is rinsed with distilled water. Following a 5-min soak in an extract solution, it is rinsed with distilled water. It is then dipped into a solution of 1% ferrous sulfate. Tannins are present if the Gold Beater's Skin turned brown or black (Pandey & Tripathi, 2014).

### Gelatin's Test

A test tube is filled with 1 mL of the extract. After adding and shaking 1% gelatin solution containing sodium chloride. The presence of tannins is indicated by the presence of white precipitate (Pandey & Tripathi, 2014).

## *Total Phenolics and Flavonoids*

The total phenolic contents of the plant extracts such as leaves, stem bark, and roots can be determined by spectrophotometric methods as described by various researchers (Singleton & Rossi, 1965). One millilitre of 10% Folin-ciocalteu's reagent is added to 1 mL crude extract (1000 μg/mL) and mixed thoroughly. To the mixture, 4 mL sodium carbonate (7.5%) and 10 mL of distilled water are added and thoroughly mixed. The mixture is allowed to stand for 2 h at room temperature and the absorbance can be read at 760 nm. The tests are performed in three replicates for better accuracy. A standard curve is also obtained using various concentrations of gallic acid (10–100 μg/mL) and the total phenols are expressed as mg of GAE gallic acid equivalent/g of extract. Total flavonoid contents in plant extracts such as leaves, stem bark, and roots can be determined by spectrophotometric methods developed by Christel et al. (2000). The crude extract (1 mL) is mixed with an equal volume of 2% methanolic solution of Aluminum chloride. Quercetin equivalents at different concentrations are employed to prepare a standard curve. All the mixtures are incubated at room temperature for 15 min and absorbance is recorded at 430 nm.

Each test is performed in three replicates. Total flavonoids content can be expressed in mg/mL of quercetin equivalent.

## Preliminary Phytochemical Characters of Different Extracts of Some Members of the Family Apocynaceae (*C. carandas* and *T. divaricate, C. procerra*)

The phytochemical analysis of various extracts extract is carried out for the screening of major phytoconstituents. These methods are widely published with their advantages and drawbacks (Kokate, 1994). The phytochemicals present in the extracts of Apocynaceae plants are widely studied. Apocynaceae members such as *C. carandas, T. divaricata and C. procerra* have been investigated for the presence of phytochemicals like alkaloids, flavonoids glycoside, steroids and triterpenoids, saponins, tannins, etc.

Preliminary phytochemical investigation different extracts of *T. divaricata* revealed that alkaloids and tannins were present in higher concentrations, tannins and glycosides were in moderate amounts and flavonoids were found present in fewer amounts. The ethyl acetate extract of this plant showed the presence of a wider group of phytochemicals in it. Table 5.1 represents the results of the phytochemical

**Table 5.1** Phytochemical analysis of different extracts of *T. divaricata*

| Phytochemicals | Test | *T. divaricata* extracts | | | | |
|---|---|---|---|---|---|---|
| | | Td (L)E | Td (L)M | Td (S)E | Td (S)M | Td Callus |
| Alkaloids | Mayer's test | +++ | +++ | +++ | +++ | ++ |
| | Wagner's test | ++ | ++ | ++ | ++ | + |
| | Dragendorff's test | +++ | +++ | +++ | +++ | ++ |
| | Hager's test | ++ | ++ | ++ | ++ | + |
| Flavonoids | Shinoda test | + | − | − | − | − |
| | Zinc HCl test | + | − | − | − | − |
| Glycosides | Keller-Killiani test | ++ | − | + | + | − |
| | Borntrager's test | + | − | − | − | − |
| Steroids and triterpenoids | Salkowski test | +++ | +++ | +++ | +++ | ++ |
| Tannins | Ferric chloride test | ++ | ++ | ++ | ++ | ++ |
| | Gelatin test | ++ | ++ | ++ | ++ | ++ |
| Saponins | Froth formation test | − | + | − | − | + |

(−) Not detected; (+) present in low concentration; (++) present in moderate concentration; (+++) present in high concentration. **Extract codes:** (Td(L)H *T. divaricata* leaves hexane extract, Td(L)E *T. divaricata* leaves ethyl acetate extract, Td(L)M *T. divaricata* leaves methanol extract, Td(S)H *T. divaricata* stem bark hexane extract, Td(S)E *T. divaricata* stem bark ethyl acetate extract, Td (S)M, *T. divaricata* stem bark methanol extract)

**Table 5.2** Phytochemical analysis of different extracts of *C. carandas*

| Phytochemicals | Test | *C. carandas* extracts | | | | |
|---|---|---|---|---|---|---|
| | | Cc (L)E | Cc (L)M | Cc (S)E | Cc (S)M | Cc Callus |
| Alkaloids | Mayer's test | ++ | +++ | +++ | +++ | ++ |
| | Wagner's test | ++ | ++ | ++ | ++ | + |
| | Dragendorff's test | +++ | +++ | +++ | +++ | ++ |
| | Hager's test | ++ | ++ | ++ | ++ | + |
| Flavonoids | Shinoda test | + | + | + | + | – |
| | Zinc HCl test | + | + | + | + | – |
| Glycosides | Keller-Killiani test | + | – | – | – | – |
| | Borntrager's test | + | – | – | – | – |
| Steroids and triterpenoids | Salkowski test | +++ | +++ | +++ | ++ | ++ |
| Tannins | Ferric chloride test | +++ | +++ | ++ | ++ | ++ |
| | Gelatin test | ++ | ++ | ++ | ++ | ++ |
| Saponins | Froth formation test | – | +++ | – | ++ | + |

(–) Not detected; (+) present in low concentration; (++) present in moderate concentration; (+++) present in high concentration. **Extract codes:** Cc(L)H *C. carandas* leaves hexane extract, Cc(L)E *C. carandas* leaves ethyl acetate extract, Cc(L)M *C. carandas* leaves methanol extract; Cc(S)H *C. carandas* stem bark hexane extract; Cc(S)E *C. carandas* stem bark ethyl acetate extract; Cc(S)M *C. carandas* stem bark methanol extract)

analysis. The cells of *T. divaricata* retain their ability to synthesize phytochemicals in an *in vitro* condition also when cultivated using plant tissue culture techniques (Bhadane et al., 2018b). The total phenolic content (TPC) of Td(L)E extract was found to be 97.67 mg GAE/g of extract. The number of phenolics was comparatively less (64.50 mg GAE/g of extract) in Td(C) extract.

Preliminary phytochemical investigation of *C. carandas* is also widely studied and well documented. The phytochemical screening revealed that alkaloids and steroids were present in higher concentrations in Cc(L)E and Cc(L)M. Tannins were present in moderate amounts and flavonoids as well as glycosides were reported at least concentration. However, saponins were dominantly present in Cc (L)M and Cc(S)M only. The maximum number of phytochemicals is found to be present in ethyl acetate and methanol extract of leaves of this plant. Table 5.2 represents the results of the phytochemical analysis. Callus extract showed a moderate presence of the same phytochemicals except for flavonoids and glycosides in comparison with wild plant extracts. It indicates that cells of *C. carandas* retain their ability to synthesize phytochemicals in an *in vitro* condition also.

The TPC of different extracts of *C. carandas*. The TPC of Cc(L)E was found to be 106.67 mg GAE/g of extract. The amount of phenolics was comparatively less (72.50 mg GAE/g of extract) in Cc(S)M.

*C. procera*, another member of this family is screened for the phytochemical investigation (Patil et al., 2015). The total ash was found to be 27.10% whereas, the water and acid-soluble ash was found to be 2.95% and 1.85%, respectively. Preliminary phytochemical screening of acetone and methanol crude extract revealed the presence of alkaloids, triterpenoids, phenols, flavonoids, tannins and saponins (Mainasara et al., 2012). Phytochemical investigation showed dominant presence of triterpenoids in both the methanol and acetone extract. The methanolic and acetone crude extracts showed the presence of polyphenols at 582.94 and 823.20 μg GAE/g of extracts, respectively (Patil et al., 2015).

## Chromatographic Separation of Compounds from a Mixture or Extracts

The fractionation process carried out using solvents of increasing polarity separates the plant extracts into various fractions. Fractionation techniques are classified into physical or chemical methods. The physical fractionation is carried out using various chromatographic techniques until a pure compound is obtained. Chromatographic techniques are used for separating compounds based on size, shape, and charge from complex biological mixtures. The idea of chromatography involves the use of a stationary phase, such as silica gel or Sephadex combined with calcium sulfate as a binder, and a mobile phase, which is the extraction solvent (Sasidharan et al., 2011). The macromolecules such as amino acids, carbohydrates, fatty acids, lipids, and alkaloids can be separated using silica gel. Proteins and amino acids can be isolated using Sephadex. Aluminum is helpful in the separation of phenols, steroids, alkaloids, vitamins, and carotenes. Separating amino acids, food colors, and alkaloids all employ cellulose powder. Steroids and organic cations can be separated using celite as a matrix. Chromatographic separation of the biomolecules is based on adsorption, partition, affinity, ion exchange, or size exclusion (Sasidharan et al., 2011; Ingle et al., 2017). The chromatographic techniques include paper chromatography (PC), Thin layer chromatography (TLC), column chromatography (CC), liquid chromatography (LC), GC, and HPLC (Smith, 2013). These techniques work on various mechanisms that are responsible for the separation of phytoconstituents (Table 5.3).

## Application of TLC and HPTLC in the Separation of Plant Extracts

TLC uses an adsorption mechanism for the separation of phytoconstituents from crude extracts. Separation is based on the interaction between the compounds in a mixture and stationary phase (usually an inert material. TLC is used for the separation of compounds with low molecular weight (Ingle et al., 2017). The silica gel is dissolved in distilled water to make slurry. The slurry of gel is then poured into a glass/aluminum plate with dimensions 20 cm × 20 cm to produce a uniform layer of

**Table 5.3** Separation mechanisms in different chromatographic techniques

| Sr. no. | Chromatography | Mechanism of separation | Reference |
|---|---|---|---|
| 1. | Partition chromatography | Separation occurs by the addition of two or more immiscible solvents into the extract. The compound partitions itself based on its solubility. The immiscible liquids are separated using a separating funnel to afford the individual compounds | Ingle et al. (2017) |
| 2. | Adsorption chromatography | Separation occurs based on the interaction between compounds to be separated and the stationary phase. It involves noncovalent interactions such as hydrophobic, non-covalent Van der Waals forces. The compound that interacts loosely with the stationary phase gets eluted early | Smith (2013) |
| 3. | Affinity chromatography | It uses a ligand positioned in a separating column as a stationary phase. The mobile phase separates compounds that have low or no affinity for the stationary phase therefore the compounds with high affinity for the stationary phase get attracted and separated | Smith (2013) |
| 4. | Size exclusion chromatography. | Separation takes place based on the molecular size of the compounds. It is also known as molecular sieving or gel filtration A small, medium and larger pores size mesh is used as a stationary phase | Sasidharan et al. (2011) |
| 5. | Ion exchange chromatography | It is used for the separation of polar compounds. Separation occurs based on the electrical charge present in the biomolecules. The charged substances are attracted to each other and get separated | Sasidharan et al. (2011) |

gel having a thickness of 1.5 mm. It is then kept for 1 h at 105 °C to get activated. The TLC analysis of bioactive metabolites from different crude extracts can also be performed on precoated 20 × 20 cm silica gel 60 F254 (0.25 mm thick) thin-layer chromatography plates (Merck, Germany). The TLC plates of desired size are cut and approximately 10–15 μL extracts are spotted on the TLC plates using the capillary tube, needle, or by using an automatic sample applicator (Fig. 5.1). The spots are air-dried and plates developed at room temperature in ascending manner. Chamber saturation time is usually 20 min or it may vary depending on the mobile phase composition. The plate development is carried out in a vertical chromatographic chamber (Fig. 5.1) with suitable dimensions Different combinations of mobile phases can be used for separation. Table 5.4 represents mobile phases used for TLC separation. Wagner and Bladt (1996) described the separation of major phytochemicals like alkaloids, terpenoids, flavonoids, glycosides, and saponins.

After 2/3 development, the plates are removed from the chamber and the mobile phase associated with the plates is evaporated using heat treatment in a hot air oven at 80 °C for 1–2 min or by using the hair drier. The detection of the separated phytoconstituents can be carried out by observing the plats under UV light at 254 nm and 365 nm wavelengths. Derivatization can be carried out using different reagents

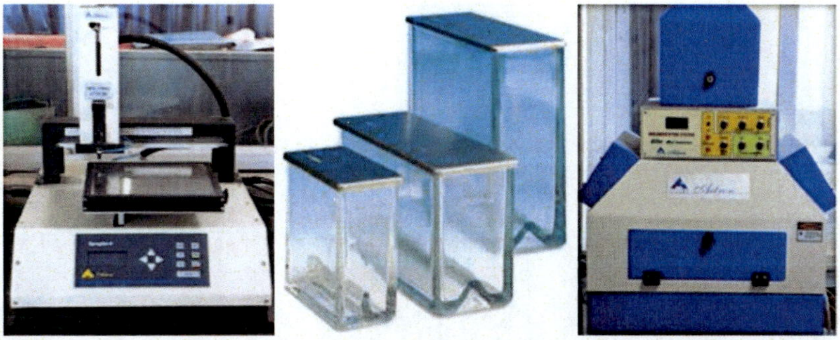

**Fig. 5.1** HPTLC automatic sample applicator, Chromatography chambers and documentation system for TLC visualization and analysis

**Table 5.4** Composition of mobile phases (v/v) used for separation of bioactive compounds of plant extracts

| Sr. no. | Mobile phase | Combinations |
|---|---|---|
| 1. | Chloroform:Methanol:$NH_4OH$:Water | 70:20:2:2 |
| 2. | Ethyl acetate:Methanol:Water | 77:15:8 |
| 3. | Hexane:Ethyl acetate | 60:40, 80:20, 50:50 |
| 4. | Toluene:Methanol | 80:20, 90:10 |
| 5. | Chloroform:Ethyl acetate | 70:30 |
| 6. | Ethyl acetate:Methanol | 100:12 |
| 7. | Toluene:Ethyl acetate | 70:10, 60:20 |
| 8. | Toluene:Ethyl acetate:Diethylamine | 70:20:10 |
| 9. | Dichloromethane:Ethyl acetate | 90:10 |
| 10. | Toluene:Ethyl acetate:Acetic acid | 70:15:5 |
| 11. | Ethyl acetate:Formic acid:Glacial acetic acid:Water | 100:11:11:26 |
| 12. | Chloroform:Glacial acetic acid:Methanol:Water | 65:32:12:8 |
| 13. | Toluene:Chloroform:Ethanol | 40:40:10 |

like 10% ethanolic sulfuric acid or anisaldehyde sulphuric acid to visualize the colored bands. The imaging and $R_f$ (Relative flow) value of separated compounds is determined using an automatic luminous documentation system and software. $R_f$ values of observed compounds were also determined by given formulae.

$$R_f = \frac{\text{Distance traveled by solute}}{\text{Distance traveled by solvent}}$$

The separation of metabolites in crude extract using HPTLC is more or less similar to that of TLC. The application of the extract on the TLC/HPTLC plate is carried out using an automatic sample applicator that applies a specific quantity of sample using a thin glass syringe with the help of pressured nitrogen gas (Fig. 5.2).

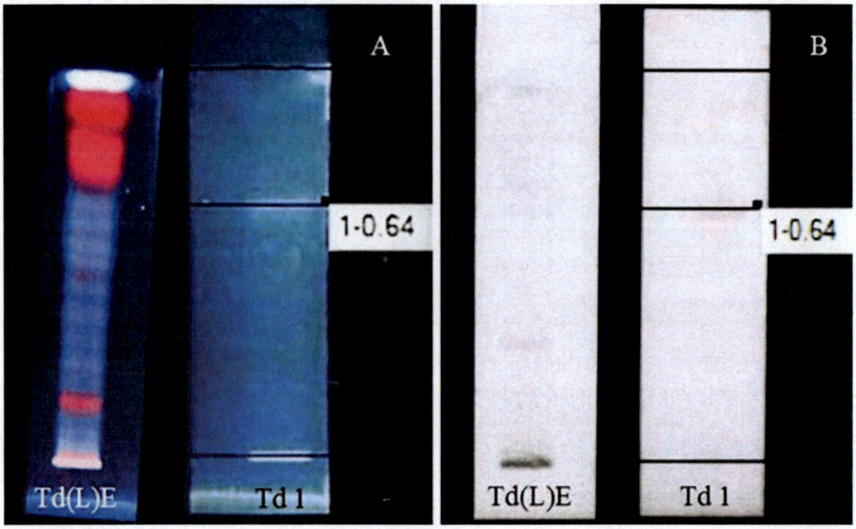


**Fig. 5.2** TLC profile of ethanol extract of *Tabarnamontana divaricate* leaves and the purified fraction showing the $R_f$ 0.64 under UV 365 nm; (**b**) and in visible light (Bhadane et al., 2018a)

## Column Chromatography

Purification of the bioactive principles can also be carried out using column chromatography. The column of varying dimensions is used for his purpose. The glass columns of 30 cm in length and 1.8 cm internal diameter equipped with the sintered disc are normally preferred used for separation (Fig. 5.3). The selection of the mobile phase for separation is generally based on the separation results achieved in TLC/HPTLC. The solvents that gave better separation in TLC are used for the elution of the column. For column preparation, Silica gel of 60–120 mesh size is poured into a clean beaker and mixed with mobile phase (1:1) to make a slurry. Before preparation of the column bed, the empty column is filled with elution solvent. Silica gel slurry is then poured gradually into the column by tapping on the outer wall of the column. For better column packing, the outer side of the column is usually tapped using silicon tubing and fingers. During column packing care should be taken that air bubbles and gaps are not formed in the column. After column packing, the mobile phase is poured on the top of the column bed and eluted several times, and the column is left to wet overnight in the mobile phase for uniform packing and saturation.

The plant extract to be separated is loaded by the dry loading method. Samples are mixed with a small amount of silica 60–120 mesh and mobile phase. The mobile phase is allowed to evaporate in rota vapor to get dry silica powder containing plant extract. The resulting sample containing plant extract is then subjected to column chromatographic separation using mobile phases as an eluent for the respective plant

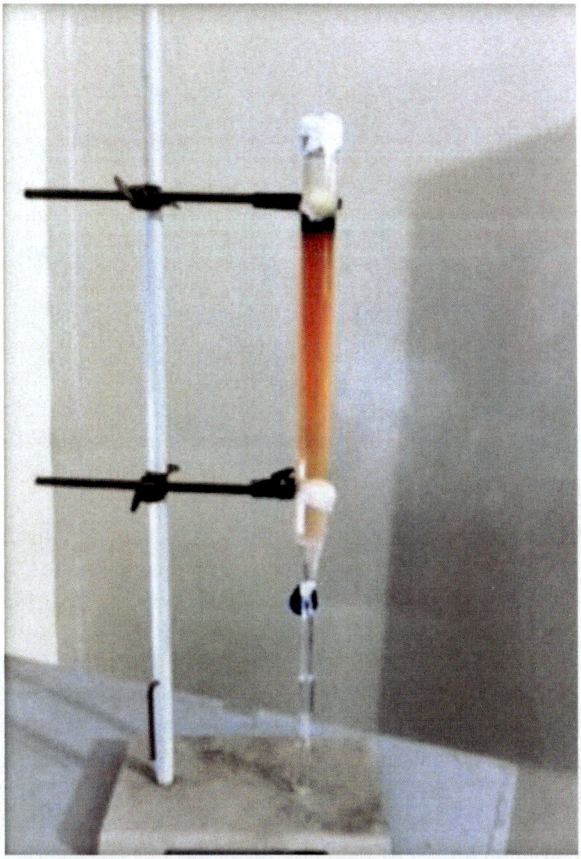

**Fig. 5.3** Silica gel column used for activity-guided separation

extract. The sample is poured carefully on the top of the column bed to get an even layer. Simultaneously, the mobile phase is added using a Pasteur pippet to wet the sample with care so that the pippet does not touch to column bed. After loading the sample, the mobile phase is poured through the side wall of the column and fractions are collected in 5 mL quantity. During elution, the constant flow rate of eluent is maintained for all fractions. Simultaneously, TLC of the eluted fraction can be carried out to check the presence of a single compound. The fractions that show the same pattern on the TLC plate with the same Rf value are mixed and used for further analysis (Patil et al., 2015).

## High-Performance Liquid Chromatography (HPLC)

HPLC is widely used for the isolation and characterization of natural products. It is a flexible, reliable, and preferred method. HPLC is a chromatographic method that can separate a mixture of compounds and is used in phytochemistry to identify, measure, and purify the individual components of the mixture (Cannell, 1998). Being the preferred method for fingerprinting research for the quality monitoring of herbal plants, this methodology is currently widely used among other analytical techniques (Fan et al., 2006). In order to completely describe a natural product's qualities, it is usually separated after a biological assay evaluates a relatively crude extract. Such multi-component samples can be processed quickly on an analytical and preparative scale because of the HPLC's resolution power. Due to its wide range of applications, reversed-phase chromatography is the separation method employed most frequently in HPLC. According to estimates, the reversed-phase mode is used for more than 65% of all HPLC separations. This is due to the ease of use, adaptability, and range of the reversed-phase approach, which can handle molecules with various polarities such as secondary plant metabolites. Several reports describe HPLC-based isolation and characterization of secondary metabolites of Apoocynaceae plants, mainly phenol compounds, terpenoids, saponins, steroids, flavonoids, and alkaloids (Patil et al., 2015).

The HPLC characterization protocols vary for different metabolites. The partially purified extracts or the bioactive fractions eluted from the silica gel column or the HPTLC eluted bands can be subjected to HPLC analysis. HPLC is carried out using a number of chromatography systems of varying configurations of pumps, detectors, columns etc. For example, LC 20 Ad liquid chromatography system (Shimadzu, Japan) equipped with SPD-M20A PDA detector and C18 120 (250 × 4.60) mm column with 5 μm internal diameter. For loading the sample, the partially purified fractions are filtered using 0.45 μ filter and sonicate prior to analysis. 20–30 μL sample is injected and the chromatogram is allowed to develop using various mobile phases in different combinations. Mobile phases commonly used for plant secondary metabolites consist of acetonitrile: water (70:30, 85:15), methanol: water (80:20, 60: 40, 50:50), and 90% acetonitrile: ammonium acetate (100 mM, pH 4.2). The flow rate is in the range of 0.5–1 mL/min. Detection of eluting compounds can be monitored in the range of 280–350 nm (Dubber & Kanfer, 2004; Hwang et al., 2011). A chromatogram (Fig. 5.4) obtained from Shimadzu LC solution software is indicated absorbance intensities and retention time of the eluted compound. Usually, the number of cycles is repeated with the same parameters to check the reproducibility in analysis.

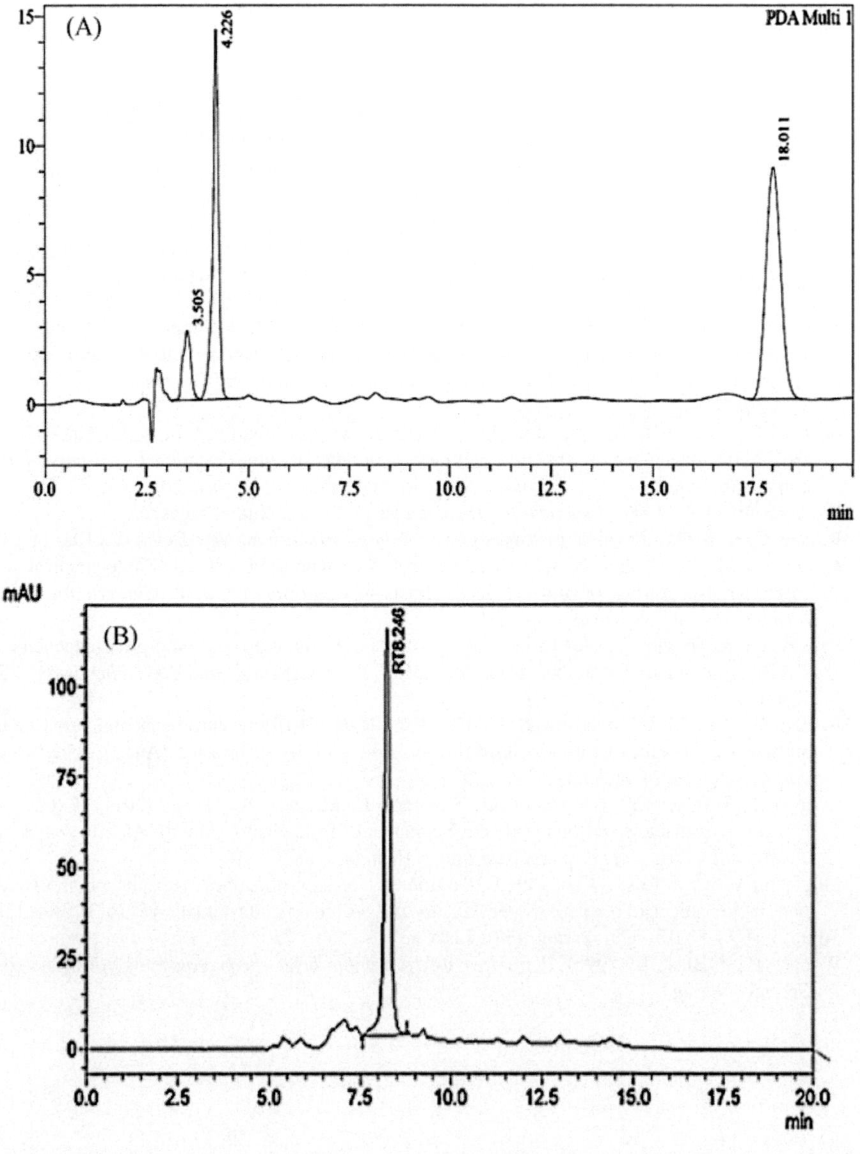


**Fig. 5.4** A typical chromatogram of partially purified extract of leaf of *T. divaricate* (**a**). Chromatogram of pure compound (**b**)

# References

Bhadane, B. S., Patil, M. P., Maheshwari, V. L., & Patil, R. H. (2018a). Ethnopharmacology, phytochemistry, and biotechnological advances of family Apocynaceae: A review. *Phytotherapy Research, 32*(7), 1181–1210.

Bhadane, B., Maheshwari, V., & Patil, R. (2018b). Quercetin and silver nitrate modulates organogenesis in Carissa carandas (L.) by inhibiting auxin transport and ethylene action. *In Vitro Cellular & Developmental Biology-Plants, 54*(6), 600–605.

Cannell, R. J. P. (1998). *Natural products isolation* (pp. 165–208). Human Press.

Christel, Q. D. (2000). Phenolic compounds and antioxidant activities of buckwheat (Fagopyrum esculentum Moench) hulls and flour. *Journal of Ethnopharmacology, 72*, 35–42.

Dubber, M. J., & Kanfer, I. (2004). High-performance liquid chromatographic determination of selected flavonols in Ginkgo biloba solid oral dosage forms. *Journal of Pharmacy & Pharmaceutical Sciences, 7*, 303–309.

Fan, X. H., Cheng, Y. Y., Ye, Z. L., Lin, R. C., & Qian, Z. Z. (2006). Multiple chromatographic fingerprinting and its application to the quality control of herbal medicines. *Analytica Chimica Acta, 555*, 217–224.

Hwang, W. T., Kim, Y. S., Hong, C. K., Lee, J., & Park, D. O. (2011). Simple stability-indicating validated HPLC method for diosgenin in cosmeceutical formulations with long term stability application for light and heat. *International Journal of Pharmacy and Pharmaceutical Sciences, 3*(5), 466–471.

Ingle, K. P., Deshmukh, A. G., Padole, D. A., Dudhare, M. S., Moharil, M. P., & Khelurkar, V. C. (2017). Phytochemicals: Extraction methods, identification, and detection of bioactive compounds from plant extracts. *Journal of Pharmacognosy and Phytochemistry, 6*(1), 32–36.

Khandelwal, K. R. (2006). *Practical pharmacognosy* (12th ed.). Nirali Prakashan.

Kokate, C. K. (1994). Practical pharmacognosy. *Vallabh Prakashan, New Delhi, 4*, 110–111.

Mainasara, M. M., Aliero, B. L., Aliero, A. A., & Yakubu, M. (2012). Phytochemical and antibacterial properties of root and leaf extracts of *Calotropis procera. Nigerian Journal of Basic and Applied Sciences, 20*, 1–6.

Pandey, A., & Tripathi, S. (2014). Concept of standardization, extraction and pre phytochemical screening strategies for herbal drug. *Journal of Pharmacognosy and Phytochemistry, 2*(5), 115–119.

Patil, S. G., Patil, M. P., Maheshwari, V. L., & Patil, R. H. (2015). In vitro lipase inhibitory effect and kinetic properties of di-terpenoid fraction from *Calotropis procera* (Aiton). *Biocatalysis and Agricultural Biotechnology, 4*(4), 579–585.

Sasidharan, S., Chen, Y., Saravanan, D., Sundram, K. M., & Latha, L. Y. (2011). Extraction, isolation and characterization of bioactive compounds from plants' extracts. *African Journal of Traditional, Complementary and Alternative Medicines, 8*(1), 1–10.

Singleton, V. L., & Rossi, J. A. (1965). Colorimetry of total phenolics with phosphomolybdic-phosphotungstic acid reagents. *American Journal of Enology and Viticulture, 16*(3), 144–158.

Smith, I. (Ed.). (2013). *Chromatography*. Elsevier.

Wagner, H., & Bladt, S. (1996). *Plant drug analysis a thin layer chromatography atlas* (2nd ed.). Springer-Verlag.

# Chapter 6
# Identification of Bioactive Secondary Metabolites of Apocynaceae Members

## Introduction

The partially purified extracts or the isolated fractions containing the single compound obtained using chromatographic procedures are subjected to compound identification. Chromatographic techniques can be used for the separation as well as for the identification of the active components in a mixture. Large-scale separation and purification of metabolites can be carried out using these techniques. Several methods were used for the identification of isolated compounds from Apocynaceae members. These methods are based on the detection of functional groups, the presence of multiple bonds and rings, hydrogen and carbon arrangement and full structural elucidation. Modern techniques such as ultraviolet spectroscopy (UV), infrared spectroscopy (IR), nuclear magnetic resonance spectroscopy (NMR), and mass spectrometry (MS) are commonly used for complete identification of the compound.

## UV–Visible Spectroscopy

Spectroscopy deals with the study of the interaction between electromagnetic radiation and matter. UV–vis spectrophotometry is one of the most commonly employed techniques in phytochemical and pharmaceutical analysis. UV–vis visible spectrophotometry measures the amount of UV or visible light absorbed by an absorbing substance in a medium. Ultraviolet–visible spectrophotometers measure the ratio of the intensity of two beams of light in the UV–vis visible region. The principle of ultraviolet–visible spectrophotometers is based on the Beer–Lambert law, which states that when a beam of parallel monochromatic light is passed through an absorbing material, the intensity of light may decrease. The absorbance is proportional to the path length of the cell through the sample and the concentration

 
R. H. Patil et al., *Apocynaceae Plants*,
https://doi.org/10.1007/978-981-99-5406-3_6

**Table 6.1** Wavelength ranges representing specific secondary metabolites

| Sr. no. | Absorption wavelength (nm) | Secondary metabolites |
|---|---|---|
| 1. | 234–676 | Flavonoids, alkaloids, phenolic compounds |
| 2. | 230–285 | Flavonoids and their derivatives |
| 3. | 230–290 | Flavonoids |
| 4. | 300–350 | Derivatives of flavonoids |
| 5. | 350–500 | Tannins, terpenes, saponins |
| 6. | 400–450 | Carotenoids |
| 7. | 400–550 | Terpenoids |
| 8. | 600–700 | Chlorophyll |

of the absorbing material. In UV analysis, a sample is placed between a light source and a photodetector, and the intensity of a beam of light is measured before and after passing through the sample. Each spectrum is background corrected using a "blank"—a cuvette filled with only the dispersing medium to guarantee that spectral features from the solvent are not included in the sample extinction spectrum (Markham, 1982). UV–visible spectroscopy can be performed for qualitative analysis and for the identification of diverse chemical compounds in both pure and biological mixtures. For UV analysis, the crude portion of the partially purified extract is centrifuged at 5000 rpm for 15 minutes to remove debris and impurities. The supernatant liquid is diluted to 1:10 when the crude extract is to be measured. The extract is scanned/spectrum is measured in wavelengths ranging from 200 to 800 nm in a spectrophotometer. The distinctive peaks of the UV–Vis can be seen at multiple wavelengths. Typical values of wavelength ranges representing specific secondary metabolites are shown in Table 6.1.

The UV spectrophotometric analysis of *Tabarnamontana divaricta* revealed maximum absorption at 280 nm when scanning was performed in the range of 200–800 nm. Some additional peaks were also noted in this range (Fig. 6.1).

Flavones and flavonols are closely related in their structures, but they can be differentiated on the basis of their UV–vis visible spectrum. Flavones show two absorption peaks: peak I between 330–350 nm and peak II between 250–270 nm. On the other hand, flavonols show peak I between 350–390 nm and peak II between 250–270 nm. The disadvantage of UV–Vis spectroscopy is that it has limitations in assigning peaks to some constituents present in the extract. Therefore, the results of UV–Vis must be supplemented with other analytical techniques, such as GC–MS or LC–MS, for exact phytochemical profiling of the sample and identification of the compound.

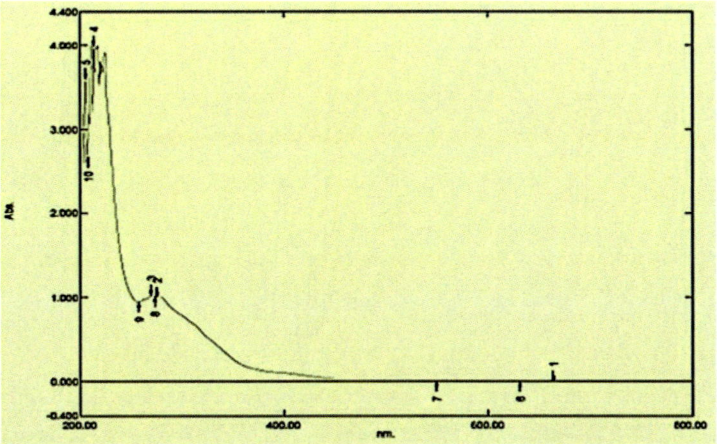


**Fig. 6.1** The UV spectrophotometric analysis of *Tabarnamontana divaricta*

## Fourier Transform Infrared (FT-IR) Spectroscopy

Fourier transform infrared spectroscopy (FT-IR) is a physicochemical analytical technique that provides a clear picture of the functional groups of the active components based on the peaks observed in the infrared region. FT-IR is widely used to elucidate the structure of unknown metabolites and the intensity of absorption spectra based on the molecular composition of the specific chemical functional groups present in the molecule (Bobby et al., 2012). Using FTIR, it has been possible to identify the complicated structures of plant secondary metabolites as well as the characterization of bacterial, fungal and plant species (Hori & Sugiyama, 2003; Yang & Yen, 2002). The dried powdered plant material or the column eluted purified fractions or the callus tissue extract, etc., is encapsulated in a 100-mg potassium bromide (KBr) pellet to prepare the sample for analysis. Usually, a thin film is prepared by applying pressure. The FT-IR scanning of the sample is carried out in the spectral range of 4000–400 $cm^{-1}$ using various models/makes of IR spectroscopes (e.g., Shimadzu 8202 PC, Kyoto, Japan, IFS 25, Bruker, Germany), and the FT-IR spectrum was analysed (Liu et al., 2014; Zohdi et al., 2015).

The FT-IR spectrophotometric analysis of *Tabarnamontana divaricate* leaf extract revealed the presence of common bonds (Fig. 6.2) in the functional group region; stretching vibrations at 3336 $cm^{-1}$ represented the presence of OH alcohol or NH amine, and stretching vibration of $CH_2$ was revealed by two bands at 2916 and 2848 $cm^{-1}$. Presence of C–H stretch vibrations in the fingerprint region at 1390 and 1107 $cm^{-1}$. Another band analysed at 1268–1037 $cm^{-1}$ in the fingerprint region arises from $CH_2$ wagging and twisting vibrations.

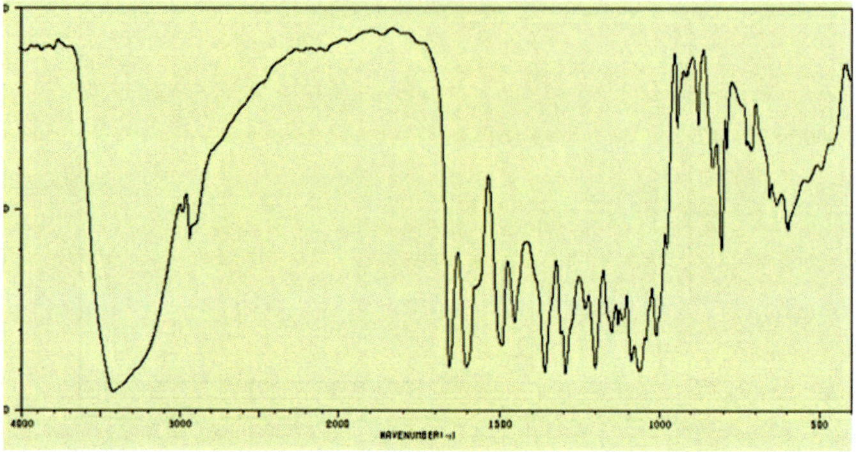

**Fig. 6.2** FT-IR spectrum of partially purified leaf extract of *Tabarnamontana divaricta* (Bhadane, 2017)

## Nuclear Magnetic Resonance (NMR) Spectroscopy

Nuclear magnetic resonance spectroscopy is a nondestructive analytical method for identifying and locating carbon and hydrogen atoms in biomolecules. It studies the radiofrequency of radiation absorption by nuclei in a magnetic field. Nuclear magnetic resonance techniques often fall into one of two categories: $^{1}$H-NMR and $^{13}$C-NMR. In NMR, the interaction between radiofrequency electromagnetic radiation and atom nuclei is studied. Nuclear spin exists in nuclei with odd numbers of protons or neutrons, such as $^{1}$H, $^{13}$C, $^{15}$N, and $^{19}$F, and such nuclei can only be studied by NMR. It pays more attention to the physical properties of the bioactive molecule, such as the number and array of carbon atoms and the presence of isotopes of carbon, hydrogen atoms, and protons. The principle of NMR is based on the fact that all nuclei have electrical charges and therefore exhibit spin. NMR spectroscopy is based on electromagnetic radiation's ability to absorb energy between the radiofrequency range of 3 kHz and 300 GHz. As a result of their electrical charge and spin, atom nuclei produce magnetic fields. Atomic nuclei will align themselves in one of two directions when an external magnetic field is present: either in the direction of the external magnetic field or in the opposite direction. Energy is transferred from the ground state to the excited state when an external magnetic field is present. When an electron transitions from an excited state to the ground state, it occurs at a wavelength that coincides with radio frequencies, and when it does, it generates a radio wave with the same frequency. The NMR spectrum is provided by this radio frequency. The intensity of the applied external magnetic field directly relates to the radiofrequency that is being radiated (Doughari, 2012; Ingale et al., 2017).

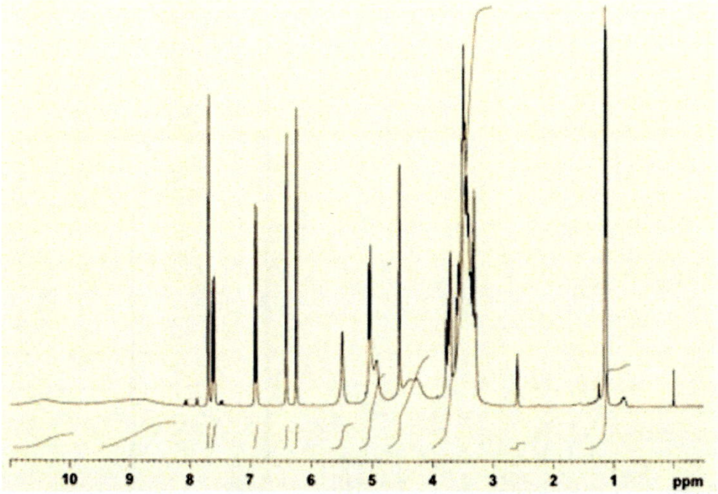


**Fig. 6.3** $^1$H NMR spectrum of a flavonoid isolated from *C. procera*

**Table 6.2** The assigned functional groups to various signals in $^1$H – NMR spectrometry (CDCl3, δ ppm) of the isolated flavonoid from *C. procera*

| δ (ppm) | Protons |
|---|---|
| 1.12, d, 3H | H6$^{iii}$ |
| 3.25–3.74, m, 4H | H5$^{ii}$ and H5$^{iii}$ |
| 4.20–4.60, m, 6H | Sugar - OH |
| 4.80–5.20, m, 8H | H2$^{ii}$ H2$^{iii}$, H3$^{ii}$ H3$^{iii}$, Ar - OH |
| 5.50, s, 2H | H1$^{ii}$ and H1$^{iii}$ |
| 6.41, d, 1H | H8 |
| 6.92, d, 1H | H5$^{i}$ |
| 7.60, d, 1H | H6$^{i}$ |
| 7.70, s, 1H | H2$^{i}$ |

A number of flavonoids have been identified in Apocynaceae plants using NMR spectroscopy. Tatsuzawa et al. (2021) isolated and successfully characterized anthocyanins, flavonols, and chlorogenic acid from the flowers of *Adenium obesum*, *Mandevilla sanderi*, and *Nerium oleander* using nuclear magnetic resonance (NMR) and mass spectrometry (MS). Cyanidin 3-*O*-[6-*O*-(rhamnosyl)-galactoside] and cyanidin 3-*O*-(galactoside) were identified as the major and minor anthocyanins, respectively, in three *A. obesum* seedlings that had red and red–purple flowers (Tatsuzawa et al., 2021).

The $^1$H NMR spectrum (Fig. 6.3) of the flavonoids isolated from *C. procera* showed the presence of two doublets at δ 6.92 and δ 7.60, evidencing a *p*-hydroxylated B ring, and two broad singlets at δ 5.50 and δ 6.24 for the 5.7.9-trioxygenated A ring of a flavonoid (Table 6.2). One doublet at δ 1.12 of a methyl group, one doublet at δ 5.20, and signals at δ 3–4 suggested the presence of a rhmanosyl moiety linked to the aglycone.

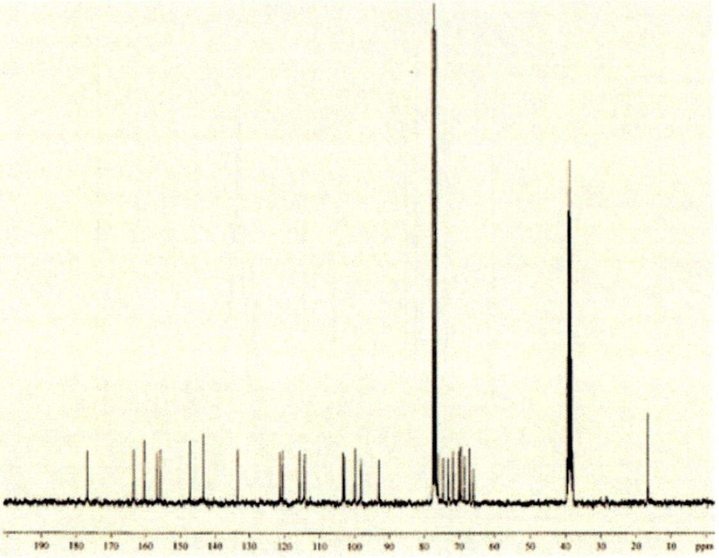


**Fig. 6.4** $^{13}$C NMR spectrum of a flavonoid isolated from *C. procera*

The $^{13}$C NMR spectrum (Fig. 6.4) showed signals at δ 69.51, 70.11, 73.09 and 74.60 for hydroxymethine carbons, one signal at δ 38.80 for one methyl carbon and at δ 102.93 for anomeric carbon, which confirmed the presence of a rhamnosyl unit linked to the aglycone. The $^{13}$C NMR data of the compound were as follows: 176 (C4), 163.4 (C8a), 160.44 (C4a), 156.71, 155.80 (d, C1$^{ii}$, C1$^{iii}$), 147 (C3$^{i}$), 143.45 (C4$^{i}$), 133.35 (C5), 121.27 (d, C5$^{ii}$), 115.83 (C7), 103.35, 102.93 (d, C2$^{ii}$, C2$^{iii}$), 99.91 (C2), 98.20 (C1$^{i}$, C5$^{i}$), 93.02 (C6), 76.56, 77.42 (t, C6$^{ii}$), 74.60, 73.0, (C3$^{ii}$, C3$^{iii}$), 70.11, 69.51 (C4$^{ii}$, C4$^{iii}$), 38.80 (q, C6$^{ii}$).

This spectrum also shows signals at δ 176, 155.80 and 133.35, suggesting the presence of an α,β-unsaturated ketone moiety of a flavonol. Comparison with literature data for flavonol aglycones evidenced a downfield shift for C-2 and C-4 and an upfield shift for C-3 for the compound, suggesting the location of the sugar moiety at C-31. The remaining signals of this spectrum confirmed the presence of and oxygenation pattern of the A and B rings (Salunke, 2006).

## Mass Spectrometry

Mass spectrometry is the method of choice for the identification of compounds based on their chemical structure and molecular mass. It aims to sequence and identify the constituents in the complex mixture. Mass spectrometry is mainly used to identify oligonucleotides and peptides (Doughari, 2012; Ingale et al., 2017). The mass spectrometry reaction involves bombarding an organic molecule with a high-energy

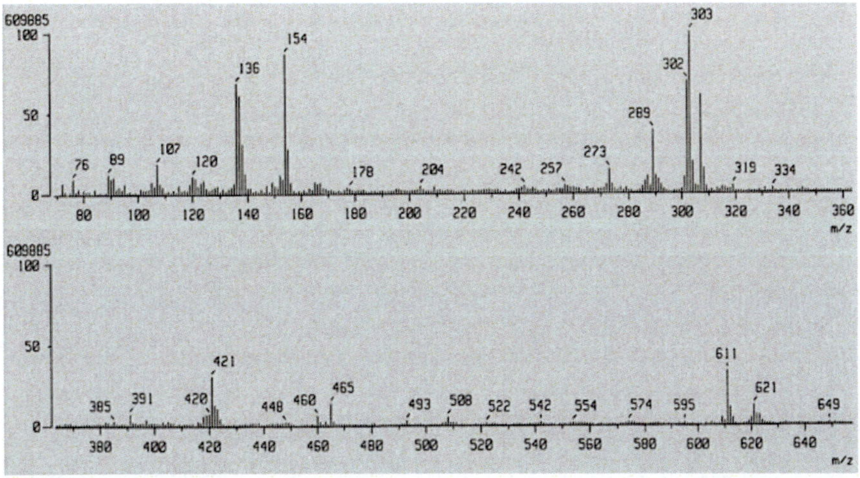


**Fig. 6.5** Mass spectrum of flavonoid (rutin) isolated from *C. procera* (Salunke, 2006)

electron that converts it into high-energy charged ions. The signals are then detected using electron ionization energy of 70 eV, and the spectra of the compound are detected and recorded as percentage peaks. The identification of the unknown constituent is based on their relative molecular mass and molecular weight. It is interpreted by plotting the mass of the fragmented ions against the charges of individual ions (Doughari, 2012; Ingale et al., 2017; Makita et al., 2016). The advantage of mass spectroscopic data is that they provide abundant and meaningful information on organic molecules. Therefore, one of the standard procedures for identifying metabolites from medicinal plants is the combination of HPLC–MS (high-performance liquid chromatography–mass spectrometry) or GC–MS (gas chromatography–mass spectrometry), which are also popular hyphenated techniques.

The fast-atom bombardment mass spectrum of several metabolites of Apocynaceae plants is well documented (Fig. 6.5) (Salunke, 2006).

The bioflavonoid rutin exhibits peaks at (*m*/*z*) 611, 421, 303 (base peak), 302, 289, 273, 257, 178, 154, and 136 as per the fragmentation pattern in MS (Fig. 6.6). Mass spectral data displayed the fragmentation pattern typical of glycosyl derivatives (losing the terminal sugars), with ions at *m*/*z* 465 (M-H-Rha) and m/z 302 $(\text{M-H-Rha-Glu})^-$; this latter ion again corresponds to the $(\text{M-H})^-$ ion of aglycon quercetin or rutin.

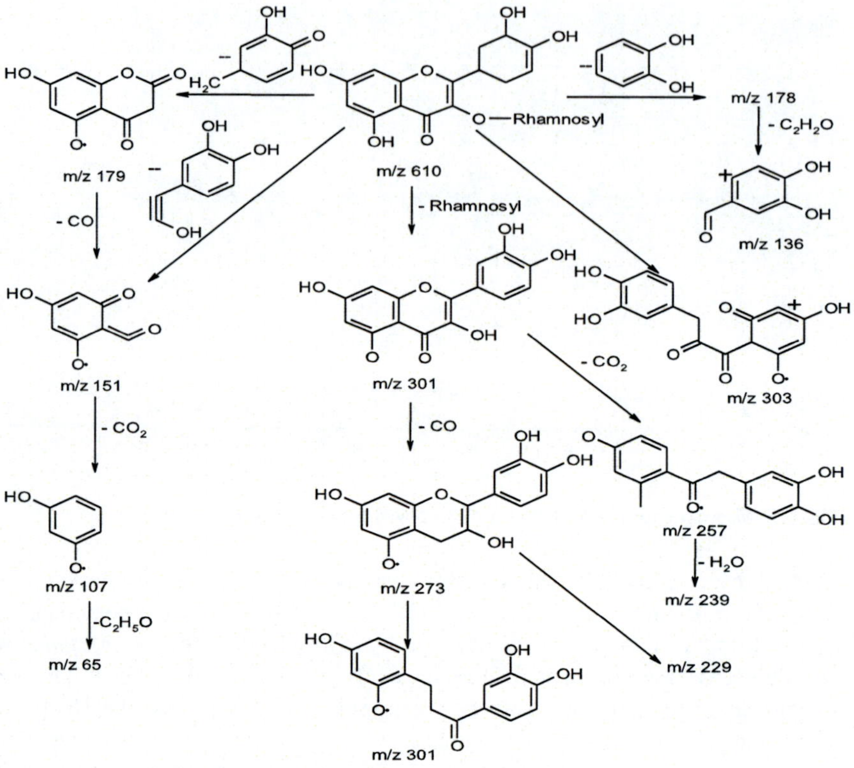


**Fig. 6.6** Mass fragmentation pattern of rutin (Salunke, 2006)

## Hyphenated Techniques in Phytoconstituent Identification

Hirschfield first introduced the term "hyphenation", which refers to the online combination of a separation technique and spectroscopic detection system. The hyphenated techniques involve a combination or coupling of a separation technique and an online modern spectroscopy-based detection technology. In the first step, chromatography produces single or nearly single compound fractions in a mixture. Spectroscopy produces selective information for identification using standards or searchable library spectra (Fig. 6.7). The continuous improvements in hyphenated techniques in the last few years have significantly increased their applications in the discovery of phytoconstituents and natural products. A number of hyphenated techniques, such as GC–MS, HPLC–MS, LC-PDA, LC–MS, LC-FTIR, LC-NMR, LC-NMR-MS, ESI-MS, and DESI-MS, have been widely used for the analysis of crude and partially purified extracts, fractions from plants and fermentation extracts. Particular emphasis is given to the hyphenated techniques that involve gas chromatography as the separation tool.

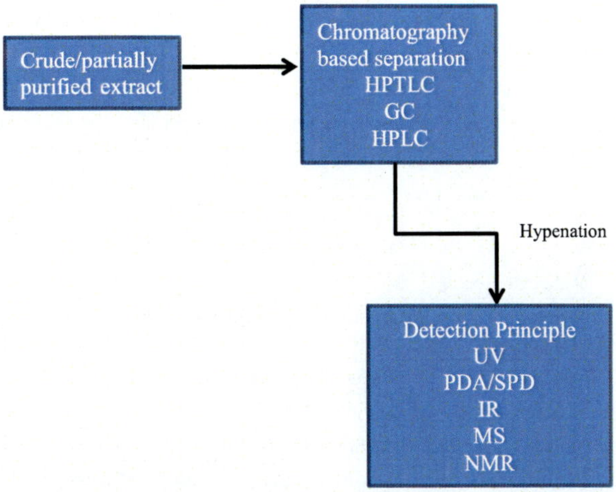


**Fig. 6.7** Block diagram of the hyphenated technique involving the separation and detection system

## Gas Chromatography–Mass Spectrometry (GC–MS)

GC/MS is a robust and very widely used technique that combines high sensitivity and specificity for a diverse group of compounds. GC–MS makes use of single- and triple-quadrupole, ion trap and time-of-flight (TOF) mass spectrometers as a detection system. It was developed by coupling GC and MS and was the first of its kind that became most useful for the analysis of volatile phytochemicals and pharmaceuticals. Mass spectra obtained by GC–MS provide more structural information based on the interpretation of the pattern of fragmentation. The fragment ions generated after ionization has different relative abundances, which can be compared with the standard compounds using the library spectra. Phytochemicals that are volatile, of low molecular weight, and sufficiently stable at high temperatures in GC conditions can be easily analysed by GC–MS. Polar compounds, especially those with a number of hydroxyl groups, are usually derivatized for GC–MS analysis. The common derivatization process involves the conversion of the metabolite to its trimethylsilyl derivative. In GC–MS, a sample is injected into the injection port of the GC device, vaporized, separated in the GC column, and analysed by an MS detector, and the spectra are recorded.

The analysis provides the "retention time''(RT), which is the time elapsed between injection and elution. GC–MS consists of an injection port at one end of a metal column packed with a sand-like material, a stationary phase, to promote maximum separation and a mass detector at the other end of the column. A number of carrier gases, such as argon, helium, nitrogen, hydrogen, etc., are used as mobile phases to propel the samples down the column. The separation of the components occurs based on their affinity to the stationary and mobile phases. The eluted compounds are ionized in a mass spectrometer, and the detector provides information that aids in the structural identification of each component (Tang et al., 2020).

**Table 6.3** Commercially available standard databases for GC–MS analysis (Halket et al., 2005)

| Online database | Class | Spectra | Website |
|---|---|---|---|
| Ehrenstorfer | Pesticides, agrochemicals | 1450 | www.ehrenstorfer.com |
| NIST20 | Organic compounds/general | 123,434 | www.hdscience.com |
| PMW | Toxins, drugs | 6300 | Instrument manufacturers |
| Stan | Pesticides, agrochemicals | 300 | www.chem.agilent.com |
| Wiley | Organic compounds/general | 444,000 | www.hdscience.com |
| Wiley/de Leeuw | Minerals | 1100 | www.hdscience.com |
| Wiley/Hennerberg | Organic compounds/general | 40,000 | www.hdscience.com |
| Wiley/Makin | Steroids, saponins | 2500 | www.hdscience.com |
| Wiley/Roesner | Designer drugs | 1700 | www.hdscience.com |
| Wiley/Yarkov | Organic compounds | 37,055 | www.hdscience.com |
| Wiley/Zeist | Volatile compounds from food and plants | 1620 | www.hdscience.com |

Wedge et al. (2009) described a typical GC-HRMS analysis method for volatile metabolites of *Tabarnamontana divaricata.* The partially purified and column-eluted fractions were separated using an Agilent 7890–5 GC (Santa Clara, CA, USA) fused with an HP-5 silica capillary column (30 m × 0.20 mm, phase thickness 0.25 μm) coupled with a mass spectrometer (Jeol Accu TOF GCV) and FID detector. The inlet temperature was 250 °C. The injection volume was 10 μL. The system was carried out in split mode with a ratio of 20:80, and the flow rate of the carrier gas was maintained at 1 ml per minute. The identification of compounds present in column-eluted fractions as well as callus extracts is carried out using the NIST 05 library search database. A number of integrated online MS databases are available with several thousand reference compounds that are digitally searchable and hence offer spectral matches for the identification of separated components. Table 6.3 provides the details of some commercially available databases for a different group of organic compounds, pesticides, volatiles, etc. (Halket et al., 2005).

## *Practical Approaches for GC-HRMS Characterization of Extracts of* C. Procera

The partially purified extracts of *C. procera* have been fully characterized by GC-HRMS (Patil et al., 2015). The GC-HRMS spectra and the bioactive metabolites present in the purified fraction are represented in Fig. 6.8 and Table 6.4. The spectra revealed six identified compounds in order of their elution from the column: phenol, 2,4-bis (1,1-dimethylethyl) ester; disisobutyl phthalate; 1,2-benzedenedicarboxylic acid, bis (2-methylpropyl) ester; dibutyl phthalate (two peaks) and phthalic acid pentyl 2-pentylester. These findings indicate that the purified fractions of *C. procera*

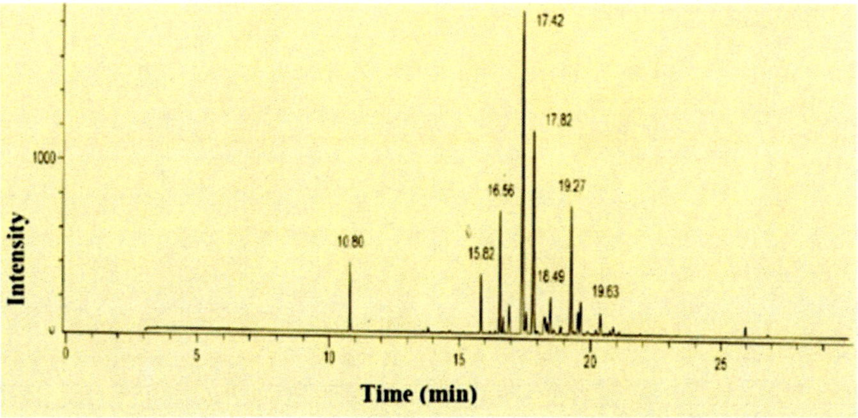


**Fig. 6.8** Gas chromatogram of the purified fraction from *C. procera* (Bhadane, 2017)

**Table 6.4** The GC-HRMS profile of the compounds present in the bioactive fraction of *C. procera*

| Retention time | Molecular weight | Molecular formula | Compound name as per NIST library |
|---|---|---|---|
| 10.80 | 206 | $C_{14}H_{22}O$ | Phenol, 2, 4 –bis (1,1-dimethylethyl) ester. |
| 15.82 | 278 | $C_{16}H_{22}O_4$ | Disisobutyl phthalate |
| 16.56 | 223 | $C_{16}H_{22}O_4$ | 1,2 benzedenedicarboxylic acid, bis (2-methylpropyl) ester |
| 17.42 | 278 | $C_{16}H_{22}O_4$ | Dibutyl phthalate |
| 17.82 | 278 | $C_{16}H_{22}O_4$ | Dibutyl phthalate |
| 19.63 | 306 | $C_{18}H_{26}O_4$ | Phthalic acid pentyl 2- pentyl ester |

contain diterpenes as major phytoconstituents, in agreement with a previous report (Velmurugan et al., 2012).

Other cyclic di-terpenoids, such as 1,2-benzenedicarboxylic acid bismethyl propyl ester ($C_{16}H_{22}O_4$), were identified as major (57.95%) compounds. The other compound identified is pentyl-2-pentyl ester ($C_{18}H_{26}O_4$) with a molecular weight of 306 using GC-HRMS.

## *Practical Approaches for GC-HRMS Characterization of Extracts of C. Carandas*

*Carissa carandas* is another member of the family Apocynaceae that is chemically characterized by GC-HRMS. Bhadane et al. (2018) revealed a major peak with a retention time of 30.97 min and a minor peak with a retention time of 19.54 min (Fig. 6.9). The chromatogram revealed the presence of two compounds, identified as benzenepropanoic acid, 3,5-bis (1,1-dimethylethyl)-4-hydroxy-octadecyl ester

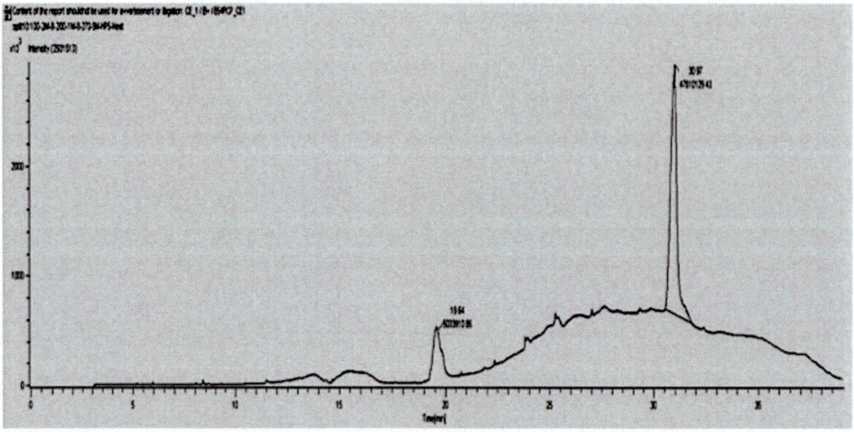

**Fig. 6.9** Gas chromatogram of purified fraction from *Carissa carandas* (Bhadane, 2017)

($C_{35}H_{62}O_3$) with molecular mass 530 and 4-tert-butyl calix [4] arene ($C_{44}H_{56}O_4$) with molecular mass 648.

Among the compounds identified by GC-HRMS analysis, a purified fraction of *Carissa carandas* revealed the presence of benzenepropanoic acid, 3,5-bis (1,1-dimethylethyl)-4-hydroxy-octadecyl ester as a major compound. This compound belongs to a class of saturated fatty acids (Orole, 2016) and was previously reported in *Catharanthus pusillus* (Subbaiyan et al., 2014). More than 30 different compounds have been reported in various parts of *Catharanthus pusillus,* and benzenepropanoic acid ester was reported for its antioxidant activity. A derivative of benzenepropanoic acid ester with hypocholesterolaemic and anti-atherosclerotic effects was isolated and characterized from the methanolic leaf extract of *Amaranthus viridis* (Salvamani et al., 2016). Another compound, 4-tert-butyl calix arene, with a minor peak shown in the GC–MS spectra, was previously reported by Vanitha (2011) in an ethanolic extract of *Annona sqamosa* leaves. Nutan and Baikar (2013) and Devi et al. (2015) reported 4-tert-butyl calix arene in *Psoralea corylifolia* stem extract and in *Zanthoxylum acanthopodium* leaf extract, respectively.

The metabolites present in the callus extract of *C. carandas* (mainly fatty acids and steroids) were identical to metabolites present in the crude extract as well as in the purified fractions. Therefore, it can be concluded that the biological activities shown by the callus extract may be attributed to the phytochemicals present in it.

The profile of phytoconstituents present in *C. carandas* calli with respect to their RT, name, molecular formula, and molecular weight is presented in Table 6.5.

Androstane-11,17-dione, 3[trimethylsilyl 10 oxy]-17-O-phenylmethyl (RT 35.73) was previously reported in *Anacardium occidentale* with antiproliferative and antiandrogenic activity (Fadeyi et al., 2015). Steroidal saponin (RT 35.18) was also previously reported in the extracts of *Calophyllum inophyllum* and *Terminalia arjuna* with antimicrobial, antioxidant and anti-inflammatory activity (Malarvizhi & Ramakrishnan, 2011; Chaudhari & Mahajan, 2015).

**Table 6.5** Phytoconstituents identified by GC-HRMS in *C. carandas* callus extract

| Sr. no. | RT | Compound name | Molecular formula | M. W |
|---|---|---|---|---|
| 1. | 5.18 | 6-acetyl-β-d-mannose | $C_8H_{14}O_2$ | 222 |
| 2 | 7.49 | Dihydrojasmone | $C_{11}H_{18}O$ | 166 |
| 3. | 10.36 | Desulphosinigrin | $C_{10}H_{17}NO_6S$ | 279 |
| 4. | 13.05 | 2-propen-1-ol, 3-(2,6,6-trimethyl-2-cyclohexen-1-yl)- | $C_{12}H_{20}O$ | 180 |
| 5. | 17.43 | 9,12-Octadecadienoyl chloride,(z,z)- | $C_{18}H_{31}ClO$ | 298 |
| 6. | 20.93 | 9,12,15 Octadecatrienoic acid, 2-[(trimethylsilyl)oxy]-1-[(trimethylsilyl)oxy)methyl] ethyl ester, (z,z,z)- | $C_{27}H_{52}O_4Si_2$ | 496 |
| 7. | 23.29 | Benzoxazol, 2,3 dihydro-2-thioxo | $C_{14}H_{16}N_2OS$ | 260 |
| 8. | 23.78 | 10 Undecenoic acid | $C_{12}H_{22}O_2$ | 198 |
| 9. | 35.18 | Spirost-8-en-11-one, 3-hydroxy-, (3β,5α,14β,20β,22β,25R) | $C_{27}H_{40}O_4$ | 428 |
| 10. | 35.73 | Androstane-11,17-dione,3[(trimethylsilyl0oxy]-17-[O9phenylmethyloxime],[3α,5α] | $C_{29}H_{43}NO_3Si$ | 481 |
| 11. | 36.72 | 1-Monolinoleoylglycerol trimethylsilyl ether | $C_{27}H_{54}O_4Si_2$ | 498 |

Glucosinolate-desulfogirin at RT 10.36 min. is reported in *Brassica oleracea* leaf extract for antioxidant activity (Fernandez-Leon et al., 2014). The alkaloid benzoxazole (RT 23.29 min) was isolated from the sea whip *Pseudopterogorgia elisabethae* (Rodriguez et al., 1999) and reported to have antituberculosis and antioxidant activity (Padalkar et al., 2011). Among the fatty acids, 10-undecenoic acid is previously isolated from the seed oil of *Cassia sophera* and reported to have insecticidal, antifungal and antitumour activities (Mostafa et al., 2007). The other fatty acids present in the extract were previously reported in the extracts of *Cynodon dactylon* (Jananie et al., 2011), *Andrographis paniculata* (Kalaivani et al., 2012) and *Cassia italica* (Sermakkani & Thangapandian, 2012).

## References

Bhadane, B. S. (2017). *Studies on biological activities and in vitro propagation of some members of family Apocynaceae*. PhD thesis. KBC North Maharashtra University.

Bhadane, B., Maheshwari, V., & Patil, R. (2018). Quercetin and silver nitrate modulates organogenesis in Carissa carandas (L.) by inhibiting auxin transport and ethylene action. In vitro Cellular and Developmental Biology-Plants.

Bobby, M. N., Wesely, E. G., & Johnson, M. (2012). FT-IR studies on the leaves of *Albizia lebbeck* Benth. *International Journal of Pharmacy and Pharmaceutical Sciences, 4*, 293–296.

Chaudhari, G. M., & Mahajan, R. T. (2015). Comprehensive study on pharmacognostic, physico and phytochemical evaluation of *Terminalia arjuna* Roxb. Stem bark. *Journal of Pharmacognosy and Phytochemistry, 4*(3), 186.

Devi, O. Z., Rao, K. S., Bidalia, A., Wangkheirakpam, R., & Singh, O. M. (2015). GC–MS analysis of phytocomponents and antifungal activities of Zanthoxylum acanthopodium DC. Collected from Manipur, India. *European Journal of Medicinal Plants, 10*(1), 1–9.

Doughari, J. H. (2012). Chapter 1: Phytochemicals: Extraction methods, basic structures and mode of action as potential chemotherapeutic agents, phytochemicals. In V. Rao (Ed.), *A global perspective of their role in nutrition and health. A global perspective of their role in nutrition and health: London* (pp. 1–32). INTECH Open Access Publisher.

Fadeyi, O. E., Olatunji, G. A., & Ogundele, V. A. (2015). Isolation and characterization of the chemical constituents of *Anacardium occidentale* cracked bark. Natural Products Chemistry and Research, 3(5), 1–8.

Fernandez-Leon, A. M., Lozano, M., González, D., Ayuso, M. C., & Fernández-Leon, M. F. (2014). Bioactive compounds content and total antioxidant activity of two savoy cabbages. *Czech Journal of Food Sciences, 32*, 549–554.

Halket, J. M., Waterman, D., Przyborowska, A. M., Patel, R. K., Fraser, P. D., & Bramley, P. M. (2005). Chemical derivatization and mass spectral libraries in metabolic profiling by GC/MS and LC/MS/MS. *Journal of Experimental Botany, 56*(410), 219–243.

Hori, R., & Sugiyama, J. (2003). A combined FTIR microscopy and principal component analysis on softwood cell walls. *Carbohydrate Polymers, 52*, 449–453. https://doi.org/10.1016/S0144-8617(03)00013-4

Ingale, K. P., Deshmukh, A. G., Padole, D. A., Dudhare, M. S., Moharil, M. P., & Khelurkar, V. C. (2017). Phytochemicals: Extraction methods, identification and detection of bioactive compounds from plant extracts. *Journal of Pharmacognosy and Phytochemistry, 6*(1), 32–36.

Jananie, R. K., Priya, V., & Vijayalakshmi, K. (2011). Determination of bioactive components of Cynodon dactylon by GC–MS analysis. *New York Science Journal, 4*, 16–20.

Kalaivani, C. S., Sathish, S. S., Janakiraman, N., & Johnson, M. (2012). GC–MS studies on Andrographis paniculata (Burm. F.) wall. Ex Nees—A medicinally important plant. *International Journal of Medicinal and Aromatic Plants, 2*(1), 69–74.

Liu, Y., Hu, T., Wu, Z., Zeng, G., Huang, D., Shen, Y., He, X., Lai, M., & He, Y. (2014). Study on biodegradation process of lignin by FTIR and DSC. *Environmental Science and Pollution Research, 21*, 14004–14013.

Makita, C., Chimuka, L., Steenkamp, P., Cukrowska, E., & Madala, E. (2016). Comparative analyses of flavonoid content in Moringa oleifera and Moringa ovalifolia with the aid of UHPLC-qTOF-MS fingerprinting. *South African Journal of Botany, 105*, 116–122.

Malarvizhi, P., & Ramakrishnan, N. (2011). GC–MS analysis of biologically active compounds in leaves of Calophyllum inophyllum L. *International Journal of ChemTech Research, 3*(2), 806–809.

Markham, K. R. (1982). *Techniques of flavonoid identification*. Academic.

Mostafa, M., Ahmed, M., Jahan, I. A., & Choudhury, J. U. (2007). Composition of oil from the seeds of Cassia Sophera Linn. *Bangladesh Journal of Scientific and Industrial Research, 42*(1), 75–78.

Nutan, M., & Baikar, S. (2013). *Diversity analysis of Psoralea corylifolia Linn Syn Cullen corylifolia and screening for DNA polymerase I inhibitor activity*. Ph.D Thesis.

Orole, O. O. (2016). GC–MS evaluation, phytochemical and antinutritional screening of Ganoderma lucidum. *Journal of Advances in Biology & Biotechnology, 5*(4), 1–10.

Padalkar, V. S., Borse, B. N., Gupta, V. D., Phatangare, K. R., Patil, V. S., Umape, P. G., & Sekar, N. (2011). Synthesis and antimicrobial activity of novel 2-substituted benzimidazole, benzoxazole and benzothiazole derivatives. *Arabian Journal of Chemistry, 9*, S1125–S1130.

Patil, S. G., Patil, M. P., Maheshwari, V. L., & Patil, R. H. (2015). In vitro lipase inhibitory effect and kinetic properties of di-terpenoid fraction from *Calotropis procera* (Aiton). *Biocatalysis and Agricultural Biotechnology, 4*(4), 579–585.

Rodriguez, A. D., Ramírez, C., Rodríguez, I. I., & González, E. (1999). Novel antimycobacterial benzoxazole alkaloids, from the west Indian Sea whip Pseudopterogorgia elisabethae. *Organic Letters, 1*(3), 527–530.

Salunke, B. K. (2006). *Biomedical basis of pulse beetle management by plant secondary metabolites*. Ph.D Thesis. North Maharashtra University.

Salvamani, S., Gunasekaran, B., Shukor, M. Y., Bakar, M. Z. A., & Ahmad, S. A. (2016). Phytochemical investigation, hypocholesterolemic and anti-atherosclerotic effects of

Amaranthus viridis leaf extract in hypercholesterolemia-induced rabbits. *RSC Advances, 6*(39), 32685–32696.
Sermakkani, M., & Thangapandian, V. (2012). GC–MS analysis of Cassia italica leaf methanol extract. *Asian Journal of Pharmaceutical and Clinical Research, 5*(2), 90–94.
Subbaiyan, B., Samydurai, P., KarthikPrabu, M., & Thangapandian, V. (2014). Gas chromatography and mass spectrum analysis of Catharanthus pusillus (Murray) g. Don (Apocyanaceae). *International Research Journal of Pharmaceutical and Applied Science, 4*(2), 48–52.
Tang, J., Ma, S., Liu, R., Yue, C., Li, G., Yu, Y., Yang, Y., & An, T. (2020). The pollution profiles and human exposure risks of chlorinated and brominated PAHs in indoor dusts from e-waste dismantling workshops: Comparison of GC–MS, GC–MS/MS and GC× GC–MS/MS determination methods. *Journal of Hazardous Materials, 394*, 122573.
Tatsuzawa F, Yoshikoshi A, Takehara A, Suzuki S. Flavonoids from the flowers of Adenium obesum (Forssk.) Roem. & Schult., Mandevilla sanderi (Hemsl.) Woodson, and Nerium oleander L.(Apocynaceae). Biochemical Systematics and Ecology, 2021;99:104347.
Vanitha, V., Umadevi, K. J., & Vijayalakshmi, K. (2011). Determination of bioactive components of *Annona squamosa* L leaf by GC-MS analysis. *International Journal of Pharmaceutical Sciences and Drug Research, 3*(4), 309–312.
Velmurugan, S., Thanga, V., Babu, M., Josephine, M., & Citarasu, T. (2012). Antimicrobial effect of *Calotropisprocera*active principles against aquatic microbial pathogens isolated from shrimp and fishes. *Asian Pacific Journal of Tropical Biomedicine, 2*, S812–S817.
Wedge, D. E., Klun, J. A., Tabanca, N., Demirci, B., Ozek, T., Baser, K. H., Liu, Z., Zhang, S., Cantrell, C. L., & Zhang, J. (2009). Bioactivity-guided fractionation and GC/MS fingerprinting of Angelica sinensis and Angelica archangelica root components for antifungal and mosquito deterrent activity. *Journal of Agricultural and Food Chemistry, 57*(2), 464–470.
Yang, J., & Yen, H. C. E. (2002). Early salt stress effects on the changes in chemical composition in leaves of ice plant and Arabidopsis—A Fourier transform infrared spectroscopy study. *Journal of Plant Physiology, 130*, 1032–1042.
Zohdi, V., Whelan, D. R., Wood, B. R., Pearson, J. T., Bambery, K. R., & Black, M. J. (2015). Importance of tissue preparation methods in FTIR microspectroscopical analysis of biological tissues:'traps for new users'. *PLoS One, 10*(2), e0116491.

# Chapter 7
# Phytochemistry of Apocynaceae Members

## Introduction

Secondary metabolites are substances produced by living cells either in response to environmental stress or as biproducts of metabolism. The cells of plants, animals and microorganisms are known to produce secondary metabolites at varying concentrations (Baker & Sardari, 2021). Apocyanaceae plants are reported to be inexpensive and sources of bioactive secondary metabolites through subsidiary pathways. These metabolites are responsible for mediating environmental interactions and are crucial for survival and reproduction. Plants are known to produce more than 50,000 secondary metabolites of medicinal and therapeutic importance. The discovery of novel secondary metabolites is a continuous process intended to discover novel molecules with better therapeutic value. Plants have evolved strong defences against herbivorous pests and other phytopathogens over the course of evolution. These species-specific metabolites have diverse effects on both plants and other living things. All living things share the basic metabolic building blocks of carbohydrates, lipids, nucleotides, and peptides, which are essential to all life processes. In contrast, secondary metabolites, which are formed from primary metabolites, serve as the plant's defense against pathogens, pests, herbivores, environmental stress, etc.

Apocynaceae is the largest family of the plant kingdom, with over 400 genera and more than 4600 species. It also has five subfamilies: Rauvolfioideae, Apocynoideae, Periplocoideae, Secamonoideae, and Asclepiadoideae. Most of the members of this family are poisonous. Apocynaceae plants are rich in cardiac glycosides, terpenoids, alkaloids and tannins and therefore are widely used for various medicinal applications. A number of plants from this family are cultivated for ornamental purposes. Apocynaceae plants are abundantly distributed in India, Pakistan, China, Bangladesh and Sri Lanka (Mahmood et al., 2011). Members of the Apocynaceae family are trees and shrubs or herbaceous plants with milky latex. The majority of the plants in this family are abundant in alkaloids and have significant therapeutic value. Additionally, many species are planted for decorative purposes. People in

 
R. H. Patil et al., *Apocynaceae Plants*,
https://doi.org/10.1007/978-981-99-5406-3_7

rural regions utilize some Apocynaceae species as food and some of them as poisons (Aiyambo, 2010). Due to the widespread use of Apocynaceae plants for medicinal purposes, members of this family are found in many different traditional medical practices, including Indian, Chinese, and Thai medicine (Van Beek et al., 1984; Bhat et al., 2012). The Indian subcontinent has a dominant presence of Apocynaceae plants. Approximately 84 species or 30 genera are present in India. Our previous report (Bhadane et al., 2018) provided a detailed account of the phytochemistry and pharmacological properties of members from this family. The important medicinal plants from this family include *C. roseus*, *A. scholaris*, *R. serpentina*, *H. antidysentriaca*, *N. indicum* and *T. Peruvian.* The plants and their parts described in this section are used in indigenous systems of medicine for the treatment of various diseases and disorders.

## Phytochemistry

Phytochemistry involves the study of chemicals derived from medicinal plants. Apocynaceae plants have been used for centuries in traditional medicine systems worldwide, and their therapeutic properties and ethnomedicinal applications are often attributed to the presence of various bioactive compounds. Phytochemical analysis involves the identification, isolation, and characterization of these chemical constituents present in plants.

Some common classes of phytochemicals found in Apocynaceae plants include alkaloids, glycosides, tannins, terpene lactones, proteins, steroids, proanthocyanidins, pigments, ecdysones, and saponins. Based on the research carried out in our laboratory and the extensive literature search on ethnomedicinal uses, recent updates on the phytochemistry of members of the family Apocynaceae are presented in the enseuing sections. Table 7.1 summarizes the phytoconstituents isolated from various parts of Apocynaceae plants with necessary details.

### *Alkaloids*

Alkaloids are a diverse group of nitrogen-containing compounds. They often possess strong physiological effects and are commonly found in Apocynaceae plants with medicinal properties. Examples of alkaloids include morphine from opium poppy (Papaver somniferum), caffeine from coffee (Coffea spp.), and quinine from cinchona bark (Cinchona spp.

The common alkaloids isolated from Apocynaceae plants are indole, Iboga and vinca alkaloids (Fig. 7.1). Vincristine, vinblastine and tabernaemontanine **(2–4)** are alkaloids derived from the Madagascar periwinkle plant, *Catharanthus roseus* and *T. divaricate.* Kam et al. (1993) isolated conophylline and conophyllidine ibogan (Conolabine A and B) **(15)** and vallesamine alkaloid derivatives. Irofoline B was

**Table 7.1** List of phytochemicals isolated from selected members of Apocynaceae family (Reproduced with permission: Bhadane, 2017; Bhadane et al., 2018).

| Phytochemical class/ subclass | Sr. no. | Name of compound | PubChem CID | Name of plant species | Source | References |
|---|---|---|---|---|---|---|
| **1. Alkaloids** Indole alkaloid | 1 | Voacangine | 73,255 | *T. divaricata* | Flowers | Arambewela and Ranatunge (1991) |
| | 2 | Voacristine | 196,982 | *T. divaricata* | Leaves, flowers | Arambewela and Ranatunge (1991) |
| | 3 | Vobasine | 5,374,722 | *T. divaricata* | Stem bark | Arambewela and Ranatunge (1991) |
| | 4 | Tabernaemontanine | 45,268,651 | *T. divaricata* | Stem bark | Arambewela and Ranatunge (1991) |
| | 5 | Conophylline | 15,226,696 | *T. divaricata* | Leaves | Kam et al. (1993) |
| | 6 | Conophyllidine | 44,566,831 | *T. divaricata* | Leaves | Kam et al. (1993) |
| Pentacyclic indole alkaloid | 7 | Lirofolines | 46,184,733 | *T. divaricata* | Stem bark | Low et al. (2010) |
| Terpenoidal indole alkaloid | 8 | Vinblastine | 241,903 | *C. roseus* | Whole plant | Van der Heijden et al. (2004) |
| | 9 | Serpentine | 73,391 | *C. roseus* | Leaves, roots | Van der Heijden et al. (2004) |
| | 10 | Vincristine | 5978 | *C. roseus* | Whole plant, leaf | Van der Heijden et al. (2004) |
| | 11 | Ajmalicine | 441,975 | *C. roseus* | Whole plant, leaf, flower, roots | Van der Heijden et al. (2004) |
| | 12 | 3′,4′-anhydrovinblastine | 443,324 | *C. roseus* | Leaves, shoot | Van der Heijden et al. (2004) |
| | 13 | Picrinine | 46,229,104 | *A. scholaris* | Leaves | Xiang-Hai et al. (2008) |
| | 14 | E and Z- alstoscholarine | 16,215,339 and 16,215,340 | *A. scholaris* | Leaves | Xiang-Hai et al. (2008) |
| Ibogan alkaloid | 15 | Conolabine A and B | – | *T. divaricata* | Stem bark | Kam et al. (1993) |
| | 16 | Vallesamine | 13,783,712 | *T. divaricata* | Stem bark | Kam et al. (1993) |
| Vinca alkaloids | 17 | Leurosine | 442,111 | *C. roseus* | Whole plant | Svoboda et al. (1959) |

(continued)

**Table 7.1** (continued)

| Phytochemical class/ subclass | Sr. no. | Name of compound | PubChem CID | Name of plant species | Source | References |
|---|---|---|---|---|---|---|
| | 18 | Vincaleukoblastine | 241,902 | *C. roseus* | Whole plant | Svoboda et al. (1959) |
| Steroidal alkaloid | 19 | Regholarrhenine | 101,529,337 | *H. antidysenterica* | Stem bark | Bhutani et al. (1988) |
| | 20 | Holadysenterine | 16,742,955 | *H. antidysenterica* | Seeds | Kumar et al. (2007) |
| | 21 | Conessine | 441,082 | *H. antidysenterica* | Seeds | Kumar et al. (2007) |
| | 22 | Isoconessimine | 11,772,257 | *H. antidysenterica* | Seeds | Kumar et al. (2007) |
| | 23 | Kurchessine | 442,979 | *H. antidysenterica* | Seeds | Kumar et al. (2007) |
| **2. Flavonoids** | 24 | Rutin | 5,280,805 | *W. tinctoria,* | Fresh leaves | Muruganandam et al. (2000) |
| | | | | *W. tomentosa,* | | |
| | | | | *W. coccinea* | | |
| | | | | *C. carandas* | Fruits | Patil et al. (2012) |
| | 25 | Epicatechin | 72,276 | *C. carandas* | Fruits | Patil et al. (2012) |
| | 26 | Quercetin | 5,280,343 | *C. carandas* | Fruits | Patil et al. (2012) |
| | | | | *A. scholaris* | Leaves | Hui et al. (2009) |
| | | | | *T. divaricata* | Leaves | Jain et al. (2013) |
| | 27 | Kaemferol | 5,483,905 | *C. carandas* | Fruits | Patil et al. (2012) |
| | | | | *A. scholaris* | Leaves | Hui et al. (2009) |
| | 28 | Isorhamnetin | 5,281,654 | *A. scholaris* | Leaves | Hui et al. (2009) |
| | 29 | Quercetin 3–0-beta-D-galactopyranoside | – | *A. scholaris* | Leaves | Hui et al. (2009) |
| | 30 | Kaemferol 3–0-beta-D-galactopyranoside | – | *A. scholaris* | Leaves | Hui et al. (2009) |
| | 31 | Isorhamnetin 3–0-beta-D-Galactopyranoside | – | *A. scholaris* | Leaves | Hui et al. (2009) |
| | 32 | Calycosin | 5,280,448 | *T. divaricata* | Root and stem | Liang et al. (2006) |
| | 33 | Formononetin | 5,280,378 | *T. divaricata* | Root and stem | Liang et al. (2006) |

| | | | | | | |
|---|---|---|---|---|---|---|
| | 34 | Farnisin | 5,378,220 | *T. divaricata* | Root and stem | Liang et al. (2006) |
| | 35 | Ellagic acid | 5,281,855 | *T. divaricata* | Leaves | Jain et al. (2013) |
| | | | | *C. carandas* | Fruits | Patil et al. (2012) |
| | 36 | Hyperoside | 5,281,643 | *T. divaricata* | Leaves | Jain et al. (2013) |
| **3. Terpenes** Monoterpenes | 37 | 3-carene | 26,049 | *C. carandas* | Flowers | Zaki et al. (1981) |
| | 38 | Camphene | 6616 | *C. carandas* | Flowers | Zaki et al. (1981) |
| | 39 | Menthol | 16,666 | *C. carandas* | Flowers | Zaki et al. (1981) |
| | 40 | ρ-Cymene | 7463 | *C. carandas* | Flowers | Zaki et al. (1981) |
| | 41 | α-Terpineol | 17,100 | *C. carandas* | Flowers | Zaki et al. (1981) |
| | 42 | Piperitone | 6987 | *C. carandas* | Flowers | Zaki et al. (1981) |
| | 43 | Neryl acetate | 1,549,025 | *C. carandas* | Flowers | Zaki et al. (1981) |
| | 44 | Geranyl acetate | 1,549,026 | *C. carandas* | Flowers | Zaki et al. (1981) |
| | 45 | β-Ionone | 638,014 | *C. carandas* | Flowers | Zaki et al. (1981) |
| Sesquiterpenes | 46 | Carindone | 101,316,738 | *C. carandas* | Root | Singh and Rastogi (1972) |
| | 47 | (+)-Carrisone | 5,086,419 | *C. carandas* | Root | Singh and Rastogi (1972) |
| | 48 | Nerolidol | 5,284,507 | *C. carandas* | Flowers | Zaki et al. (1981) |
| | 49 | Farnesol | 445,070 | *C. carandas* | Flowers | Zaki et al. (1981) |
| Triterpenes | 50 | Lupeol | 259,846 | *C. carandas* | Roots | Pakrashi et al. (1968) |
| | | | | *H. antidysenterica* | Seeds | Bhattachartya et al. (2009) |
| | 51 | Carrisic acid | – | *C. carandas* | Leaves | Naim et al. (1988) |
| | 52 | Carrisol | – | *C. carandas* | Fruits | Naim et al. (1985) |
| | 53 | Betulinic acid | 64,971 | *C. carandas* | Leaves | Begum et al. (2013) |
| | | | | *N. oleander* | Leaves | Siddiqui et al. (1986), Fu et al. (2005) |
| | | | | *H. antidysenterica* | Seeds | Bhattachartya et al. (2009) |
| | 54 | 16β-hydroxybetulinic acid | – | *C. carandas* | Root | Singh and Uppal (2015) |
| | 55 | Ursolic acid | 64,945 | *C. carandas* | Roots | Pakrashi et al. (1968) |

(continued)

**Table 7.1** (continued)

| Phytochemical class/ subclass | Sr. no. | Name of compound | PubChem CID | Name of plant species | Source | References |
|---|---|---|---|---|---|---|
| | | | | *N. oleander* | Leaves | Siddiqui et al. (1986), Fu et al. (2005) |
| | 56 | α-Amyrin | 73,170 | *C. carandas* | Root | Singh and Uppal (2015) |
| | | | | *T. divaricata* | Root and stem | Liang et al. (2006) |
| | 57 | α-Amyrin acetate | 92,842 | *T. divaricata* | Root and stem | Liang et al. (2006) |
| | 58 | β-Amyrin | 73,145 | *W. tinctoria* | Seed pods | Ramchandra et al. (1993) |
| | 59 | Betulin | 72,326 | *N. oleander* | Leaves | Siddiqui et al. (1986), Fu et al. (2005) |
| | 60 | Oleanderol | – | *N. oleander* | Leaves | Siddiqui et al. (1986) |
| | 61 | Oleandric acid | – | *N. oleander* | Leaves | Fu et al. (2005) |
| | 62 | Oleanerolide | 11,113,483 | *N. oleander* | Leaves | Fu et al. (2005) |
| | 63 | Epoxydammarane 3β, 25-diol | – | *N. oleander* | Leaves | Fu et al. (2005) |
| | 64 | Wrightial | – | *W. tinctoria* | Seed pods | Ramchandra et al. (1993) |
| | 65 | Taraxasterol acetate | 13,889,352 | *T. divaricata* | Root and stem | Liang et al. (2006) |
| | 66 | Alstonic acid A and B | 91,895,416 and 91,895,417 | *A. scholaris* | Leaves | Wang et al. (2009) |
| | 67 | 2,3-Secofernane | – | *A. scholaris* | Leaves | Wang et al. (2009) |
| | 68 | *Cis* and *trans* Karenin | 6,440,661 and 101,921,671 | *N. oleander* | Leaves | Siddiqui et al. (1995) |
| | 69 | 3β,27-dihydroxy-urs-18-en-13,28-olid | – | *N. oleander* | Uncrushed leaves | Begum et al. (1999) |
| | 70 | 3β,22α,28-trihydroxy-25-nor- | – | *N. oleander* | Uncrushed leaves | Begum et al. (1999) |

| | | | | | | |
|---|---|---|---|---|---|---|
| | | Lup-1(10),20(29)-dien-2-one | | | | |
| | 71 | Oleanolic acid | 10,494 | *C. carandas* | Root | Itankar et al. (2011) |
| | | | | *N. oleander* | Leaves | Siddiqui et al. (1986) |
| | 72 | Betulinaldehyde | 99,615 | *H. antidysenterica* | Seeds | Bhattachartya et al. (2009) |
| Isohopane triterpenoid | 73 | Carandinol | 102,202,376 | *C. carandas* | Leaves | Begum et al. (2013) |
| **4. Glycosides** Cardiac glycosides | 74 | Kaneroside | – | *N. oleander* | Leaves | Siddiqui et al. (1986) |
| | 75 | Neriumoside | – | *N. oleander* | Leaves | Siddiqui et al. (1986) |
| | 76 | Oleandrin | 11,541,511 | *N. indicum* | Fresh leaves | Dai et al. (2011) |
| | 77 | Neriantin | 12,313,293 | *N. indicum* | Fresh leaves | Dai et al. (2011) |
| | 78 | Adynerin | 441,840 | *N. indicum* | Fresh leaves | Dai et al. (2011) |
| | 79 | Deacetyloeandrin | 37,401 | *N. indicum* | Fresh leaves | Dai et al. (2011) |
| | 80 | Odoroside-H | 205,840 | *C. carandas* | Roots | Hegde and Joshi (2009) |
| | | | | *N. oleander* | Leaves | Begum et al. (1999) |
| | 81 | Rhamnoside | – | *C. carandas* | Roots | Hegde and Joshi (2009) |
| | 82 | Rhamnoglycoside | – | *C. carandas* | Roots | Hegde and Joshi (2009) |
| | 83 | Evomoside | – | *C. carandas* | Roots | Hegde and Joshi (2009) |
| | 84 | Thevetin | 159,331 | *T. peruviana* | Seeds | Kohls et al. (2012) |
| | 85 | Acetylthevetin | 90,479,481 | *T. peruviana* | Seeds | Kohls et al. (2012) |
| Cardenolides | 86 | Neridiginoside | – | *N. oleander* | Leaves | Begum et al. (1999) |
| | 87 | Nerizoside | 10,816,404 | *N. oleander* | Leaves | Begum et al. (1999) |
| | 88 | Neritaloside | 44,566,654 | *N. oleander* | Leaves | Begum et al. (1999) |
| | 89 | Neriifolin | 441,867 | *T. peruviana.* | Stem bark | Miyagawa et al. (2009) |
| | 90 | Peruvoside | 12,314,120 | *T. peruviana.* | Stem bark | Miyagawa et al. (2009) |
| | 91 | Thevefolin | 44,537,896 | *T. peruviana.* | Stem bark | Miyagawa et al. (2009) |
| | 92 | | – | *T. peruviana.* | Stem bark | Miyagawa et al. (2009) |

(continued)

**Table 7.1** (continued)

| Phytochemical class/ subclass | Sr. no. | Name of compound | PubChem CID | Name of plant species | Source | References |
|---|---|---|---|---|---|---|
| | | (20S)-18,20 epoxydigitoxigenin R-L-thvetoside | | | | |
| Terpenoids glycosides (Sesquiterpene glucoside) | 93 | Carandoside | 44,179,851 | *C. carandas* | Stem | Wangteeraprasert and Likhitwitayawuid (2009) |
| (Anthocynidin glycosides) | 94 | 7-O-Methylcyanidin 3-robinobioside | – | *C. roseus* | Flowers | Toki et al. (2008) |
| | 95 | Rosinidin 3-robinobioside | – | *C. roseus* | Flowers | Toki et al. (2008) |
| (Secoiridoid glycoside) | 96 | Alstonoside | – | *A. scholaris* | Stem | Thomas et al. (2008) |
| Flavonoid glycosides (Isoflavone glycosides) | 97 | Formononetin 7-O-β-D-apiofuranosyl -(1–6)-β-D-glucopyranoside | – | *A. scholaris* | Stem | Thomas et al. (2008 |
| | 98 | Biochanin A 7-O-β-D-apiofuranosy l-(1–6)-β-D-glucopyranoside | – | *A. scholaris* | Stem | Thomas et al. (2008)) |
| (Flavonol glycosides) | 99 | Syringetin-3-O-robinobioside | – | *C. roseus* | Stem | Brun et al. (1999) |
| | 100 | Quercetin-3-*O*-(2,6-di-*O*-rhamnosyl-galactoside) | – | *C. roseus* | Stem, leaves, seed, petals | Ferreres et al. (2008) |
| | 101 | Quercetin –3-*O*-(2,6-di-*O*-rhamnosyl-glucoside) | – | *C. roseus* | Petals | Ferreres et al. (2008) |
| | 102 | Kaempferol-3-*O*-(2,6-di-*O*-rhamnosyl-galactoside) | – | *C. roseus* | Stem, leaves, seed, petals | Ferreres et al. (2008) |

| | | | | | | |
|---|---|---|---|---|---|---|
| | 103 | Kaempferol-3-*O*-(2,6-di-*O*-rhamnosyl-glucoside) | – | *C. roseus* | Stem, seed, petals | Ferreres et al. (2008) |
| | 104 | Isorhamnetin-3-*O*-(2,6-di-*O*-rhamnosylgalactoside) | – | *C. roseus* | Stem, leaves | Ferreres et al. (2008) |
| | 105 | Isorhamnetin-3-*O*-(2,6-di-*O*-rhamnosyl-glucoside) | – | *C. roseus* | Seeds | Ferreres et al. (2008) |
| | 106 | Quercetin 3-O-{β-D-glucopyranosyl-(1–2)-[α-L-rhamnopyranosyl-(1–6)]-β-D-galactopyranoside} | – | *T. peruviana* | Leaves | Tewtrakul et al. (2002) |
| (flavanone glycosides) | 107 | (2S)-5-O-β-D-glucopyranosyl-7,4′-dihydroxy-3′,5′-dimethoxyflavanone | – | *T. peruviana* | Leaves | Tewtrakul et al. (2002) |
| | 108 | (2R)-5-O-β-D-glucopyranosyl-7,4′-dihydroxy-3′,5′-dimethoxyflavanone | – | *T. peruviana* | Leaves | Tewtrakul et al. (2002) |
| **5. Sterols** | 109 | Stigmasterol | 5,280,794 | *A. scholaris* | Root, root bark | Adotey et al. (2012) |
| | | | | *W. tinctoria* | Stem | Ramchandra et al. (1993), Jain and Bari (2010) |
| | | | | *C. carandas* | Roots | Galipalli et al. (2015) |
| | 110 | β-Sitosterol | 222,284 | *A. scholaris* | Stem | Adotey et al. (2012) |
| | | | | *W. tinctoria* | Stem | Ramchandra et al. (1993) |
| | | | | *C. carandas* | Roots | Pakrashi et al. (1968), Bhaskar and Balakrishnan (2015) |

(continued)

**Table 7.1** (continued)

| Phytochemical class/ subclass | Sr. no. | Name of compound | PubChem CID | Name of plant species | Source | References |
|---|---|---|---|---|---|---|
| | 111 | Campesterol | 173,183 | *A. scholaris* | Root, root bark | Adotey et al. (2012) |
| | | | | *W. tinctoria* | Stem | Ramchandra et al. (1993), Jain and Bari (2010) |
| | 112 | 14α-methylzymosterol | – | *W. tinctoria* | Seeds | Akihisa et al. (1988) |
| | 113 | Desmosterol | 439,577 | *W. tinctoria* | Seeds | Akihisa et al. (1988) |
| | 114 | Clerosterol | 5,283,638 | *W. tinctoria* | Seeds | Akihisa et al. (1988) |
| | 115 | 24-methylene-25-methylcholesterol | 5,283,645 | *W. tinctoria* | Seeds | Akihisa et al. (1988) |
| | 116 | 24-dehydropollinastanol | – | *W. tinctoria* | Seeds | Akihisa et al. (1988) |
| | 117 | 7-dihydositosterol | 5,283,634 | *R. serpentine* | Roots | Karmakar and Chakraborty (1983) |
| | 118 | Sitosterol glucoside | 70,699,351 | *C. carandas* | Roots | Hegde et al. (2012) |
| | 119 | Cholest-5-en-3- β-ol | – | *C. carandas* | Roots | Balakrishnan et al. (2011) |
| | 120 | 20-Hydoxypregnan-18-oic acid | – | *C. carandas* | Leaves | Bhadane and Patil (2017) |
| | 121 | Sitosta-5,23-dien-3β-ol | – | *H. antidysenterica* | Stem bark | Usmani (1995) |
| **6. Simple phenolic compounds** | 122 | Syringic acid | 10,742 | *C. carandas* | Fruits | Patil et al. (2012) |
| | 123 | Vanillic acid | 8468 | *C. carandas* | Fruits | Patil et al. (2012) |
| | | | | *C. roseus* | Leaves | Proestos et al. (2005) |
| | 124 | ρ-Coumaric acid | 637,542 | *C. carandas* | Fruits | Patil et al. (2012) |
| | 125 | Caffeic acid | 689,043 | *C. carandas* | Fruits | Patil et al. (2012) |
| | 126 | Chlorogenic acid | 1,794,427 | *C. carandas* | Fruits | Patil et al. (2012) |
| | | | | *C. roseus* | Leaves | Proestos et al. (2005) |
| | 127 | Scopoletin | 5,280,460 | *C. carandas* | Root | Galipalli et al. (2015) |

| | | | | | | |
|---|---|---|---|---|---|---|
| | 128 | Gallic acid | 370 | *C. roseus* | Leaves | Proestos et al. (2005) |
| | 129 | Hydroxytyrosol | 82,755 | *C. roseus* | Leaves | Proestos et al. (2005) |
| | 130 | Ferulic acid | 445,858 | *C. roseus* | Leaves | Choi et al. (2004) |
| | 131 | 3-O-Caffeoylquinic acid (3-CQA) | – | *C. roseus* | Leaves | Ferreres et al. (2011), Wong et al. (2014) |
| | | | | *N. oleander* | Leaves | Wong et al. (2014) |
| | 132 | 4-O-Caffeoylquinic acid (4-CQA) | – | *C. roseus* | Leaves | Ferreres et al. (2011), Wong et al. (2014) |
| | | | | *N. oleander* | Leaves | Wong et al. (2014) |
| | 133 | 5-O-Caffeoylquinic acid (5-CQA) | – | *C. roseus* | Leaves | Ferreres et al. (2011), Wong et al. (2014) |
| | | | | *N. oleander* | Leaves | Wong et al. (2014) |
| **7. Lignans** | 134 | Carinol | – | *C. carandas* | Roots | Pal et al. (1975) |
| | 135 | (6R,7S,8S)-7a-[(b-glucopyranosyl) oxy] lyoniresinol | – | *C. carandas* | Stem | Wangteeraprasert and Likhitwitayawuid (2009) |
| | 136 | Carissanol | 340,931 | *C. carandas* | Stem | Wangteeraprasert and Likhitwitayawuid (2009) |
| | 137 | (−)-nortrachelogenin | 394,846 | *C. carandas* | Stem | Wangteeraprasert and Likhitwitayawuid (2009) |
| | 138 | Syringaresinol 4-O-b-glucopyranoside | | *V. major* | Leaves | Sohretoglu et al. (2013) |
| **8. Others** | 139 | Eicosanoic acid | 10,467 | *C. carandas* | Seeds | Shrivastava and Bakodia (1979) |
| | | | | *T. divaricata* | Flower | Ali Khan et al. (2013) |
| | 140 | Hexadecanoic acid, 15-methyl ester | 985 | *C. carandas* | Seeds | Shrivastava and Bakodia (1979) |
| | | | | *W. tinctoria* | Flower | Ramalakshmi et al. (2012) |
| | | | | *C. roseus* | Leaves | Brun et al. (2001) |

(continued)

**Table 7.1** (continued)

| Phytochemical class/ subclass | Sr. no. | Name of compound | PubChem CID | Name of plant species | Source | References |
|---|---|---|---|---|---|---|
| | | | | *T. divaricata* | Flower | Ali Khan et al. (2013) |
| | | | | *H. antidysenterica* | Flower | Brun et al. (2001) |
| | 141 | Octadecanoic acid, 9Z,12Z-octadecadienoic acid | – | *C. carandas* | Seeds | Shrivastava and Bakodia (1979) |
| | | | | *T. divaricata* | Flower | Ali Khan et al. (2013) |
| | 142 | 9Z-octadecenoic acid | 445,639 | *C. carandas* | Seeds | Shrivastava and Bakodia (1979) |
| | 143 | Pentadecanoic acid ester | 38,762 | *W. tinctoria* | Flower | Ramalakshmi et al. (2012) |
| | | | | *C. roseus* | Leaves | Brun et al. (2001) |
| | 144 | Caproic acid | 8892 | *C. roseus* | Leaves | Brun et al. (2001) |
| | 145 | Capric acid | – | *C. roseus* | Leaves | Brun et al. (2001) |
| | 146 | Caprylic acid | 379 | *C. roseus* | Leaves | Brun et al. (2001) |
| | 147 | Nonanoic acid | 8158 | *C. roseus* | Leaves | Brun et al. (2001) |
| | 148 | Tridecanoic acid | 12,530 | *C. roseus* | Leaves | Brun et al. (2001) |
| | 149 | Arabinogalactun | 24,847,856 | *N. oleander* | Leaves | Dong and Fang (2001) |
| | 150 | L-ascorbic acid | 54,670,067 | *C. carandas* | Fruits | Zaki et al. (1981), Patil et al. (2012) |
| | 151 | Dihydrojasmone | 62,378 | *C. carandas* | Flowers | Rastogi et al. (1966) |
| | 152 | 6-Decaprenylphenol | – | *C. carandas* | Roots | Zaki et al. (1981), Patil et al. (2012) |
| | 153 | 4-Amino-1-(4-amino-2-oxo-1(2H) -pyrimidinyl)-1,4-dideoxy -b-D-glucopyranuronic acid | – | *C. carandas* | Roots | Zaki et al. (1981), Patil et al. (2012) |
| | 154 | Vitamin E | 14,985 | *T. divaricata* | Flower | Ali Khan et al. (2013) |

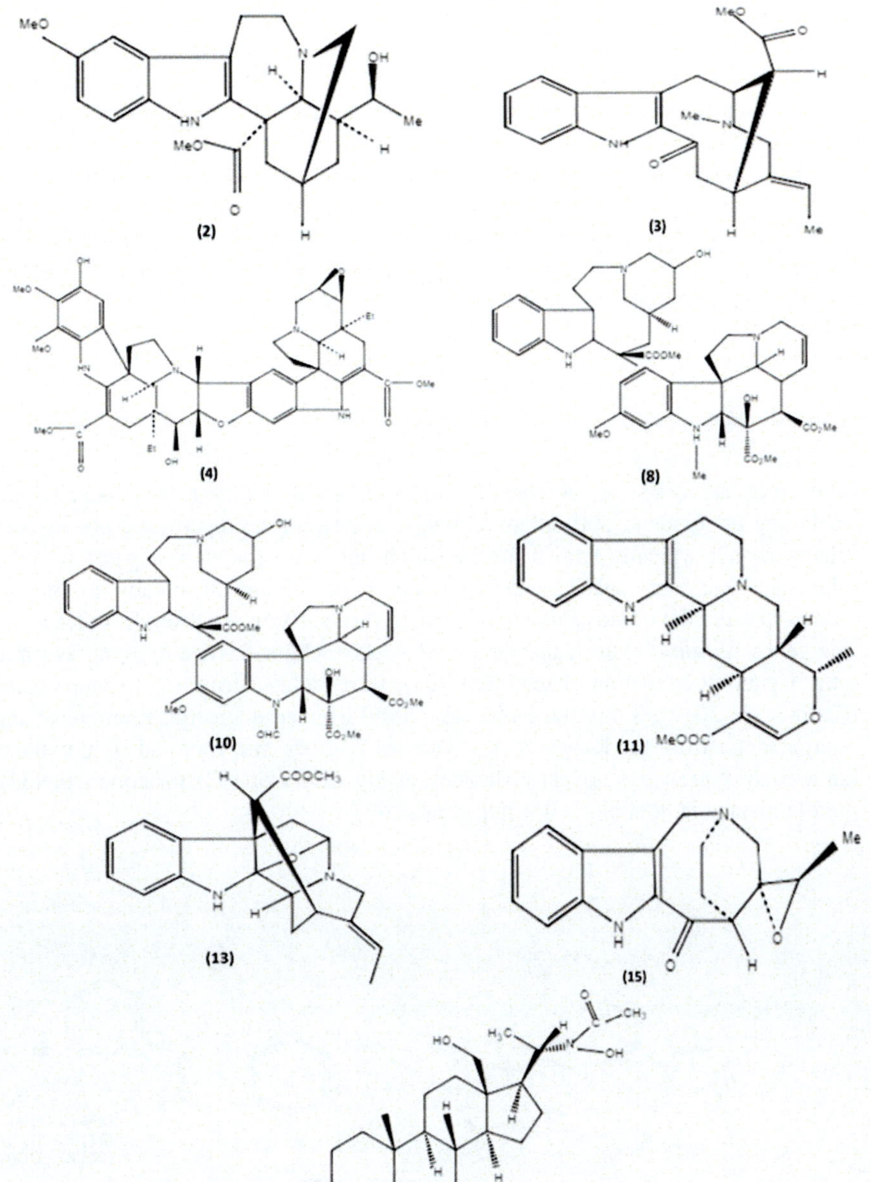


**Fig. 7.1** Alkaloids isolated from Apocynaceae plants (Bhadane, 2017)

reported from *T. divaricata* (Low et al., 2010). Indole alkaloids (vinblastine, serpentine, vincristine ajmalicine, and 3′,4′-anhydrovinblastine) **(8,10,11)** were reported in *C. roseus* (Van der Heijden et al., 2004). Svoboda et al. (1959) reported leurosine and vincaleukoblastine. Picrinine **(13)** and alstoscholarine were isolated from *A. scholaris* (Xiang-Hai et al., 2008; Jagetia). Bhutani et al. (1988) isolated regholarrhenine from *H. antidysenterica.* Holadysenterine**,** conessine, isoconessimine and kurchessine **(20)** were extracted from *H. antidysenterica* (Kumar et al., 2007).

These are just a few examples of alkaloids found in plants of the Apocynaceae family. Each alkaloid has unique properties and potential medicinal applications. However, it is important to note that some alkaloids can be highly toxic or have significant side effects, so their use should be strictly controlled and monitored by healthcare professionals.

## *Flavonoids*

Flavonoids are a class of polyphenolic compounds with diverse biological activities, including antioxidant, anti-inflammatory, and anticancer properties. Apocynaceae plants are not typically known for their high flavonoid content compared to other plant families; some species within this family do contain flavonoids. Figure 7.2 shows isolated flavonoids from Apocynaceae plants. Rutin was from fresh leaves of the genus *Wrightia* (Muruganandam et al., 2000). Other flavonoids, such as rutin, epicatechin, quercetin and kaempferol **(24),** were isolated from *C. carandas* fruits (Patil et al., 2012). Similarly, Hui et al. (2009) isolated quercetin, kaempferol and isorhamnetin from the leaves of *A. scholaris. T. divaricata* root and stem extracts showed the presence of novel flavonoids, namely, calycosin **(32)**, formononetin **(33)**, and farnisin (Liang et al., 2006; Jain et al., 2013).

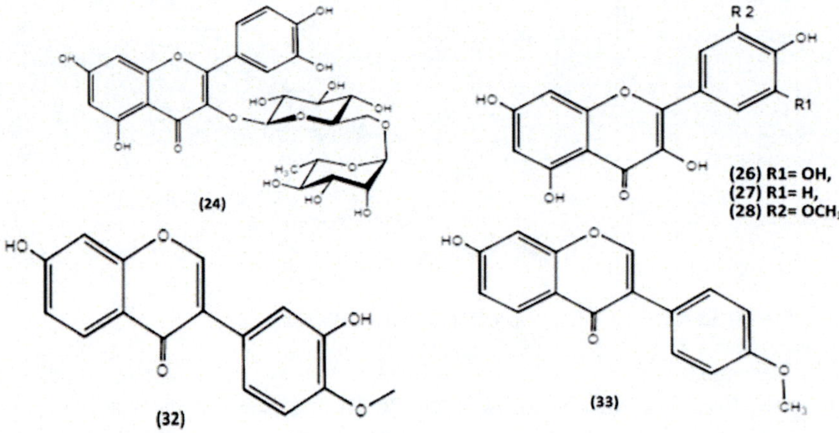


**Fig. 7.2** Some examples of flavonoids isolated from Apocynaceae plants (Bhadane, 2017)

## *Terpenes*

Apocynaceae plants are rich sources of terpenes, which are a large and diverse class of natural compounds derived from the isoprene unit. Terpenes are responsible for the aromatic and medicinal properties of many plants. Terpenes are largely present in plants from the family Asteraceae (such as daisies and sunflowers). Apocynaceae members generally contain all four classes of terpenes.

*Monoterpenes:* Monoterpenes are composed of two isoprene units and are known for their pleasant aromas. Some Apocynaceae plants may contain monoterpenes such as limonene, which is responsible for the citrusy aroma found in citrus fruits and some species of Apocynaceae.

*Diterpenes*: Diterpenes consist of four isoprene units and are commonly found in various plant families, including Apocynaceae. One well-known diterpene from Apocynaceae is Taxol (paclitaxel), which is derived from the bark of the Pacific yew tree (Taxus brevifolia). Taxol is used as a chemotherapy agent for various cancers.

*Triterpenes:* Triterpenes are composed of six isoprene units and are widespread in the plant kingdom. Some Apocynaceae plants may contain triterpenes such as betulinic acid, which has demonstrated potential anticancer properties and is found in plants such as white birch (Betula spp.). Triterpenoids such as betulinic acid, betulin oleanolic acid and ursolic acid were isolated by Siddiqui et al. (1995).

*Sesquiterpenes:* Sesquiterpenes are composed of three isoprene units and are known for their strong and diverse biological activities. While they are more commonly found in plants from other families, certain Apocynaceae plants may contain sesquiterpenes such as caryophyllene and humulene.

Triterpenoids **(50, 53, 55)** were found to be evenly distributed in leaves, roots and fruits (Pakrashi et al., 1968; Naim et al., 1985, 1988; Singh & Uppal, 2015). The sesquiterpene carrisone, triterpene lupeol and oleanolic acid **(71)** were found to exhibit anti-inflammatory activity against proinflammatory mediators such as nitric oxide (NO), TNF-α and IL-1b (Itankar et al., 2011). Recently, a total of 15 triterpenoids, including previously known and newly isolated oleandric acid and oleanerolide **(61, 64),** from *N. oleander* were evaluated for their anti-inflammatory and anticancer potential (Fu et al., 2005). *W. tinctoria* of this family contains Wrightial **(64)** as a new triterpenoid (Ramchandra et al., 1993). Taraxasterol acetate **(65)** was isolated from the roots and stem of *T. divaricata* (Liang et al., 2006). Alstonic acids A and B as well as 2,3-secofernane **(66, 71)** triterpenoids were reported from the leaves of *A. scholaris* (Wang et al., 2009) (Fig. 7.3).

## *Glycosides*

Phenolic glycosides are compounds containing both a phenolic group and a sugar molecule and are present in some Apocynaceae plants. These compounds often

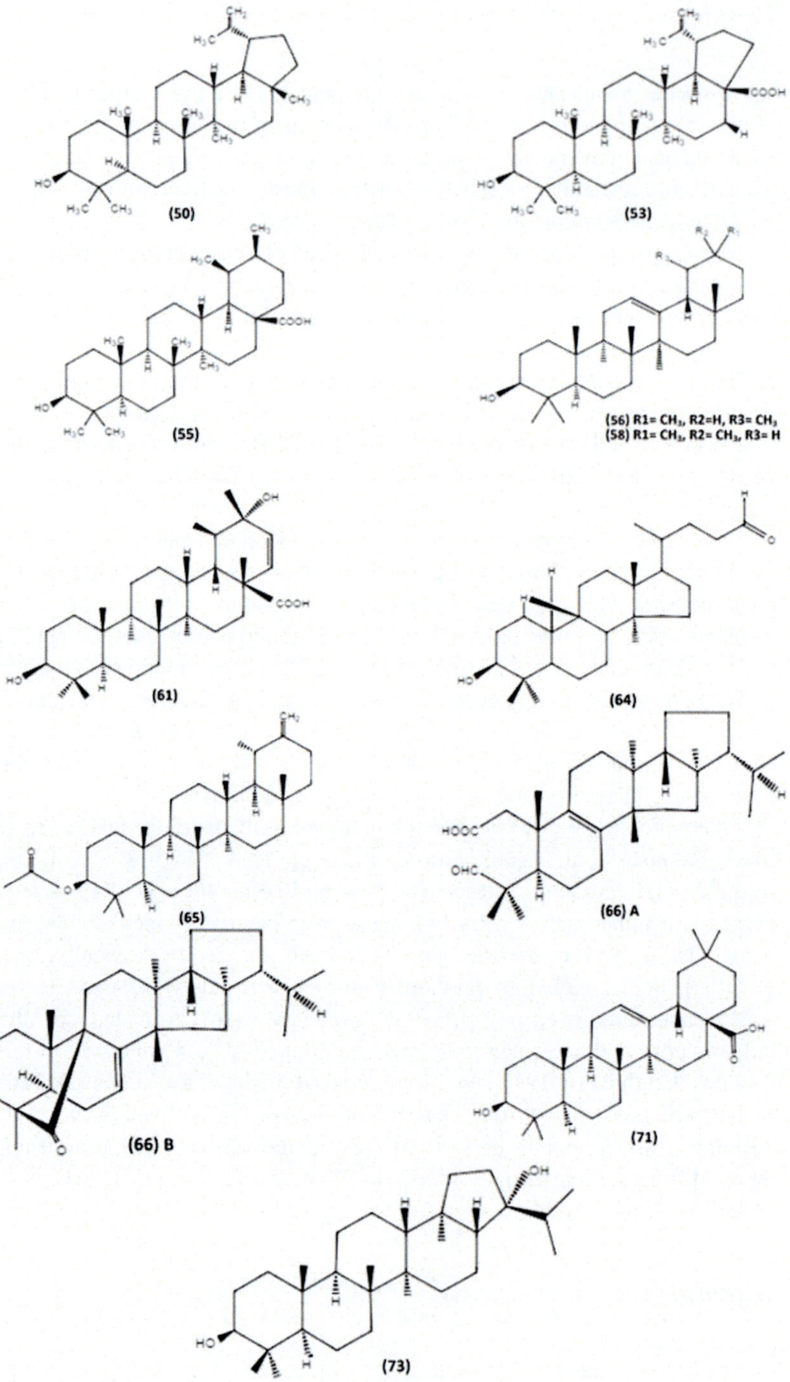


**Fig. 7.3** Chemical structures of common terpenes present in Apocynaceae plants (Bhadane, 2017)

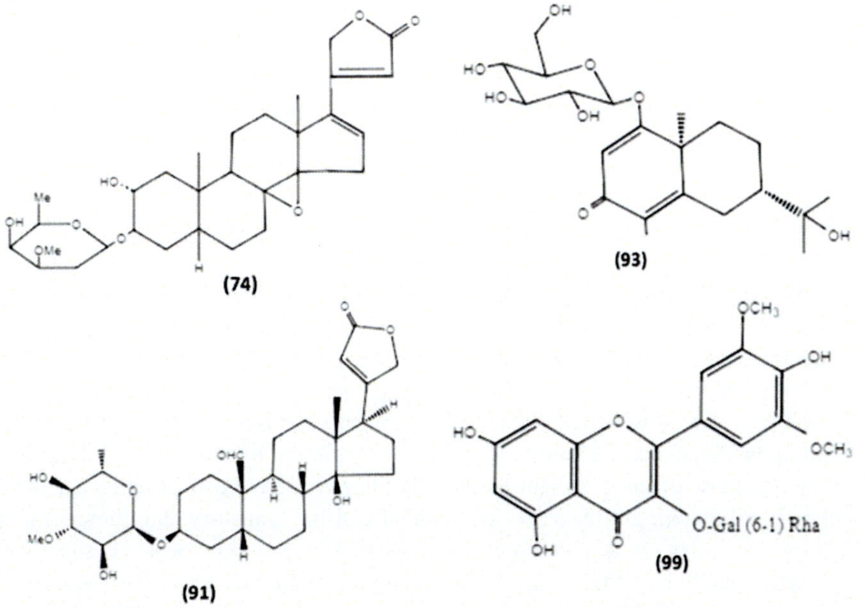


**Fig. 7.4** Chemical structure of glycosides isolated from Apocynaceae plants (Bhadane, 2017)

possess various biological activities and contribute to the medicinal properties of plants. While the prevalence and diversity of phenolic glycosides in Apocynaceae may not be as extensive as in other plant families, here are a few examples of phenolic glycosides that have been reported in certain Apocynaceae plants. The composition of phenolic glycosides may vary among different species within the Apocynaceae family. Cardiac glycosides are a major class of glycosides and are widely used in the treatment of heart failure. The genera *Thevetia* and *Nerium* are reported to possess the highest glycosides (Fig. 7.4). Siddiqui et al. (1986) isolated two cardiac glycosides, kaneroside and neriumoside **(74),** from the leaves of *N. oleander. A* sesquiterpene glucoside, carandoside **(91),** was extracted from *C. carandas* stem (Wangteeraprasert & Likhitwitayawuid, 2009). Toki et al. (2008) isolated and characterized anthocyanidin glycosides **(93, 99)** from orange–red flowers of *C. roseus.*

Oleuropein is a phenolic glycoside present in some Apocynaceae members, such as *Olea europaea* (olive tree). It is known for its antioxidant and anti-inflammatory properties and is associated with numerous health benefits, such as cardiovascular protection and immune system modulation. Another phenolic glycoside, secologanoside, is found in several Apocynaceae plants, particularly in species of *Securidaca.* It has been investigated for its potential analgesic and anti-inflammatory effects. Verbascoside has been isolated and identified in some Apocynaceae plants, including species of *Nerium* and *Vinca.* It possesses antioxidant and anti-inflammatory properties and has been studied for its potential applications in skin health and wound healing. Calotropin is a toxic cardiac glycoside that contains both a phenolic group and a sugar molecule. It is found in several members of the genera *Calotropis* and *Catharanthus*. Calotropin and related cardiac glycosides have been used traditionally in certain medicinal preparations but should be used with caution due to their potential toxicity.

## *Sterols*

Sterols are bioactive compounds that are known for their hepatoprotective, anti-inflammatory and antihyperlipidemic activities (Zaki et al., 1981). Sterols play important roles in plant physiology and can have various biological activities. One well-known sterol found in Apocynaceae plants is β-sitosterol. β-Sitosterol is a phytosterol that is widely distributed in the plant kingdom, including some species within the Apocynaceae family. It is chemically similar to cholesterol and has been found to have potential health benefits. β-Sitosterol has been investigated for its cholesterol-lowering effects and its potential anti-inflammatory and immunomodulatory properties. It is commonly used as a dietary supplement for supporting prostate health and managing urinary symptoms. Stigmasterol, β-sitosterol and campesterol **(109–110)** are present in various parts of these plants (Fig. 7.5) (Ramchandra et al., 1993; Adotey et al., 2012). Stigmasterol is another common sterol found in the Apocynaceae family. It is structurally similar to β-sitosterol and has been studied for its potential cholesterol-lowering and anti-inflammatory effects. Stigmasterol has also been investigated for its anticancer properties and its ability to support cardiovascular health (Jain & Bari, 2010). Akihisa et al. (1988) reported some uncommon sterols **(113–114)** from unsaponifiable lipids of *W. tinctoria* seeds.

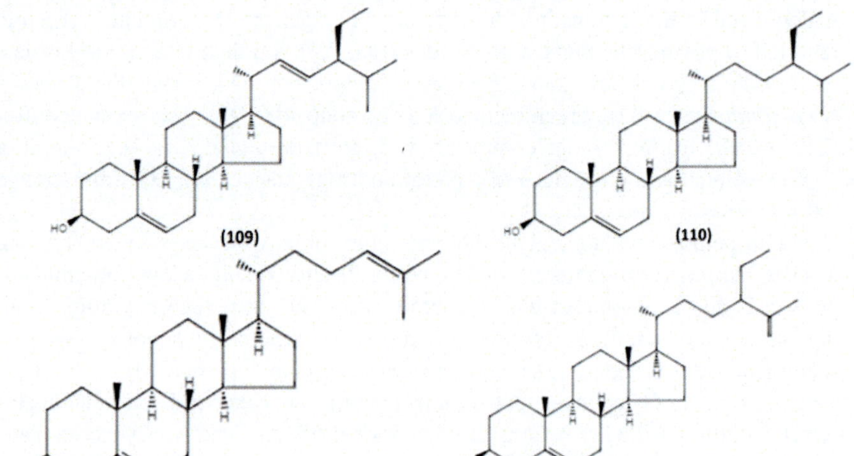


**Fig. 7.5** Chemical structure of sterols isolated from Apocynaceae plants (Bhadane, 2017)

Studies on phytochemical aspects of Apocynaceae plants offer potential benefits for human health, as well as their applications in industries such as agriculture, cosmetics, and pharmaceuticals. Phytochemistry helps us identify and isolate the bioactive compounds in Apocynaceae plants that have medicinal properties, which can then be used to develop new drugs or improve existing ones. Understanding the phytochemical aspects of Apocynaceae plants is important in the field of agriculture because it helps us understand the chemical makeup of plants, which can be used to improve crop yields, increase resistance to pests and diseases, and develop new plant varieties. Apocynaceae plants harbor several compounds that have beneficial effects on the skin, such as moisturizing, antiaging, and anti-inflammatory properties. Phytochemistry studies help us identify these compounds and develop new cosmetic products that can improve skin health. Phytochemistry can be used to study the chemical makeup of plants in different ecosystems, which can help us understand the effects of environmental factors such as pollution and climate change on plant growth and health.

## References

Adotey, J. P. K., Adukpo, G. E., Opoku Boahen, Y., & Armah, F. A. (2012). A review of the ethnobotany and pharmacological importance of *Alstonia boonei* De wild (Apocynaceae). *ISRN Pharmacology, 2012*, 1–9.

Aiyambo, D. (2010). Traditional uses of selected members of the Apocynaceae family in Namibia. *Ministry of agriculture, water and forestry, Windhock.*

Akihisa, T., Ahmad, I., Singh, S., Tamura, T., & Matsumoto, T. (1988). 14α-Methylzymosterol and other sterols from *Wrightia tinctoria* seeds. *Phytochemistry, 27*(10), 3231–3234.

Ali Khan, M. S., Mat Jais, A. M., & Afreen, A. (2013). Prostaglandin analogous and antioxidant activity mediated gastroprotective action of *Tabernaemontana divaricata* (L.) R. Br. Flower methanolic extract against chemically induced gastric ulcers in rats. *BioMed Research International, 2013*, 1–18.

Arambewela, L. S., & Ranatunge, T. (1991). Indole alkaloids from *Tabernaemontana divaricata. Phytochemistry, 30*(5), 1740–1741.

Baker, Z. K., & Sardari, S. (2021). Molecularly imprinted polymer (MIP) applications in natural product studies based on medicinal plant and secondary metabolite analysis. *Iranian Biomedical Journal, 25*(2), 68–77.

Balakrishnan, N., Balasubaramaniam, A., Sangameswaran, B., & Bhaskar, V. (2011). Hepatoprotective activity of two Indian medicinal plants from Western Ghats, Tamil Nadu. *Journal of Natural Pharmaceuticals, 2*(2), 92–98.

Begum, S., Siddiqui, B. S., Sultana, R., Zia, A., & Suria, A. (1999). Bioactive cardenolides from the leaves of *Nerium oleander. Phytochemistry, 50*(3), 435–438.

Begum, S., Syed, S. A., Siddiqui, B. S., Sattar, S. A., & Choudhary, M. I. (2013). Carandinol: First isohopane triterpene from the leaves of *Carissa carandas* L. and its cytotoxicity against cancer cell lines. *Phytochemistry Letters, 6*(1), 91–95.

Bhadane, B. S. (2017). *Studies on biological activities and in vitro propagation of some members of family Apocynaceae.* PhD thesis,. KBC North Maharashtra University.

Bhadane, B. S., Patil, M. P., Maheshwari, V. L., & Patil, R. H. (2018). Ethnopharmacology, phytochemistry, and biotechnological advances of family Apocynaceae: A review. *Phytotherapy Research, 32*(7), 1181–1210.

Bhadane, B. S., & Patil, R. H. (2017). Isolation, purification and characterization of antioxidative steroid derivative from methanolic extract of *Carissa carandas* (L.) leaves. *Biocatalysis and Agricultural Biotechnology, 10*, 216–223.

Bhaskar, V. H., & Balakrishnan, N. (2015). Analgesic, anti-inflammatory and antipyretic activities of *Pergularia daemia* and *Carissa carandas*. *DARU Journal of Pharmaceutical Sciences, 17*(3), 168–174.

Bhat, P., Hegde, G., & Hegde, G. R. (2012). Ethnomedicinal practices in different communities of Uttara Kannada district of Karnataka for treatment of wounds. *Journal of Ethnopharmacology, 143*(2), 501–514.

Bhattachartya, S., Tarafdar, S., & Saha, C. N. (2009). Triterpenoids and steroids from *Holarrhena pubescens* seeds. *Pharmacognosy Magazine, 5*(20), 407–411.

Bhutani, K. K., Ali, M., Sharma, S. R., Vaid, R. M., & Gupta, D. K. (1988). Three new steroidal alkaloids from the bark of *Holarrhena antidysenterica*. *Phytochemistry, 27*(3), 925–928.

Brun, G., Bessière, J. M., Dijoux-Franca, M. G., David, B., & Mariotte, A. M. (2001). Volatile components of *Catharanthus roseus* (L.) G. Don (Apocynaceae). *Flavour and Fragrance Journal, 16*(2), 116–119.

Brun, G., Dijoux, M. G., David, B., & Mariotte, A. M. (1999). A new flavonol glycoside from *Catharanthus roseus*. *Phytochemistry, 50*(1), 167–169.

Choi, Y. H., Tapias, E. C., Kim, H. K., Lefeber, A. W., Erkelens, C., Verhoeven, J. T. J., & Verpoorte, R. (2004). Metabolic discrimination of *Catharanthus roseus* leaves infected by phytoplasma using 1H-NMR spectroscopy and multivariate data analysis. *Plant Physiology, 135*(4), 2398–2410.

Dai, L., Wang, W., Dong, X., Hu, R., & Nan, X. (2011). Molluscicidal activity of cardiac glycosides from *Nerium indicum* against *Pomacea canaliculata* and its implications for the mechanisms of toxicity. *Environmental Toxicology and Pharmacology, 32*(2), 226–232.

Dong, Q., & Fang, J. N. (2001). Structural elucidation of a new arabinogalactan from the leaves of *Nerium indicum*. *Carbohydrate Research, 332*(1), 109–114.

Ferreres, F., Figueiredo, R., Bettencourt, S., Carqueijeiro, I., Oliveira, J., Gil-Izquierdo, A., & Barceló, A. R. (2011). Identification of phenolic compounds in isolated vacuoles of the medicinal plant *Catharanthus roseus* and their interaction with vacuolar class III peroxidase: An $H_2O_2$ affair? *Journal of Experimental Botany, 62*(8), 2841–2854.

Ferreres, F., Pereira, D. M., Valentão, P., Andrade, P. B., Seabra, R. M., & Sottomayor, M. (2008). New phenolic compounds and antioxidant potential of *Catharanthus roseus*. *Journal of Agricultural and Food Chemistry, 56*(21), 9967–9974.

Fu, L., Zhang, S., Li, N., Wang, J., Zhao, M., Sakai, J., & Kiuchi, M. (2005). Three new triterpenes from Nerium oleander and biological activity of the isolated compounds. *Journal of Natural Products, 68*(2), 198–206.

Galipalli, S., Patel, N. K., Prasanna, K., & Bhutani, K. K. (2015). Activity-guided investigation of *Carissa carandas* (L.) roots for anti-inflammatory constituents. *Natural Product Research, 29*(17), 1670–1672.

Hegde, K., & Joshi, A. B. (2009). Hepatoprotective effect of *Carissa carandas* Linn root extract against CCl 4 and paracetamol induced hepatic oxidative stress. *Indian Journal of Experimental Biology, 47*, 660–667.

Hegde, K., Satyanarayana, D., & Joshi, A. B. (2012). Phytochemical investigation of root extract of the plant *Carissa spinarum*. *RGUHS Journal of Pharmaceutical Sciences, 2*(1), 55–59.

Hui, T., Sun, Y., Zhu, L., Guo, W., & Rao, G. (2009). Flavonoids in leaves of Alstonia scholaris. *Zhongguo Zhong yao za zhi Zhongguo zhongyao zazhi. China Journal of Chinese Materia Medica, 34*(9), 1111–1113.

Itankar, P. R., Lokhande, S. J., Verma, P. R., Arora, S. K., Sahu, R. A., & Patil, A. T. (2011). Antidiabetic potential of unripe *Carissa carandas* Linn. Fruit extract. *Journal of Ethnopharmacology, 135*(2), 430–433.

Jain, P. S., & Bari, S. B. (2010). Isolation of lupeol, stigmasterol and campesterol from petroleum ether extract of woody stem of *Wrightia tinctoria*. *Asian Journal of Plant Sciences, 9*(3), 163–167.

Jain, S., Sharma, P., Ghule, S., Jain, A., & Jain, N. (2013). In vivo anti-inflammatory activity of *Tabernaemontana divaricata* leaf extract on male albino mice. *Chinese Journal of Natural Medicines, 11*(5), 472–476.

Kam, T. S., Loh, K. Y., & Wei, C. (1993). Conophylline and conophyllidine: New dimeric alkaloids from *Tabernaemontana divaricata*. *Journal of Natural Products, 56*(11), 1865–1871.

Karmakar, T., & Chakraborty, D. P. (1983). 7-dehydrositosterol from *Rauwolfia serpentina*. *Phytochemistry, 22*(2), 608–609.

Kohls, S., Scholz-Böttcher, B. M., Teske, J., Zark, P., & Rullkötter, J. (2012). Cardiac glycosides from Yellow Oleander (Thevetia peruviana) seeds. *Phytochemistry, 75*, 114–127.

Kumar, N., Singh, B., Bhandari, P., Gupta, A. P., & Kaul, V. K. (2007). Steroidal alkaloids from *Holarrhena antidysenterica* (L.) WALL. *Chemical and Pharmaceutical Bulletin, 55*(6), 912–914.

Liang, S., Chen, H. S., Du, J. L., Wang, H. P., Shen, Y., & Huang, M. Z. (2006). Study on chemical constituents of *Tabernaemontana divaricata*. *Academic Journal of Second Military Medical University, 8*, 023.

Low, Y. Y., Lim, K. H., Choo, Y. M., Pang, H. S., Etoh, T., Hayashi, M., & Kam, T. S. (2010). Structure, biological activity, and a biomimetic partial synthesis of the lirofolines, novel pentacylic indole alkaloids from *Tabernaemontana divaricata*. *Tetrahedron Letters, 51*(2), 269–272.

Mahmood, T., Meer, F., Munir, F., Nazar, N., & Naveed, I. (2011). Genetic diversity of selected Apocynaceae species based on chloroplast gene rps11. *Journal of Medicinal Plant Research: Planta Medica, 5*, 4382–4387.

Miyagawa, T., Ohtsuki, T., Koyano, T., Kowithayakorn, T., & Ishibashi, M. (2009). Cardenolide glycosides of Thevetia peruviana and triterpenoid saponins of *Sapindus emarginatus* as TRAIL resistance-overcoming compounds. *Journal of Natural Products, 72*(8), 1507–1511.

Muruganandam, A. V., Bhattacharya, S. K., & Ghosal, S. (2000). Indole and flavanoid constituents of *Wrightia tinctoria, W. tomentosa and W. coccinea*. *Indian Journal of Chemistry, 39*, 125–131.

Naim, Z., Khan, M., & Nizami, S. (1985). Isolation of a new triterpenic alcohol from of *Carissa carandas*. *Pakistan Journal of Science and Industrial Research, 28*(6), 378–381.

Naim, Z., Khan, M. A., & Nizami, S. S. (1988). Isolation of a new isomer of ursolic acid from fruits and leaves of *Carissa carandas*. *Pakistan Journal of Scientific and Industrial Research, 31*, 753–755.

Pakrashi, S. C., Datta, S., & Ghosh-Dastidar, P. P. (1968). Indian medicinal plants—XVII. 3 phytochemical examination of *Carissa* SPP. *Phytochemistry, 7*(3), 495–496.

Pal, R., Kulshreshtha, D. K., & Rastogi, R. P. (1975). A new lignan from *Carissa carandas*. *Phytochemistry, 14*(10), 2302–2303.

Patil, R. P., Pai, S. R., Pawar, N. V., Shimpale, V. B., Patil, R. M., & Nimbalkar, M. S. (2012). Chemical characterization, mineral analysis, and antioxidant potential of two underutilized berries (*Carissa carandus* and *Eleagnus conferta*) from the Western Ghats of India. *Critical Reviews in Food Science and Nutrition, 52*(4), 312–320.

Proestos, C., Chorianopoulos, N., Nychas, G. J., & Komaitis, M. (2005). RP-HPLC analysis of the phenolic compounds of plant extracts. Investigation of their antioxidant capacity and antimicrobial activity. *Journal of Agricultural and Food Chemistry, 53*(4), 1190–1195.

Ramalakshmi, S., Edaydulla, N., Ramesh, P., & Muthuchelian, K. (2012). Investigation on cytotoxic, antioxidant, antimicrobial and volatile profile of *Wrightia tinctoria* (Roxb.) R. Br. Flower used in Indian medicine. *Asian Pacific Journal of Tropical Disease, 2*, S68–S75.

Ramchandra, P., Basheermiya, M., Krupadanam, G. L. D., & Srimannarayana, G. (1993). Wrightial, a new terpene from *Wrightia tinctoria*. *Journal of Natural Products, 56*(10), 1811–1812.

Rastogi, R. C., Vohra, M. M., Rastogi, R. P., & Dhar, M. L. (1966). Studies on Carissa carandas Linn. I. Isolation of cardiac active principles. *Indian Journal of Chemistry, 4*(3), 132–138.

Shrivastava, R. M., & Bakodia, M. M. (1979). Studies in vegetable oils: Composition of seeds oil of *Carissa carandas*. *Journal of Scientific Research, 1*, 57–60.
Siddiqui, B. S., Begum, S., Siddiqui, S., & Lichter, W. (1995). Two cytotoxic pentacyclic triterpenoids from *Nerium oleander*. *Phytochemistry, 39*(1), 171–174.
Siddiqui, S., Hafeez, F., Begum, S., & Siddiqui, B. S. (1986). Isolation and structure of two cardiac glycosides from the leaves of *Nerium oleander*. *Phytochemistry, 26*(1), 237–241.
Singh, A., & Uppal, G. K. (2015). A review on carissa carandas phytochemistry, ethnopharmacology, and micropropagation as conservation strategy. *Asian Journal of Pharmaceutical and Clinical Research, 8*(3), 26–30.
Singh, B., & Rastogi, R. P. (1972). The structure of carindone. *Phytochemistry, 11*(5), 1797–1801.
Sohretoglu, D., Masullo, M., Piacente, S., & Kirmizibekmez, H. (2013). Iridoids, monoterpenoid glucoindole alkaloids and flavonoids from Vinca major. *Biochemical Systematics and Ecology, 49*, 69–72.
Svoboda, G. H., Neuss, N., & Gorman, M. (1959). Alkaloids of Vinca rosea Linn. (Catharanthus roseus G. Don.) V. preparation and characterization of alkaloids. *Journal of the American Pharmaceutical Association, 48*(11), 659–666.
Tewtrakul, S., Nakamura, N., Hattori, M., Fujiwara, T., & Supavita, T. (2002). Flavanone and flavonol glycosides from the leaves of Thevetia peruviana and their HIV-1 reverse transcriptase and HIV-1 integrase inhibitory activities. *Chemical and Pharmaceutical Bulletin, 50*(5), 630–635.
Thomas, P. S., Kanaujia, A., Ghosh, D., Duggar, R., & Katiyar, C. K. (2008). Alstonoside, a secoiridoid glucoside from Alstonia scholaris. *Indian Journal of Chemistry—Section B (IJC-B), 47*, 1298–1302.
Toki, K., Saito, N., Irie, Y., Tatsuzawa, F., Shigihara, A., & Honda, T. (2008). 7-O-methylated anthocyanidin glycosides from Catharanthus roseus. *Phytochemistry, 69*(5), 1215–1219.
Usmani, S. B. (1995). *Studies on the chemical constituents of Holarrhena antidysenterica L and the β-carboline series of bases and their pharmacological activity* (doctoral dissertation, Ph. D., Karachi).
Van Beek, T. A., Verpoorte, R., Svendsen, A. B., Leeuwenberg, A. J., Bisset, N. G., & Tabernaemontana divaricata L. (1984). (Apocynaceae): A review of its taxonomy, phytochemistry, ethnobotany and pharmacology. *Journal of Ethnopharmacology, 10*(1), 1–56.
Van der Heijden, R., Jacobs, D. I., Snoeijer, W., Hallard, D., & Verpoorte, R. (2004). The catharanthus alkaloids: Pharmacognosy and Biotechnology. *Current Medicinal Chemistry, 11*, 607–628.
Wang, F., Ren, F. C., & Liu, J. K. (2009). Alstonic acids a and B, unusual 2,3-secofernane triterpenoids from Alstonia scholaris. *Phytochemistry, 70*(5), 650–654.
Wangteeraprasert, R., & Likhitwitayawuid, K. (2009). Lignans and a sesquiterpene glucoside from Carissa carandas stem. *Helvetica Chimica Acta, 92*(6), 1217–1223.
Wong, S. K., Lim, Y. Y., Ling, S. K., & Chan, E. W. C. (2014). Caffeoylquinic acids in leaves of selected Apocynaceae species: Their isolation and content. *Pharmacognosy Research, 6*(1), 67–72.
Xiang-Hai, C. A. I., Ya-Ping, L. I. U., Tao, F. E. N. G., & Xiao-Dong, L. U. O. (2008). Picrinine-type alkaloids from the leaves of Alstonia scholaris. *Chinese Journal of Natural Medicines, 6*(1), 20–22.
Zaki, A. Y., El-Tohamy, S. F., Hashem, M., & Abd El-Fattah, S. (1981). Preliminary phytochemical and pharmacological screening of Carissa carandas L. and Carissa grandiflora DC. Growing in Egypt. Egyptian. *Journal of Pharmaceutical Sciences, 22*, 113–126. (Chemical Abstracts, 100, 117843b).

# Chapter 8
# Biological and Pharmacological Properties of Apocynaceae Members

## Introduction

Medicinal plants have been used for centuries by different cultures around the world. The biological activities of these plants are attributed to various phytocoistituents present in them. Apocynaceae plants contain various active bioactive compounds with antibacterial, antifungal, antiviral, anti-inflammatory, antioxidant, and anticancer properties. Ethnomedicinal knowledge of many members of this family has led to extensive studies on their biological and pharmacological properties. The pharmacological evaluation of many plants has explored crude extracts as well as purified phytoconstituents. However, very few members of this family have been systematically studied and explored for pharmaceutical/medicinal applications despite their vast ethnomedcinal value. This chapter highlights a detailed outlook of the biological activities exhibited by Apocynaceae members with respect to crude extracts, fractions and single isolated compounds (Table 8.1).

## Antioxidant Activity

The Apocynaceae family includes a number of species with potential antioxidant properties. Antioxidants are compounds that help protect cells from oxidative damage caused by free radicals, which are highly reactive molecules that can contribute to various diseases and aging processes. While not all plants within the Apocynaceae family have been extensively studied for their antioxidant potential, several species have shown promising results.

*Catharanthus roseus* contains various alkaloids, including vincamine and vinblastine, which have demonstrated antioxidant activity. Vincamine has been shown to scavenge free radicals and inhibit lipid peroxidation, protecting cells from oxidative stress. The leaves and flowers of *Allamanda cathartica* have been investigated



R. H. Patil et al., *Apocynaceae Plants*,
https://doi.org/10.1007/978-981-99-5406-3_8

**Table 8.1** Biological and pharmacological properties of Apocynaceae members (Reproduced with permission from Bhadane, 2017; Bhadane et al., 2018)

| Plant and plant part | Biological activity | Extract/Compound | Effective concentration/Dose | Study model | Refernces |
|---|---|---|---|---|---|
| *A. scholaris* | | | | | |
| Flowers | Antioxidant (DPPH) | Methanolic and aqueous extract | $IC_{50}$ 34.2 and 50 μg/mL | *In vitro* | James et al. (2011) |
| Leaves | Anticancer | 80% methanolic extract | 4% crude extract | *In vitro* | Surendren et al. (2012) |
| | Anti-HIV | Methanol and butanol extract | 10 and 15 μg/mL, respectively | *In vitro* (using CEM-GFP cell lines infected with HIV-1NL4.3) | Sabde et al. (2011) |
| | Antiarthritic | Ethanolic extract | 400 mg/kg | *In vivo* (Freund's adjuvant induced rat) | |
| Stem bark | Cardioprotective | Aqueous extract | 300 mg/kg | *In vivo* (Streptozotocin induced rat) | Bandawane et al. (2011) |
| | Antidiabetic | Aqueous extract | 300 mg/kg | *In vivo* (Streptozotocin induced rat) | Bandawane et al. (2011) |
| | Anti-inflammatory | Methanolic extract | 400 mg/kg | *In vivo* (carrageenan induced rat paw) | Subraya and Gupta (2012) |
| *C. carandas* | | | | | |
| Leaves | Antioxidant (DPPH and $OH^-$ radical scavenging) | Methanolic extract | $EC_{50}$ 546 and 498 μg/mL | *In vitro* | Bhadane and Patil (2017) |
| | Anti-inflammatory | Methanolic extract | 1.3 mg/mL | *In vitro* | Bhadane and Patil (2017) |
| | Cytotoxic | Carandinol from ethyl acetate fraction of methanolic extract | $IC_{50}$ 6.87 μg/mL | *In vitro* (using HeLa cell line) | Begum et al. (2013) |
| | Hepatoprotective | Methanol and acetone (3:1) extract | 100 mg/kg | *In vivo* ($CCl_4$ induced liver injury in albino rat) | Bhati et al. (2014) |
| Fruit | Antidiabetic | | | | |

| | | | | | |
|---|---|---|---|---|---|
| Roots | Gastroprotective | | | | |
| | Anti-inflammatory | Carissone and Scopoletin from methanolic extract | $IC_{50}$ 20.1 and 24.6 μg/mL, respectively | *In vitro* (using J774A.1 cell lines) | Galipalli et al. (2015) |
| | Hepatoprotective | Ethanolic extract | 400 mg/kg | *In vivo* ($CCl_4$ and paracetamol induced liver injury in albino rat) | Hegde and Joshi (2009) |
| *C. roseus* | | | | | |
| Leaves | Hepatoprotective | Ethanolic extract | 500 mg/kg | *In vivo* (simvastatin induced albino rat) | Ahmed and Srinivasa Rao (2013) |
| | Antimalarial | Ethanolic extract | $IC_{50}$ 49.63 μg/mL | *In vitro* (Antiplasmodial against *plasmodium falciparum*) | Ravikumar et al. (2012) |
| | Anti- HIV | Methanolic extract | 2.5 μg/mL | *In vitro* (using CEM-GFP cell lines infected with HIV-1NL4.3) | Sabde et al. (2011) |
| | Antibacterial | Ethanolic extract | 5 mg/mL | *In vitro* | Govindasamy and Srinivasan (2012) |
| Whole plant | Cytotoxic | Leurosine (indole alkaloid) from 95% ethanolic extract | $IC_{50}$ 0.73 μM | *In vitro* (using MDA-MB-231 human breast cancer cell lines) | Wang et al. (2012) |
| *H. antidysenterica* | | | | | |
| Leaves | Antioxidant (superoxide and $OH^-$ radical scavenging) | Methanolic extract | $IC_{50}$ 500 and 285 μg/mL, respectively | *In vitro* | Ganapathy et al. (2011) |
| Seeds | Antidiabetic | Ethanolic extract | 600 mg/kg | *In vivo* (Streptozotocin induced rat) | Keshri et al. (2012) |
| | Antiurolithic | 70% ethanolic extract | $IC_{50}$ 0.27 mg/mL | *In vitro* (inhibition of CaOx crystal aggregation) | Khan et al. (2012) |
| | | | 100 mg/kg | | |

(continued)

**Table 8.1** (continued)

| Plant and plant part | Biological activity | Extract/Compound | Effective concentration/Dose | Study model | Referernces |
|---|---|---|---|---|---|
| | | | | *In vivo* (CaOx-induced urolithiasis rat) | |
| *N. oleander* | | | | | |
| Leaves | Cardioprotective | Oleandrin from methanolic extract | 300 mg/kg | *In vivo* (using Guinea pigs) | Botelho et al. (2017) |
| | | | 0.3 mg/mL | *In vitro* (using ventricular myocytes of mice) | |
| Flower | Antioxidant (DPPH, ABTS, superoxide and $OH^-$ radical scavenging) | Methanolic extract | $IC_{50}$ 193, 156, 210 and 211 μg/mL, respectively | *In vitro* | Singhal and Gupta (2012) |
| | Hepatoprotective | Methanolic extract | 400 mg/kg | *In vivo* ($CCl_4$-induced liver injury in rats) | Singhal and Gupta (2012) |
| | Anticonvulsant | 50% hydroalcoholic extract | 200 mg/kg | *In vivo* (Pentylenetetrazol induced convulsion in Swiss albino mice) | Singhal and Gupta (2011) |
| Leaves | Anticandida | Methanolic extract | MIC 1 mg/mL | *In vitro* | Wankhede et al. (2013) |
| *R. serpentina* | | | | | |
| Roots | Anti-HIV | Methanolic extract | 10 μg/mL | *In vitro* (using CEM-GFP cell lines infected with HIV-1NL4.3) | Sabde et al. (2011) |
| Seeds | Antimalarial | Petroleum ether extract | 100 ppm | *In vivo* (mosquito larvicidal) | Das and Chandra (2012) |
| | Cytotoxic | Reserpine (indole alkaloid) | $IC_{50}$ 14.52 μM | *In vitro* (drug resistant tumour cell lines) | Abdelfatah and Efferth (2015) |
| Leaves | Antidiarrheal | Methanolic extract | 400 mg/kg | *In vivo* (Castor oil-induced diarrhea in mice) | Ezeigbo et al. (2012) |

| | | | | | |
|---|---|---|---|---|---|
| *T. divaricata* | | | | | |
| Leaves | Antioxidant (metal chelating, $OH^-$, DPPH and superoxide radical scavenging) | Ethanolic extract | 68% metal chelation by extract ($IC_{50}$ 5.3, 5.2 and 5.8 mg/mL for radical scavenging) | *In vitro* | Anbukkarasi et al. (2016) |
| | Anti-inflammatory | Hexane of methanolic extract | 0.7 $\mu g{\cdot}cm^{-2}$ | *In vivo* (croton oil induced edema in albino mice) | Jain et al. (2013) |
| | Anticandida | Methanolic extract | MIC 1 mg/mL | *In vitro* | Wankhede et al. (2013) |
| | Antidiarrheal | Hydroalcoholic and aqueous extract | 300 mg/kg | *In vivo* (Castor oil induced diarrhea in rats) | Raj et al. (2013) |
| Flower | Antidiabetic | Methanolic extract | 300 mg/kg | *In vivo* (Alloxan-induced mice) | Rahman et al. (2011) |
| | Gastroprotective | Methanolic extract | 500 mg/kg | *In vivo* (aspirin and ethanol-induced gastric ulcer in rats) | Ali Khan et al. (2013) |
| | Antifertility and abortifacient activity | Methanolic extract | 500 mg/kg | *In vivo* (using female Wistar rats) | Mukhram et al. (2012) |
| Whole plant | Hepatoprotective | Ethanolic extract | 400 mg/kg | *In vivo* (DEN and Fe NTA-induced albino rats) | Poornima et al. (2014) |
| | Anticonvulsant | Ethanolic extract | 400 mg/kg | *In vivo* (maximal electroshock, pentylenetetrazol, isoniazid induced convulsion in albino mice) | Basavaraj et al. (2011) |
| *T. peruviana* | | | | | |
| Leaves | Wound healing | Hexane extract | 200 mg/kg | *In vivo* (dead space wound rat model) | Rahman et al. (2017) |
| Fruits | Wound healing | Water extract | 200 mg/kg | *In vivo* (dead space wound rat model) | Rahman et al. (2017) |
| | Anticancer | Methanolic extract | 100 mg/kg | *In vivo* (EAC cell line in albino mice) | Haldar et al. (2015) |

(continued)

**Table 8.1** (continued)

| Plant and plant part | Biological activity | Extract/Compound | Effective concentration/Dose | Study model | Referernces |
|---|---|---|---|---|---|
| Seeds | Antimalarial | Ethanolic extract | $IC_{50}$ 58.83 µg/mL | *In vitro* (Antiplasmodial against *plasmodium falciparum*) | Ravikumar et al. (2012) |
| Flowers | Gastroprotective | Linalool and 1,8-cineole from volatile oil | 33 and 36 mg/kg, respectively | *In vivo* (ethanol-induced gastric ulcer in mice) | Kumar et al. (2015) |
| Whole plant | Anticancer | Cardenolides from methanolic extract | 200 µg/mL | *In vitro* (against Hep $G_2$, HL-60 and PC 62 cancerous cells) | Barbon et al. (2012) |
| ***W. tinctoria*** | | | | | |
| Leaves | Antimalarial (Larvicidal) | Petroleum ether and aqueous extract | $LC_{50}$ 0.37 and 0.11% | *In vitro* (against filarial larvae *Culex quinquefasciatus*) | Sakthivadivel et al. (2014) |
| Fruit | Antimalarial (Larvicidal) | Petroleum ether and aqueous extract | $LC_{50}$ 1.31 and 0.09% | *In vitro* (against filarial larvae *Culex quinquefasciatus*) | Sakthivadivel et al. (2014) |
| | Antidiabetic | Methanolic and ethyl acetate extract | 300 mg/kg | *In vivo* (Alloxan-induced mice) | Rani et al. (2012) |
| Flower | Antioxidant (metal chelating, $OH^-$ and DPPH radical scavenging) | Ethanolic extract | $IC_{50}$ 43.16 µg/mL | *In vitro* | Ramalakshmi et al. (2012) |

for their antioxidant properties. Extracts from *Allamanda cathartica* exhibited significant antioxidant activity in various *in vitro* studies, attributed to the presence of phenolic compounds. *Nerium oleander* (oleander) has been shown to possess antioxidant potential. Its leaf extract has been found to scavenge free radicals and inhibit lipid peroxidation, suggesting a protective effect against oxidative stress. *Plumeria spp.* have been investigated for their antioxidant activity. The flowers and leaves of *Plumeria* contain various phenolic compounds, which contribute to their antioxidant potential. *Tabernaemontana divaricata* is a highly explored member of this family for obtaining natural antioxidants. The antioxidant potential of *Tabernaemontana divaricata* is attributed to the presence of phenolic compounds. Various extracts of *Tabernaemontana divaricate have* demonstrated the ability to scavenge free radicals and inhibit lipid peroxidation in a number of studies (Bhadane, 2017).

The unique property of phenolic compounds is that they act as excellent reducing agents, metal chelators, and free radical quenchers (Rice-Evans et al., 1996). The antioxidant activity of *N. oleander* was evaluated by various *in vitro* methods, such as DPPH and reducing power assays (Singhal & Gupta, 2012). The antioxidative potential of ethanol extracts of *W. tinctoria* and *T. divaricata* was reported using DPPH scavenging activity (Ramalakshmi et al., 2012; Anbukkarasi et al., 2016; Rumzhum et al., 2012). The *in vitro* antioxidant activity of *H. antidysenterica* was attributed to the higher total phenolic content of the plant extract (Ganapathy et al., 2011). James et al. (2011) explored the antioxidative potential of *A. scholaris* fruit extract. The DNA damage inhibition and antioxidative potential of *C. carandas* were studied by Verma et al. (2015).

## Anti-Inflammatory/Analgesic Activity

Inflammation is a natural response of the body's immune system to injury, infection, or tissue damage. It is a complex process involving various immune cells, signaling molecules, and biochemical pathways. Inflammatory diseases can be treated with a variety of medications, including NSAIDs (nonsteroidal anti-inflammatory drugs), but their extended use might have negative side effects. Plants have long been a valuable source of anti-inflammatory compounds, many of which have been used in traditional medicine for centuries. Numerous plant species in the Apocynaceae family have been explored for their possible anti-inflammatory properties. However, the Apocynaceae family has not been fully explored for obtaining anti-inflammatory compounds.

Members of the family, such as *Carissa carandas, Plumeria rubra, Wrightia tinctoria, A. scholaris* and *Vinca major,* are known for their anti-inflammatory activity potential (Arulmozhi et al., 2012). Subraya and Gupta (2012) studied the *in vivo* anti-inflammatory activity of *A. scholaris* using animal models. *T. divaricata* leaf extracts showed significant anti-inflammatory activity in a male albino mouse model Jain et al. (2013). The isolated phytoconstituents from *C. carandas* roots, such as lupeol, stigmasterol, oleanolic acid, carissone and scopoletin, showed inhibitory potential against pro-inflammatory mediators (Galipalli et al., 2015). Bhadane and Patil (2017) demonstrated the *in vitro* anti-inflammatory activity of a steroidal saponin of *C. carandas* leaves using RBC membrane stabilising potential.

## Anticancer/Cytotoxic Activity

Conventional anticancer compounds have played a significant role in cancer treatment, and they are not without drawbacks. The limitations and drawbacks of conventional drugs are that they exert several side effects on normal cells, and therefore, their applications are restricted. Natural sources, including plants, marine organisms, and microorganisms, have been extensively studied for their potential anticancer compounds. The anticancer and cytotoxic potential of several members of the Apocynaceae family has been explored using animal models. A few compounds from Apocynaceae plants have shown promising anticancer activity.

Alkaloids such as vinblastine and vincristine have been isolated from the Madagascar periwinkle plant (*Catharanthus roseus*) and have been successfully used as chemotherapy drugs for the treatment of various types of cancer, including Hodgkin's lymphoma, leukemia, and some solid tumors. Vinblastine and vincristine interfere with microtubule formation during cell division, leading to cell cycle arrest and cell death. A cardiac glycoside, cerberin, is found in the seeds of *Cerbera odollam*, also known as the "suicide tree." While cerberin is highly toxic, it has demonstrated potent anticancer effects. It has been shown to induce apoptosis (programmed cell death) in cancer cells and inhibit tumor growth in preclinical studies. Oleandrin is isolated from the leaves of *Nerium oleander* and has shown cytotoxic effects against various cancer cell lines, including breast, ovarian, and lung cancer. Oleandrin has been studied for its potential as an anticancer agent, primarily by inducing apoptosis and inhibiting cancer cell proliferation. Cardenolide compounds such as thevetin and thevetoxin were isolated from the seeds and other parts of *Thevetia peruviana*, commonly known as yellow oleander. They have exhibited anticancer activity by inducing apoptosis and inhibiting cell proliferation in various cancer cell lines.

Rumzhum et al. (2012) reported potent cytotoxic activity of *T. divaricata* leaves *using an in vitro* bioassay. Wang et al. (2012) showed cytotoxic activity of different alkaloids isolated from *C. roseus* against breast cancer cells with an $IC_{50}$ of 0.72 μM. Siddiqui et al. (2012) isolated compounds such as odoroside A and B, oleandrin, adynerin and doxorubicin from *N. oleander* with anticancer properties. *A. scholaris* showed antiproliferative effects on cancer/tumor cells (Surendren et al., 2012). Recently, carandinol, an isohopane triterpene from *C. carandas,* showed potent cytotoxic activity (Begum et al., 2013). The crude extract of *T. peruviana* showed considerable antitumour activity in albino mice (Haldar et al., 2015). Barbon et al. (2012) demonstrated the growth inhibition of cancer cells using cardenolides isolated from *T. peruviana.* In another study, a reserpine alkaloid from *R. serpentina* was reported for its anticancer potential (Abdelfatah & Efferth, 2015).

## Cardioprotective Activity

The two most significant species in this family that are utilised to treat cardiovascular problems are *N. oleander* and *T. peruviana*. The antiatherogenic activity of *A. scholaris* stem bark extract was assessed by Bandawane et al. (2011), and the extract's ability to reduce LDL cholesterol may have contributed to the prevention of cardiovascular problems. The cardioprotective effect of *N. oleander* (Gayathri et al., 2011) was studied in rats. Roots of *R. serpentina* (Azmi & Qureshi, 2012) exhibited antidiabetic and cardioprotective potential in mice.

## Antidiabetic Activity

Several members of this family have been studied for their antidiabetic activity. For example, *Catharanthus roseus* alkaloids, which are mainly used in cancer treatment, have also been found to possess antidiabetic properties. The leaf extract of *Catharanthus roseus* can reduce blood glucose levels and improve glucose tolerance. The bark extract of *Pterocarpus marsupium* (Indian kino tree) has been traditionally used in Ayurvedic medicine for treating diabetes. Studies have indicated that the extract of this plant can lower blood glucose levels by increasing insulin secretion from pancreatic beta cells and improving insulin sensitivity.

*Gymnema sylvestre* is a well-known plant in traditional medicine for its antidiabetic properties. It is commonly used in Ayurvedic medicine to help control blood sugar levels. Gymnemic acids are known to have antidiabetic effects. These compounds can reduce the absorption of glucose from the intestines and increase insulin production in the pancreas. *Nerium oleander* has also been studied for its antidiabetic potential. The leaf extract of *Nerium oleander* possesses hypoglycemic properties, helping to lower blood glucose levels (Gayathri et al., 2011). The bark of *A. scholaris* showed hypoglycemic activity in rats (Bandawane et al., 2011). *H. antidysenterica* seeds (Keshri et al., 2012); the methanolic extract of flowers of *T. divaricata* (Rahman et al., 2011); and a partially purified fraction of the methanol extract of *C. carandas* fruit (Itankar et al., 2011; Rani et al., 2012) have shown potential antidiabetic activities in various animal models.

## Antimalarial Activity

According to the World Health Organisation (WHO), in 2019, an estimated 229 million cases of malaria were reported worldwide, resulting in approximately 409,000 deaths. *Plasmodium falciparum* and *Plasmodium vivax* are responsible for the majority of malaria cases worldwide. Therefore, the development of antimalarial drugs and the identification of natural compounds with antimalarial properties have been crucial in the fight against malaria. Plant metabolites have played a significant role in the development of antimalarial drugs and the control of malaria. The use of plant metabolites for the treatment of malaria is well documented in the traditional system of medicine. Plant metabolites such as alkaloids, terpenoids, flavonoids, and phenolic compounds are known for their role in malaria control. For example, artemisinin, derived from the plant *Artemisia annua* (sweet wormwood), and its derivatives (e.g., artesunate, artemether) are potent antimalarial compounds. They exert their action by producing free radicals within the malaria parasite, leading to parasite death. The popular antimalarial drug quinine and its derivatives, such as quinidine, were isolated from the bark of *Cinchona* tree. They were also used to treat certain heart rhythm disorders. Curcumin, an alkaloid found in turmeric, has been demonstrated to have antimalarial activity against both *Plasmodium falciparum* (the most deadly malaria parasite) and *Plasmodium vivax* (Miettinen et al., 2014). Other metabolites, such as berberine and cryptolepine, found in various plants, including *Berberis* species, have been shown to possess antimalarial activity by inhibiting the growth of *Plasmodium* parasites. Saponins derived from plants such as *Quillaja saponaria* or *Sapindus spp*. have shown antimalarial activity. Flavonoids and phenolic compounds from the Apocynaceae family have been investigated for their antiplasmodial activity (Chan et al., 2016). *In vivo* antiplasmodial activity of *A. boonei* (Iyiola et al., 2011); *C. roseus* and *T. peruviana* (Ravikumar et al., 2012); the mosquito larvicidal activity of *R. serpentina* seed extracts (Das & Chandra, 2012) and the anti-malaria and chikungunia larvicidal activity of *W. tinctoria* extract are well documented (Sakthivadivel et al., 2014).

## References

Abdelfatah, S. A., & Efferth, T. (2015). Cytotoxicity of the indole alkaloid reserpine from *Rauwolfia serpentina* against drug-resistant tumor cells. *Phytomedicine, 22*(2), 308–318.

Ahmed, M. F., & Srinivasa Rao, A. (2013). Comparative hepatoprotective activities of selected Indian medicinal plants. *Global Journal of Medical Research, 13*(2).

Ali Khan, M. S., Mat Jais, A. M., & Afreen, A. (2013). Prostaglandin analogous and antioxidant activity mediated gastroprotective action of Tabernaemontana divaricata (L.) R. Br. flower methanolic extract against chemically induced gastric ulcers in rats. *BioMed Research International, 2013*, 1–18.

Anbukkarasi, M., Thomas, P. A., Sundararajan, M., & Geraldine, P. (2016). Gas chromatography–mass spectrometry analysis and *in vitro* antioxidant activity of the ethanolic extract of the leaves of *Tabernaemontana divaricata*. *Pharmacognosy Journal, 8*(5), 451–458.

Arulmozhi, S., Mazumder, P. M., Sathiyanarayanan, L., & Thakurdesai, P. A. (2012). Analgesic, anti-inflammatory and anti-ulcerogenic activities of fractions from *Alstonia scholaris*. *Pharmacologia, 3*(5), 132–137.

Azmi, M. B., & Qureshi, S. A. (2012). Methanolic root extract of *Rauwolfia serpentina* benth improves the glycemic, antiatherogenic, and cardioprotective indices in alloxan-induced diabetic mice. *Advances in Pharmacological Sciences, 2012*, 1–11.

Bandawane, D., Juvekar, A., & Juvekar, M. (2011). Antidiabetic and antihyperlipidemic effect of *Alstonia scholaris* Linn. Bark in streptozotocin induced diabetic rats. *Indian Journal of Pharmaceutical Education and Research, 45*(2), 114–120.

Barbon, T. C., da Silva, C. P., Sampaio, S. V., & Baldo, M. A. (2012). 164. Evaluation of anticancer activity promoted by molecules contained in the extracts of *Thevetia Peruviana*. *Toxicon, 60*(2), 179–180.

Basavaraj, P., Shivakumar, B., Shivakumar, H., & Manjunath, V. J. (2011). Evaluation of anticonvulsant activity of Tabernaemontana divaricata (Linn) r. br. flower extract. International. *Journal of Pharmacy and Pharmaceutical Sciences, 3*(3), 310–315.

Begum, S., Syed, S. A., Siddiqui, B. S., Sattar, S. A., & Choudhary, M. I. (2013). Carandinol: first isohopane triterpene from the leaves of Carissa carandas L. and its cytotoxicity against cancer cell lines. *Phytochemistry Letters, 6*(1), 91–95.

Bhadane, B. S. (2017). *Studies on biological activities and in vitro propagation of some members of family Apocynaceae*. PhD thesis. KBC North Maharashtra University.

Bhadane, B. S., & Patil, R. H. (2017). Isolation, purification and characterisation of antioxidative steroid derivative from methanolic extract of *Carissa carandas* (L.) leaves. *Biocatalysis and Agricultural Biotechnology, 10*, 216–223.

Bhadane, B. S., Patil, M. P., Maheshwari, V. L., & Patil, R. H. (2018). Ethnopharmacology, phytochemistry, and biotechnological advances of family Apocynaceae: A review.

Bhati, P., Shukla, A., & Sharma, M. (2014). Hepatoprotective activity of leaves extracts of Carissa carandas Linn. *American Journal of Pharm Research, 4*(11), 5185–5192.

Botelho, A. F. M., Santos-Miranda, A., Joca, H. C., Mattoso, C. R. S., de Oliveira, M. S., Pierezan, F., Cruz, J. S., Soto-Blanco, B., & Melo, M. M. (2017). Hidroalcoholic extract from *Nerium oleander* L.(Apocynaceae) elicits arrhythmogenic activity. *Journal of Ethnopharmacology, 206*, 170–177.

Chan, E. W. C., Wong, S. K., & Chan, H. T. (2016). Apocynaceae species with antiproliferative and/or antiplasmodial properties: A review of ten genera. *Journal of Integrative Medicine, 14*(4), 269–284.

Das, D., & Chandra, G. (2012). Mosquito larvicidal activity of *Rauvolfia serpentina* L. seeds against Culex quinquefasciatus say. *Asian Pacific Journal of Tropical Medicine, 5*(1), 42–45.

Ezeigbo, I. I., Ezeja, M. I., Madubuike, K. G., Ifenkwe, D. C., Ukweni, I. A., Udeh, N. E., & Akomas, S. C. (2012). Antidiarrhoeal activity of leaf methanolic extract of Rauwolfia serpentina. *Asian Pacific Journal of Tropical Biomedicine, 2*(6), 430–432.

Galipalli, S., Patel, N. K., Prasanna, K., & Bhutani, K. K. (2015). Activity-guided investigation of Carissa carandas (L.) roots for anti-inflammatory constituents. *Natural Product Research, 29*(17), 1670–1672.

Ganapathy, P. S., Ramachandra, Y. L., & Rai, S. P. (2011). *In vitro* antioxidant activity of *Holarrhena antidysenterica* wall. Methanolic leaf extract. *Journal of Basic and Clinical Pharmacy, 2*(4), 175–178.

Gayathri, V., Ananthi, S., Chandronitha, C., Ramakrishnan, G., Sundaram, R. L., & Vasanthi, H. R. (2011). Cardioprotective effect of *Nerium oleander* flower against isoproterenol-induced myocardial oxidative stress in experimental rats. *Journal of Cardiovascular Pharmacology and Therapeutics, 16*(1), 96–104.

Govindasamy, C., & Srinivasan, R. (2012). In vitro antibacterial activity and phytochemical analysis of Catharanthus roseus (Linn.) G. Don. *Asian Pacific Journal of Tropical Biomedicine, 2*(1), S155–S158.

Haldar, S., Karmakar, I., Chakraborty, M., Ahmad, D., & Haldar, P. K. (2015). Antitumour potential of *Thevetia peruviana* on Ehrlich's *ascites carcinoma*—Bearing mice. *Journal of Environmental Pathology, Toxicology and Oncology, 34*(2), 105–113.

Hegde, K., & Joshi, A. B. (2009). Hepatoprotective effect of Carissa carandas Linn root extract against CCl 4 and paracetamol induced hepatic oxidative stress, *47*, 660–667.

Itankar, P. R., Lokhande, S. J., Verma, P. R., Arora, S. K., Sahu, R. A., & Patil, A. T. (2011). Antidiabetic potential of unripe *Carissa carandas* Linn. Fruit extract. *Journal of Ethnopharmacology, 135*(2), 430–433.

Iyiola, O. A., Tijani, A. Y., & Lateef, K. M. (2011). Antimalarial activity of ethanolic stem bark extract of *Alstonia boonei* in mice. *Asian Journal of Biological Sciences, 4*(3), 235–243.

Jain, S., Sharma, P., Ghule, S., Jain, A., & Jain, N. (2013). *In vivo* anti-inflammatory activity of *Tabernaemontana divaricata* leaf extract on male albino mice. *Chinese Journal of Natural Medicines, 11*(5), 472–476.

James, J., Veettil, A. K. T., Pratyush, K., Misra, C. S., Sahadevan, L. D. M., & Thankamani, V. (2011). *In vitro* antioxidant activity of flowers and fruits of *Alstonia scholaris*. *International Journal of Phytomedicine, 3*(4), 475–479.

Keshri, U. P., Chandra, S., & Sharma, J. (2012). Antidiabetic efficacy of ethanolic extract of *Holarrhena antidysenterica* seeds in streptozotocin–induced diabetic rats and it's influence on certain biochemical parameters. *Journal of Drug Delivery and Therapeutics, 2*(4), 159–162.

Khan, A., Khan, S. R., & Gilani, A. H. (2012). Studies on the in vitro and in vivo antiurolithic activity of Holarrhena antidysenterica. *Urological Research, 40*(6), 671–681.

Kumar, P., Shukla, P., Shukla, P., & Alok, S. (2015). Antinociceptive and gastro protective effects of inhaled and orally administered thevetia peruviana pers. K. Schum. Essential oil. *International Journal of Pharmaceutical Sciences and Research, 6*(10), 4496–4502.

Miettinen, K., Dong, L., Navrot, N., Schneider, T., Burlat, V., Pollier, J., & Verpoorte, R. (2014). The seco-iridoid pathway from *Catharanthus roseus*. *Nature Communications, 5*, 1–11.

Mukhram, M. A., Shivakumar, H., Viswanatha, G. L., & Rajesh, S. (2012). Anti-fertility effect of flower extracts of Tabernaemontana divaricata in rats. *Chinese Journal of Natural Medicines, 10* (1), 58–62.

Poornima, K., Chella Perumal, P., & Gopalakrishnan, V. K. (2014). Protective effect of ethanolic extract of Tabernaemontana divaricata (L.) R. Br. against DEN and Fe NTA induced liver necrosis in Wistar Albino rats. *BioMed Research International*, 1–9.

Rahman, M., Islam, M. S., Ali, M. S., Islam, M. R., & Hossain, M. Z. (2011). Antidiabetic and cytotoxic activities of methanolic extract *of Tabernaemontana divaricata* (L.) flowers. *International Journal of Drug Development and Research, 3*(3), 270–276.

Rahman, N., Rahman, H., Haris, M., & Mahmood, R. (2017). Wound healing potentials of Thevetia peruviana: Antioxidants and inflammatory markers criteria. *Journal of Traditional and Complementary Medicine.* https://doi.org/10.1016/j.jtcme.2017.01.005

Raj, C. N., Balasubramaniam, A., & Nadeem, S. (2013). Antidiarrheal potential of Tabernaemontana divaricata. *Phytopharmacology, 4*(1), 61–68.

Ramalakshmi, S., Edaydulla, N., Ramesh, P., & Muthuchelian, K. (2012). Investigation on cytotoxic, antioxidant, antimicrobial and volatile profile of Wrightia tinctoria (Roxb.) R. Br. flower used in Indian medicine. *Asian Pacific Journal of Tropical Disease, 2*, S68–S75.

Rani, M. S., Pippalla, R. S., Mohan, G. K., & Gangaraju, M. (2012). Anti-diabetic activity of methanolic and ethyl acetate extracts of *Wrightia tinctoria* R. Br. fruit. *International Journal of Pharmaceutical Sciences and Research, 3*(10), 3861–3866.

Ravikumar, S., Inbaneson, S. J., & Suganthi, P. (2012). *In vitro* antiplasmodial activity of ethanolic extracts of south Indian medicinal plants against plasmodium falciparum. *Asian Pacific Journal of Tropical Disease, 2*(3), 180–183.

Rice-Evans, C. A., Miller, N. J., & Paganga, G. (1996). Structure-antioxidant activity relationships of flavonoids and phenolic acids. *Free Radical Biology and Medicine, 20*(7), 933–956.

Rumzhum, N. N., Rahman, M. M., & Kazal, M. K. (2012). Antioxidant and cytotoxic potential of methanol extract of *Tabernaemontana divaricata* leaves. *International Current Pharmaceutical Journal, 1*(2), 27–31.

Sabde, S., Bodiwala, H. S., Karmase, A., Deshpande, P. J., Kaur, A., Ahmed, N., et al. (2011). Anti-HIV activity of Indian medicinal plants. *Journal of Natural Medicines, 65*(3–4), 662–669.

Sakthivadivel, M., Gunasekaran, P., Annapoorani, J. T., Samraj, D. A., Arivoli, S., & Tennyson, S. (2014a). Larvicidal activity of *Wrightia tinctoria* R. BR.(Apocynaceae) fruit and leaf extracts against the filarial vector Culex quinquefasciatus say (Diptera: Culicidae). *Asian Pacific Journal of Tropical Disease, 4*, S373–S377.

Siddiqui, B. S., Khatoon, N., Begum, S., Farooq, A. D., Qamar, K., Bhatti, H. A., & Ali, S. K. (2012). Flavonoid and cardenolide glycosides and a pentacyclic triterpene from the leaves of *Nerium oleander* and evaluation of cytotoxicity. *Phytochemistry, 77*, 238–244.

Singhal, K. G., & Gupta, G. D. (2011). Some central nervous system activities of *Nerium oleander* Linn (Kaner) flower extract. *Tropical Journal of Pharmaceutical Research, 10*(4), 455–461.

Singhal, K. G., & Gupta, G. D. (2012). Hepatoprotective and antioxidant activity of methanolic extract of flowers of *Nerium oleander* against CCl4–induced liver injury in rats. *Asian Pacific Journal of Tropical Medicine, 5*(9), 677–685.

Subraya, C. K., & Gupta, D. (2012). Antioxidant anti-inflammatory activity of Alstonia scholaris R. Br. stem bark extract. *Free Radicals and Antioxidants, 2*(2), 55–57.

Surendren, P., Jayanthi, G., & Smitha, K. R. (2012). *In vitro* evaluation of the anticancer effect of methanolic extract of *Alstonia scholaris* leaves on mammary carcinoma. *Journal of Applied Pharmaceutical Science, 2*(5), 142–149.

Verma, K., Shrivastava, D., & Kumar, G. (2015). Antioxidant activity and DNA damage inhibition *in vitro* by a methanolic extract of *Carissa carandas* (Apocynaceae) leaves. *Journal of Taibah University for Science, 9*(1), 34–40.

Wang, C. H., Wang, G. C., Wang, Y., Zhang, X. Q., Huang, X. J., Zhang, D. M., & Ye, W. C. (2012). Cytotoxic dimeric indole alkaloids from *Catharanthus roseus*. *Fitoterapia, 83*(4), 765–769.

Wankhede, S. B., Routh, M. M., Rajput, S. B., & Karuppayil, S. M. (2013). Antifungal properties of selected plants of Apocynaceae family against the human fungal pathogen Candida albicans. *International Current Pharmaceutical Journal, 2*(7), 122–125.

# Chapter 9
# Agricultural Importance of Apocynaceae Members: Insecticidal Properties

## Introduction

Agricultural insects and pests cause damage to crops, livestock, or stored products. These pests include insects, mites, nematodes, rodents, birds, and plant pathogens such as fungi, bacteria, and viruses. They can cause a significant amount of damage to crops, resulting in lower yields and lower-quality produce. Insects are among the most damaging pests in agriculture. They can cause direct damage to crops by feeding on plant tissues or by transmitting plant diseases. Some of the most common insect pests in agriculture include aphids, caterpillars, beetles, and thrips. In addition to insects, mites can also cause damage to crops by feeding on plant tissues. Controlling pests in agriculture is important for maintaining healthy crops and ensuring high yields. Integrated pest management (IPM) is a sustainable approach for the control of insect pests that involves multiple control methods, including cultural practices and biological and chemical control, that involve the limited use of pesticides while maintaining effective pest control. IPM strategies may include crop rotation, use of resistant plant varieties, introduction of natural predators or parasites, and targeted use of pesticides when necessary. Biopesticides/botanical pesticides are a type of pesticide derived from natural sources such as plants, microbes, and other naturally occurring substances. They are used to control pests and diseases in agriculture, forestry, and public health. There are three main categories of biopesticides: microbial biopesticides (microorganisms such as bacteria, fungi, and viruses), plant-derived biopesticides (these are derived from natural plant extracts or oils and are used to repel, deter, or kill pests), and biochemical biopesticides (naturally occurring substances such as pheromones, enzymes, and plant growth regulators). Biopesticides have several advantages over synthetic pesticides. They are generally less toxic to humans, wildlife, and the environment and have a shorter residual life. They are also more target specific, meaning that they have a lower impact on beneficial organisms such as friendly insects and pollinators (Salunke, 2006).



R. H. Patil et al., *Apocynaceae Plants*,
https://doi.org/10.1007/978-981-99-5406-3_9

**Fig. 9.1** *Calotropis procera* (Apocynaceae), a widely used species for insecticidal purposes

Apocynaceae plants have been widely used as a source of insecticidal metabolites. Several plants have been documented to have insecticidal properties (Bhadane et al., 2018). A common medicinal plant found in the tropics of Asia and Africa is *Calotropis procera* (Ait) R. Br., often known as "Arka" in India. It is a spreading shrub or small tree that exudes abundant amounts of milky sap when cut or broken. Leaves are opposite, large, grey–green up to 15 cm long and 10 cm wide, with a pointed tip, two rounded basal lobes, and no leaf stalk; its waxy white flowers are carried in stalked clusters at the ends of the branches; its fruit is grey–green, inflated, and 8–12 cm long, and it has many seeds with tufts (Sharma & Sharma, 1999) (Fig. 9.1). External parts as well as the roots of this plant have been widely utilized in conventional medical practices. According to Mascolo et al. (1988), an ethanol extract of the plant's flower has antibacterial, anti-inflammatory, antipyretic, analgesic, anticancer (Smit et al., 1995), and antimalarial (Sharma & Sharma, 1999, 2001) properties. In addition, the extracts of C. procera in water, ethanol and other solvents show insecticidal, larvicidal, antibacterial, and antiparasitic properties (Moursy, 1997; Markouk et al., 2000). The primary components of *C. procera* latex are flavonoids, cardenolides, triterpenoids, alkaloids, resins, anthocyanins, saponins, tannins, and proteolytic enzymes (Deepak, 1995).

## *Insects and Pests*

Insects are among the highly diverse forms of life living on earth. They can be found in diverse habitats, such as the ocean, swamps, forests, and deserts, as well as in harsh environments (Imms, 1964). They are highly adaptable to diverse environmental conditions, as their total numbers are far greater than those of other animals. Insects make up over 85% of all animal species known to exist on Earth, yet it is generally believed that just a fifth of all pest species have been recognized and categorized. Since the Palaeozoic era (230–570 million years ago), insects have been prevalent on Earth, and insects that resemble modern species have been discovered in rocks from the Permian era (230–280 million years ago) (FAO, 2006). Most insects have direct benefits for both people and the environment. Despite their enormous diversity, fewer than 0.5% of all known insect species are regarded as pests, and only a small number of these potentially pose a major threat to humans.

## *Insects and Agricultural Losses*

Insects, weeds, or phytopathogens that affect the crop while it is still in the field cause preharvest losses. At different times during the life of a crop, cutworms, caterpillars, aphids, and fly maggots operate as leaf miners, stem borers, and pod borers (Atwal, 1976). Helicoverpa species, distributed worldwide, have polyphagous feeding habits (Harsulkar et al., 1998; Grover & Pental, 2003). As it is a serious pest of cotton in the USA, it is also called the American bollworm. In India, it causes serious damage to cotton, sorghum and other rabi crops, especially red gram and Bengal gram (Pradhan, 1992; Gujar et al., 2000). Cutworms (Agrotis sp.): These are a small group of polyphagous pests active in the rabi season. They are known as cutworms because they cause whole plants or crops such as gram or their twigs to fall on the ground (Pradhan, 1992). The plume moth caterpillar (*Exelastis atoms*) Walsingham (Family: Pterophoridae): This pest causes serious damage to developing seeds of red gram. Leaf miners and other *Agromyzid* flies: Leaf miners of major importance belong to the family of *Agromyzid* flies. Common examples are the pea leaf miner (*Phytomyza atricornis* Meigen), the pea stem borer (*Melanagromyza phaseoli* Coquillett) and the red gram pod fly (*Agromyza obtuse* Mallas). Aphids: A black species of aphid called *Aphis craccivora* K. is a serious pest on leguminous crops, particularly peas. Aphids damage the plant by sucking the cell sap. Other phytopathogenic fungi include *Aspergillus candidus, Aspergillus clavatus, Aspergillus fumigatus, Aspergillus parasiticus, Aspergillus restrictus, Alternaria alternata, Fusarium graminearum, Fusarium moniliforme, Fusarium roseum, Fusarium tricinctum, Helminthosporium* spp., *Scopulariopsis* spp., *Penicillium verrucosum, etc.* ICIPE, 2006.

## *Traditional Pest Control Methods and their Drawbacks*

A number of traditional pest control methods are used in practice to control these insect pests. Traditionally used materials such as inert dust are mixed with the product, which contributes to decreasing pest activity (Arthur, 1996). Fumigation, control by drying, exposure to sunlight, or exposure followed by sieving of the grains, is a well-known technique among farmers in Asia and sub-Saharan Africa. The use of synthetic is also popular among farmers. Pesticides are agrochemicals required to combat the attack of various pests (s) on agricultural and horticultural crops. Synthetic pesticides have rapid action, are target specific and have relatively high residual toxicity. However, their injudicious use has resulted in the disruption of biological control by natural enemies. It also caused the development of resistance in pests (Varma & Dubey, 1999). It resulted in a rise in pest insects that attack stored grains and had unfavorable consequences on nontarget organisms, raising worries for the environment and human health. The public is becoming increasingly concerned about synthetic/chemical pesticides due to resistant strains/insects, release hazardous residues in food and feed, human safety, irregular supply, and high prices. Finding new approaches to combat insect pests at various phases of crop production and storage is therefore essential. A biological approach may offer a practical and sustainable substitute for the management of agricultural pests. It has been successful to employ the bacteria *Bacillus thuringiensis* to combat pests of stored grain and seedlings. B. *thuringiensis* was found to be the best choice for application on grains and pulses. Thus, the importance of microbial, viral, and bacterial insecticides for crop protection is growing. The first microbial pesticides to be registered in the US were bacteria from the genus *Bacillus* (*B. popilliae* and *B. popilliae Lentimorbus*) to control the Japanese beetle, *Popillia japonica*. The most popular and effective microbial pesticide ever registered is *B. thuringiensis* (Bt). Industrial bioreactors with huge capacities are used to produce Bt.

Toxin of *B. thuringiensis* (*Bt*) Berlang. var. Kurstaki, a Gram-positive, aerobic, sporulating bacterium, is effective against lepidopteran pests. It produces a proteinaceous crystalline protoxin. This protoxin, a 130 kDa molecule, is not toxic to insects. After ingestion by the insect, due to the alkaline pH of the insect gut, it is cleaved into smaller subunits (60 and 70 kDa) called δ-endotoxins. The activated toxins bind to the receptors present on the larval midgut epithelium, disrupting the membrane and concomitant death of the insect (Khanna, 1998). *Bt israleansis* (*Bti*) was the first *Bt* subspecies discovered by Morgalit in 1976 and has activity only against Diptera (Becker & Margalit, 1993). On the other hand, entomopathogenic fungi are found to produce more than 60 insecticidal compounds that have been isolated and characterized from various fungi possessing a broad spectrum of activity. Avermectins of *Streptomyces avermitilis*, milbemycins of *S. hygroscopicus*, tetranactin from *S. aureus*, piericidins from *S. mobaraensis* and destruxins from *Metarrhizium anisopliae* are some examples of the insecticidal compounds produced by different fungi. The term "integrated pest management" (IPM) has been defined and used in a variety of ways. Previously known as "integrated control," is now expanded as IPM that encompasses all biological, cultural, and artificial practices. The most frequent risk that chemical and synthetic pesticides pose is to their natural enemies. The best strategy for decreasing the detrimental effects on natural enemies is to modify pesticide use practices in light of IPM strategies. Alternatively, IPM programs have been run with "pesticide management" goals rather than "pest management" goals (Lewis et al., 1997).

## Botanicals as Pesticides: An Overview

Plants have developed effective countermeasures to withstand herbivorous pests/ diseases through the production of highly diverse types of organic compounds and metabolites called secondary metabolites (Salunke et al., 2009). The primary metabolites carbohydrates, lipids, nucleotides and peptides are produced by living cells to control basic life processes. On the other hand, secondary metabolites are obtained from primary metabolites or produced by cells, as the biproducts of their metabolism are involved in protecting plants against various biotic and abiotic stresses, including diseases and pests. A number of plant families produce alkaloids, tannins, terpenes, and phenolics that are widely used for the control of insects and pests. There are several types of secondary plant metabolites with insecticidal/pesticidal properties, *viz.*, repellents, antifeedants, phagostimulants and toxins (Krishnayya & Rao, 1996; Nivsarkar et al., 2001). Many classes of plant proteins and secondary plant substances have been shown to have toxic or antimetabolic effects on insects and have been proposed as possible candidates (Simmonds et al., 1989; Kim et al., 2003; Kotkar et al., 2002; Upasani et al., 2003; Patil et al., 2010). The important character of plant-derived insecticidal metabolites is that they have chronic toxicity on insects, and therefore, their effects are less dramatic than those of synthetic insecticides (Estruch et al., 1997; Park et al., 2003).

Plant-derived pesticides or botanical pesticides, although natural, are not necessarily always safe, and there are a few limitations (Mendki et al., 2001). Most of them are expensive to produce compared to synthetic insecticides and require a large volume of starting material; therefore, an adequate supply of raw material may be a limiting factor in their availability. Moreover, the bioefficacy of botanical pesticides may vary even in identical preparations because of a number of factors. Although natural in origin, they are toxicity issues associated with botanical pesticides in several cases. They are toxic to pests and beneficial insects alike, and repeated use can disrupt biological control. Some commonly used botanical pesticides with their source, mode of action and target insects are shown in Table 9.1.

However, unlike many synthetic pesticides, natural compounds are usually biodegradable and are potentially environmentally safer than many of the synthetic products on the market today. The major benefits of biopesticides are that they are generally safer than conventionally used synthetic pesticides, and they are renewable. Being biodegradable, they can be metabolized, so there is no or less risk of resistance in the target pest. There is a lower risk of developing quick resistance.

**Table 9.1** Some examples of major botanicals and their pesticidal activities (Source: Weinzierl, 1998)

| Plant based molecules/ botanicals | Plant source | Probable mechanism of action | Mammalian $LD_{50}$ | Target insect(s) |
|---|---|---|---|---|
| Pyrethrum | Flowers of *chrysanthemum cinerariaefolium* | Interferes with nerve transmission | > 1000 | On fleas, ticks and lice |
| Rotenone | Roots of *derris* sp., *Lonchocarpus* and tropical legumes | Disrupts energy metabolism in mitochondria | Dermal >1000 Oral 25–3000 | Beetles |
| Sabadilla | Seeds of *Schoenocaulon officinale* and *Veratrum album* | Interferes with Na and K ion movement in nerve axons | Oral near 4000 | Bugs and thrips |
| Ryania | Stems of *Ryania speciosa* | Activates ca ion release channels | Oral near 1000 Dermal near 4000 | Caterpillars and thrips |
| Nicotine | Leaves of *Nicotiana* sp | Mimics acetylcholine to overstimulate receptor cells | Oral near 3–188 | Aphids, thrips and mites |
| Azadirachtin | Leaves, bark, and seeds of *Azadirachta indica* and *Melia azedarach* | Growth hormone analog | Oral >5000 | Soft-bodied insect pests |
| Limonene/ linalool | Citrus oils | Stimulation of sensory nerves | Oral >5000 Dermal >3500 | Fleas and ticks |

$LD_{50}$ = dose required to kill 50% of the test animal population; higher numbers indicate lower toxicity (Source: Weinzierl, 1998)

Most importantly, chances of accumulation of residues in food and fodder, disturbance of soil microflora and hazard to wild life are less compared to synthetic pesticides.

## Plants of the Apocynaceae Family as Bioresources for Pesticidal Molecules

The plant kingdom contains an estimated 500,000–600,000 different species. It is also estimated that 3000 plants belonging to more than 75 plant families are known to exhibit insecticidal activities (Dev & Koul, 1997).

For pesticidal activities, the family Solanaceae is most widely used, followed by Amaranthaceae, Apocynaceae, Fabaceae and Malvaceae. *Calotropis procera* (Ait.) R. Br. and *Calotropis gigantea* R. Br. are two species of family Apocynaceae and subfamily Asclepiadoideae (milkweeds), which are soft-wooded, evergreen, perennial shrubs. Naturally occurring plant materials or products derived from such plant materials are termed 'botanicals' (Jacobson, 1989). They are also referred to as "first generation" insecticides. They include compounds such as Derris, pyrethrum, rotenone, nicotine, sabadilla, physostigmine and quassin, as well as lead aresenate and calcium arsenate. Some of these compounds formed the basis for synthetic "second generation" insecticides. For example, pyrethrum from Chrysanthemum flowers gave rise to synthetic pyrethroids, and physostigmine from Physostigma venenosa gave rise to structurally similar synthetic carbamates. Botanical insecticides were extensively used (1800–1940) until the discovery of the synthetic insecticide DDT. Table 9.2 provides a review of Apocynaceae members with insecticidal properties.

## General Procedure Isolation of Insecticidal Metabolites(S) from *Calotropis Procera*

Parts of the identified plants were collected and carried to the laboratory. Different parts, such as leaf stems, flowers, bark, and roots, can be used for the isolation of insecticidal molecules. The plant parts can be used fresh or subjected to sun drying. Pant materials (leaf/stem/root powders of *C. procera*) were extracted in methanol using the Soxhlet apparatus at 65–70 °C for 6–8 h. The extracted material is dried in a rotary vacuum evaporator or drier, and the yield of the product is measured. This extract was dissolved in methanol and used for various experiments as well as for studying the larvicidal, insecticidal and oviposition deterrent activities (Fig. 9.2).

## Insecticidal Metabolites from Apocynaceae Plants

Secondary metabolites play an important role in the response to biotic and abiotic stress-related mechanisms in plants. The adaptive role of secondary metabolites in plant defense against insect pests is gaining importance in our understanding of plant defense. They participate in photosynthesis and cellular energy transfer mechanisms, operate as antioxidants or as enzyme inhibitors, and may be precursors to hazardous compounds or have therapeutic effects (Harborne & Williams 2000; Winkel-Shirley, 2002).

Strong adulticidal activity against the stored grain pest *C. chinensis* was observed with the various secondary metabolites from *A. squamosa, R. communis, C. procera* and *A. calamus* (Upasani et al., 2003; Salunke et al., 2005) (Table 9.3). Both dose and exposure duration had an impact on the adulticidal action. Other studies have achieved similar outcomes for the toxicity of numerous additional plant bioactive compounds (Kim & Ahn, 2001). According to El-Nahal et al. (1989), exposure

**Table 9.2** Insecticidal properties of Apocynaceae members

| Plant and plant part | Pesticidal/insecticidal property | Crude solvent extract/ purified compound | Effective concentration/ Dose | Study model | References |
|---|---|---|---|---|---|
| *Catharanthus roseus* | | | | | |
| Leaves | Larvicidal | Ethanolic extract | $IC_{50}$ 49.63 µg/ mL | In vitro (Antiplasmodial against *plasmodium falciparum*) | Salunke, (2006) |
| *Wrightia tinctoria* | | | | | |
| Leaves | Antimalarial (Larvicidal) | Petroleum ether and aqueous extract | $LC_{50}$ 0.37 and 0.11% | In vitro (against filarial larvae *Culex quinquefasciatus*) | Salunke, (2006) |
| Fruits | Antimalarial (Larvicidal) | Petroleum ether and aqueous extract | $LC_{50}$ 1.31 and 0.09% | In vitro (against filarial larvae *Culex quinquefasciatus*) | Salunke, (2006) |
| | Antidiabetic | Methanolic and ethyl acetate extract | 300 mg/kg | In vivo (Alloxan induced mice) | Prabhu et al. (2018) |
| *Calatropis procera* | | | | | |
| Milkweed juice | Larvicidal activity against *Musca domestica* larvae | Alkaloids, steroids and resinous | 80 µg/larvae | In vivo Larvicidal activity against Musca domestica larvae | Mendki et al. (2005), Khatter and Abuldahab (2012) |
| Leaves | Larvicidal activity against *Macrophomina phaseolina* | Hydrocarbon, azocompounds, aromatic carboxylic acids | 1%, 2.5%, 4%, 5.5% and 7% of | In vivo model of *Macrophomina phaseolina* | Waheed et al. (2016) |
| Leaves | Insecticidal activity against *Callosobruchus maculatus* | Methanol | 1.0, 1.5 and 5.0% | In vitro oviposition deterrent activity | Chauhan et al. (2016) |
| Leaves | Insecticidal activity against *C. chinensis* | Flavonoids: Rutin and quercetin | 5 and 10 mg $mL^{-1}$ | In vitro oviposition deterrent activity *Callosobruchus chinensis* | Salunke et al. (2005) |

| | | | | | |
|---|---|---|---|---|---|
| *Calotropis gigantea* | | | | | |
| Leaves | Larvicidal activity against *Helicoverpa armigera* | Aqueous suspensions | 2%, 4%, 10% | Leaf disc bioassay method | Prabhu et al. (2018) |
| Fruits | Larvicidal activity against *Helicoverpa armigera* | Aqueous suspensions | 2%, 4%, 10% | Leaf disc bioassay method | Prabhu et al. (2018) |
| Flower | Larvicidal activity against *Helicoverpa armigera* | Aqueous suspensions | 2%, 4%, 10% | Leaf disc bioassay method | Prabhu et al. (2018) |
| Roots | Larvicidal activity against *Helicoverpa armigera* | Aqueous suspensions | 2%, 4%, 10% | Leaf disc bioassay method | Prabhu et al. (2018) |
| Stem | Larvicidal activity against *Helicoverpa armigera* | Aqueous suspensions | 2%, 4%, 10% | Leaf disc bioassay method | Prabhu et al. (2018) |

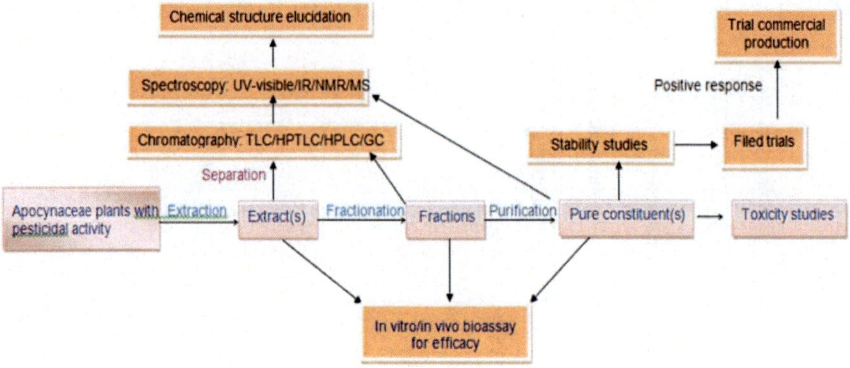


**Fig. 9.2** Isolation of insecticidal metabolites(s) from *Calotropis procera* (Salunke, 2006)

**Table 9.3** Effect of purified secondary metabolites on different stages of development of *C. chinensis* (Source: Salunke, 2006)

| Sr. no. | Secondary metabolite | Emergence after treatment | | | | | |
|---|---|---|---|---|---|---|---|
| | | 1[a] | 7[a] | 13[a] | 17[a] | 21[a] | 26[a] |
| 1 | *A. squamosa* Myricetin | 12.7 ± 0.8b | 21.8 ± 0.8 cd | 28.8 ± 0.8b | 64.7 ± 0.8c | 74.2 ± 0.8c | 81 ± 0.6bcd |
| 2 | *C. procera* quercetin | 19.2 ± 1.2c | 28.2 ± 0.8e | 49.3 ± 1d | 68.7 ± 0.8d | 77.8 ± 1.2d | 80.3 ± 0.5b |
| 3 | *A. squamosa* long chain fatty alcohol | 8.17 ± 0.8a | 12.3 ± 0.8a | 25 ± 0.9a | 54.8 ± 0.8a | 67.8 ± 0.8a | 77.8 ± 0.8a |
| 4 | *A. Calamus* Rutin | 11.5 ± 1b | 20.3 ± 0.8b | 28.3 ± 0.5b | 60.3 ± 0.5b | 71.2 ± 1b | 80.2 ± 0.8b |
| 5 | Flavone | 31 ± 0.9e | 44.5 ± 1 g | 58.3 ± 0.8f | 77 ± 0.9e | 79.5 ± 1d | 81.7 ± 0.5bcd |
| 6 | Rutin | 11.5 ± 1b | 19.2 ± 0.8b | 28.3 ± 0.8b | 60.2 ± 0.8b | 70.2 ± 1b | 80.7 ± 1.2bc |
| 7 | Quercetin | 20 ± 1.7c | 27.5 ± 1e | 48.2 ± 0.8d | 70.2 ± 1.5d | 78.5 ± 0.5d | 81.3 ± 0.8bcd |
| 8 | Myricetin | 12.7 ± 0.8b | 22.8 ± 2.6d | 30.8 ± 1.2c | 63.3 ± 0.8c | 71.5 ± 1b | 80.3 ± 1b |
| 9 | Fisetin | 37.8 ± 0.8f | 48.2 ± 0.8 h | 60 ± 0.9f | 78 ± 0.9f | 79.5 ± 1d | 81 ± 0.9bcd |
| 10 | Quercitrin | 24.8 ± 0.8d | 37.7 ± 1f | 54.7 ± 0.8e | 75.7 ± 1e | 78.3 ± 0.8d | 82.2 ± 0.8 cd |
| 11 | Solvent control | 81.7 ± 1.2 g | 81.7 ± 1.2i | 81.7 ± 1.2 g | 81.7 ± 1.2 g | 81.7 ± 1.2e | 81.7 ± 1.2 cd |
| 12 | Control without treatment | 82.3 ± 0.5 g | 82.3 ± 0.5i | 82.3 ± 0.5 g | 82.3 ± 0.5 g | 82.3 ± 0.5e | 82.3 ± 0.5d |

Means within the column followed by the same letter are not significantly different by Tukey's MRT ($P = 0.05$). Means (±SD) of the data are reported.

[a]Days after oviposition; (Dose applied at 10 mg $mL^{-1}$) (Source: PhD Thesis, Salunke et al., 2005)

duration, not dosage, proved to have the greatest impact on the toxicity of the essential oil of *A. calamus* on adults of five stored-product insect species.

Terpenoids are volatile plant-derived substances popularly used as fumigants. According to Ahn et al. (1998), the monoterpene carvacrol functions as an insecticidal fumigant and has a wide range of insecticidal actions against agricultural, stored goods, and medical insect pests. Terpenoids from *A. squamosa* were found to have fumigant action against *C. chinensis*. In closed cups, as opposed to open cups, partially purified terpenoids and long-chain fatty alcohols were significantly more effective against adults of *C. chinensis*. These findings suggest that the chemicals identified from *A. squamosa* may behave as fumigants, which may cause toxic effects to enter the insect body through the respiratory system and produce insecticidal effects. Secondary metabolites are known to reduce oviposition at a 10 mg $mL^{-1}$ dose of secondary metabolites of *C. procera,* and no egg laying occurred in most of the combined forms of partially purified secondary metabolite treatments (Salunke et al., 2005). This suggests that flavonoids act as ovipositional deterrents at higher concentrations. The reduced oviposition may be attributed to the triglycerides present in the oils (Hill & Schoonhoven, 1981; Rajapakse & Van Emden, 1997).

The application of plant secondary metabolites at higher doses (10 mg/ml) was able to reduce progeny production, similar to the effect of oil (Rajapakse & Van Emden, 1997). Credland (1992) demonstrated the ovicidal effect of oils obtained from plants that were previously reported to have insecticidal properties. The effect of plant secondary metabolites on eggs may be attributed to their chemical toxicity or physical characteristics, which may reduce the surface tension within the egg similar to those of surfactants/oils (Singh et al., 1978). Flavonoids cause lethal effects not only through direct contact but also by penetrating the cotyledons and inhibiting larval development. Different flavonoids are found to alter molting in insects, causing death (Stamp & Yang, 1996). The majority of plant secondary metabolites either act as antiestrogens or inhibit cytochrome P450 isozyme expression and activity (Mitchell et al., 1993; Tsyrlov et al., 1994). The effect of plant secondary metabolites was reduced when applied 21 and 26 days after oviposition, probably because by that time, most of the larvae became pupae and entered deep inside the grains.

## Botanical Pesticides: A Way Forward

Human, animal and environmental health concerns have prompted governments all over the world to restrict the use of chemical pesticides and other agricultural products. This has resulted in the search for new safe and sustainable alternatives for insect and pest control. Plant secondary metabolites may represent a new class of extremely physiologically active chemicals and an alternative to traditional synthetic materials. The rate of excretion or inactivation of these compounds, changes in the target receptors for these compounds and interactions with endogenous factors that may override the target responses are just a few examples of how the response to a given compound may be significantly influenced by other factors, both endogenous and exogenous. For this reason, understanding the physiological mechanism by which these active plant secondary metabolites function is crucial for their potential use in agriculture. Additionally, there is a need for the formulation of new strategies for understanding the mode of action of botanical pesticides.

Plant secondary metabolites have multiple modes of action, and the exact mechanism of their activity may need further investigation. There is an important role of plant secondary metabolites in the metamorphism and reproduction of beetles. There is a relationship between oxidative stress and membrane damage, physiology and nervous system, digestion, ecdysteroid titres and egg laying levels, but further study is needed to ascertain the mechanisms of action in these crucial life cycle events. It can be concluded that the Apocynaceae plant can be considered a reliable, cost-effective source of biopesticides, and it could be an eco-friendly strategy for the protection of crops and has the potential to be included in future insecticide development programs.

## References

Ahn, Y. J., Lee, S. B., Lee, H. S., & Kim, G. H. (1998). Insecticidal and acaricidal activity of carvacrol and β-thujaplicine derived from *Thujopsis dolabrata* var. hondai sawdust. *Journal of Chemical Ecology, 24*, 81–90.

Arthur, F. H. (1996). Grain protectants: Current status and prospects for the future. *Journal of Stored Products Research, 32*(4), 293–302.

Atwal, A.S. (Ed.) (1976). Insect pests of stored grain and other products. In *Agricultural pests of India and South–East Asia* (pp. 389–415). New Delhi: Kalyani Publishers.

Becker, N., & Margalit, J. (1993). Use of *Bacillus thuringiensis* israelensis against mosquitoes and blackflies. In *Bacillus thuringiensis, an environmental biopesticide: Theory and practice* (pp. 147–170). John Wiley & Sons.

Bhadane, B. S., Patil, M. P., Maheshwari, V. L., & Patil, R. H. (2018). Ethnopharmacology, phytochemistry, and biotechnological advances of family Apocynaceae: A review. *Phytotherapy Research, 32*(7), 1181–1210.

Chauhan, A., Ranjan, A., & Jindal, T. (2016). Insecticidal activity of Methanolic extract of Calotropis Procera against *Callosobruchus Maculatus* using moong seeds (Vigna Radiata). *Journal of Biomedical and Pharmaceutical Research, 5*(6), 75–82.

Credland, P. F. (1992). The structure of bruchid eggs may explain the ovicidal effect of oils. *Journal of Stored Products Research, 28*(1), 1–9.

Deepak, D. (1995). The taxonomy and phytochemistry of the Asclepiadaceae in tropical Asia. In Ruthkiew (Ed.), *Universiti* (p. 33). Pertanian.

Dev, S., & Koul, O. (1997). *Insecticides of natural origin* (p. 365). Harwood Academic Publishers.

El-Nahal, A. K., Schmidt, G. H., & Risha, E. M. (1989). Vapours of Acorus calamus oil—A space treatment for stored-product insects. *Journal of Stored Products Research, 25*(4), 211–216.

Estruch, J. J., Carozzi, N. B., Desai, N., Duck, N. B., Warren, G. W., & Koziel, M. G. (1997). Transgenic plants: An emerging approach to pest control. *Nature Biotechnology, 15*(2), 137–141.

FAO. (2006). *Food and Agriculture Organization*. Retrieved from http://www.fao.org (accessed on 20.02.2006).

Grover, A., & Pental, D. (2003). Breeding objectives and requirements for producing transgenics for major field crops of India. *Current Science, 84*(3), 310–320.

Gujar, G. T., Vinay, K., & Archana, K. (2000). Bioactivity of *bacillus thuringiensis* against the American bollworm, Helicoverpa armigera (Hübner). *Annals of Plant Protection Sciences, 8*(2), 125–131.

Harborne, J. B., & Williams, C. A. (2000). Advances in flavonoid research since 1992. *Phytochemistry, 55*(6), 481–504.

Harsulkar, A. M., Giri, A. P., Gupta, V. S., Sainani, M. N., Deshpande, V. V., Patankar, A. G., & Ranjekar, P. K. (1998). Characterization of *Helicoverpa armigera* gut proteinases and their interaction with proteinase inhibitors using gel X-ray film contact print technique. *Electrophoresis, 19*(8–9), 1397–1402.

Hill, J. M., & Schoonhoven, A. V. (1981). The use of vegetable oils in controlling insect infestations in stored grains and pulses. *Recent Advances in Food Science and Technology, 1*, 473–481.

ICIPE. (2006). *International Centre of Insect Physiology and Ecology*. Retrieved from http://www.icipe.org (accessed on 22.02.2006).

Imms, A. D. (1964). *Outlines of entomology* (5th ed., p. 224). Methuen.

Jacobson, M. (1989). Botanical pesticides past, present and future. In J. T. Arnason, B. J. R. Philogene, & P. Morand (Eds.), *Insecticides of plant origin* (pp. 44–58). American Chemical Society Symposium Series No. 387.

Khanna, V. (1998). *Molecular characterization of native isolates of Bacillus thuringiensis and standardization of fermenter parameters for mass production*. Ph.D. Thesis,. Punjabi University.

Khatter, N. A., & Abuldahab, F. F. (2012). Insecticidal activity of Calotropis procera extracted groups on some biochemical aspects of the house fly, *Musca domestica* vicina (Diptera: Muscidae). *Journal of American Science, 8*(7), 687–693.

Kim, D. H., & Ahn, Y. J. (2001). Contact and fumigant activities of constituents of *Foeniculum vulgare* fruit against three coleopteran stored-product insects. *Pest Management Science: Formerly Pesticide Science, 57*(3), 301–306.

Kim, S. I., Park, C., Ohh, M. H., Cho, H. C., & Ahn, Y. J. (2003). Contact and fumigant activities of aromatic plant extracts and essential oils against *Lasioderma serricorne* (Coleoptera: Anobiidae). *Journal of Stored Products Research, 39*(1), 11–19.

Kotkar, H. M., Mendki, P. S., Sadan, S. V., Jha, S. R., Upasani, S. M., & Maheshwari, V. L. (2002). Antimicrobial and pesticidal activity of partially purified flavonoids of *Annona squamosa*. *Pest Management Science: Formerly Pesticide Science, 58*(1), 33–37.

Krishnayya, P. V., & Rao, P. J. (1996). Plumbagin: A natural bioactive principle against insects. *The Andhra Agricultural Journal, 43*, 1–5.

Lewis, W. J., Van Lenteren, J. C., Phatak, S. C., & Tumlinson Iii, J. H. (1997). A total system approach to sustainable pest management. *Proceedings of the National Academy of Sciences, 94*(23), 12243–12248.

Markouk, M., Bekkouche, K., Larhsini, M., Bousaid, M., Lazrek, H. B., & Jana, M. (2000). Evaluation of some Moroccan medicinal plant extracts for larvicidal activity. *Journal of Ethnopharmacology, 73*(1–2), 293–297.

Mascolo, N., Sharma, R., Jain, S. C., & Capasso, F. (1988). Ethnopharmacology of *Calotropis procera* flowers. *Journal of Ethnopharmacology, 22*(2), 211–221.

Mendki, P. S., Maheshwari, V. L., Kothari, R. M., & Gowda, B. S. (2001). Botanical pesticides: Emerging trends, advantages and limitations. *Physiology and Molecular Biology of Plants, 7*, 107–115.

Mendki, P. S., Salunke, B. K., Kotkar, H., Maheshwari, V., Mahulikar, P., & Kothari, R. (2005). Antimicrobial and insecticidal activities of flavonoid glycosides from *Calotropis procera* L. for postharvest preservation of pulses. *Biopesticides International, 1*(3), 4.

Mitchell, M. J., Keogh, D. P., Crooks, J. R., & Smith, S. L. (1993). Effects of plant flavonoids and other allelochemicals on insect cytochrome P-450 dependent steroid hydroxylase activity. *Insect Biochemistry and Molecular Biology, 23*(1), 65–71.

Moursy, L. E. (1997). Insecticidal activity of Calotropis procera extracts of the flesh fly, *Sarcophaga haemorrhoidalis* fallen. *Journal of the Egyptian Society of Parasitology, 27*(2), 505–514.

Nivsarkar, M., Cherian, B., & Padh, H. (2001). Alpha-terthienyl: A plant-derived new generation insecticide. *Current Science*, 667–672.

Park, I. K., Lee, S. G., Choi, D. H., Park, J. D., & Ahn, Y. J. (2003). Insecticidal activities of constituents identified in the essential oil from leaves of Chamaecyparis obtusa against *Callosobruchus chinensis* (L.) and *Sitophilus oryzae* (L.). *Journal of Stored Products Research, 39*(4), 375–384.

Patil, S. V., Salunke, B. K., Patil, C. D., Salunkhe, R. B., Gavit, P., & Maheshwari, V. L. (2010). Potential of extracts of the tropical plant Balanites aegyptiaca (L) Del.(Balanitaceae) to control the mealy bug, Maconellicoccus hirsutus (*Homoptera: Pseudococcidae*). *Crop Protection, 29*(11), 1293–1296.

Prabhu, S., Priyadharshini, P., & Thangamalar, A. (2018). Evaluation of antifeedant activity of different parts of *Calotropis gigantea* against *Helicoverpa armigera*. *Journal of Pharmacognosy and Phytochemistry, 7*(2), 2919–2922.

Pradhan, S. (1992). Pests of pulse crops. In M. G. Jotwani (Ed.), *Insect pests of crops* (pp. 59–69). National Book Trust of India.

Rajapakse, R., & Van Emden, H. F. (1997). Potential of four vegetable oils and ten botanical powders for reducing infestation of cowpeas by *Callosobruchus maculatus, C. chinesis* and *C. rhodesianus*. *Journal of Stored Products Research, 33*(1), 59–68.

Salunke, B. K. (2006). *Biomedical basis of pulse beetle management by plant secondary metabolites*. Ph.D Thesis. North Maharashtra University.

Salunke, B. K., Kotkar, H. M., Mendki, P. S., Upasani, S. M., & Maheshwari, V. L. (2005). Efficacy of flavonoids in controlling *Callosobruchus chinensis* (L.) (*Coleoptera: Bruchidae*), a postharvest pest of grain legumes. *Crop Protection, 24*(10), 888–893.

Salunke, B. K., Prakash, K., Vishwakarma, K. S., & Maheshwari, V. L. (2009). Plant metabolites: An alternative and sustainable approach towards post harvest pest management in pulses. *Physiology and Molecular Biology of Plants, 15*, 185–197.

Sharma, P., & Sharma, J. D. (1999). Evaluation of in vitro schizontocidal activity of plant parts *of Calotropis procera*—An ethnobotanical approach. *Journal of Ethnopharmacology, 68*(1–3), 83–95.

Sharma, P., & Sharma, J. D. (2001). In vitro hemolysis of human erythrocytes—By plant extracts with antiplasmodial activity. *Journal of Ethnopharmacology, 74*(3), 239–243.

Simmonds, M. S., Blaney, W. M., Ley, S. V., Savona, G., Bruno, M., & Rodriguez, B. (1989). The antifeedant activity of clerodane diterpenoids from Teucrium. *Phytochemistry, 28*(4), 1069–1071.

Singh, S. R., Luse, R. A., Leuschner, K. E., & Nangju, D. (1978). Groundnut oil treatment for the control of *Callosobruchus maculatus* (F.) during cowpea storage. *Journal of Stored Products Research, 14*(2–3), 77–80.

Smit, H. F., Woerdenbag, H. J., Singh, R. H., Meulenbeld, G. J., Labadie, R. P., & Zwaving, J. H. (1995). Āyurvedic herbal drugs with possible cytostatic activity. *Journal of Ethnopharmacology, 47*(2), 75–84.

Stamp, N. E., & Yang, Y. (1996). Response of insect herbivores to multiple allelochemicals under different thermal regimes. *Ecology, 77*(4), 1088–1102.

Tsyrlov, I. B., Mikhailenko, V. M., & Gelboin, H. V. (1994). Isozyme-and species-specific susceptibility of cDNA-expressed CYP1A P-450s to different flavonoids. *Biochimica et Biophysica Acta (BBA)-Protein Structure and Molecular Enzymology, 1205*(2), 325–335.

Upasani, S. M., Kotkar, H. M., Mendki, P. S., & Maheshwari, V. L. (2003). Partial characterization and insecticidal properties of *Ricinus communis* L foliage flavonoids. *Pest Management Science: Formerly Pesticide Science, 59*(12), 1349–1354.

Varma, J., & Dubey, N. K. (1999). Prospectives of botanical and microbial products as pesticides of tomorrow. *Current Science*, 172–179.

Waheed, N., Jabeen, K., Iqbal, S., & Javaid, A. (2016). Biopesticidal activity of *Calotropis procera* L. against *Macrophomina phaseolina*. *African Journal of Traditional, Complementary, and Alternative Medicines, 13*(6), 163–167. https://doi.org/10.21010/ajtcam.v13i6.23

Weinzierl, R. A. (1998). Botanical insecticides, soaps and oils. In J. E. Rechcigl & N. A. Rechcigl (Eds.), *Biological and biotechnological control of insect pests* (pp. 101–121). Lewis Publ.

Winkel-Shirley, B. (2002). Biosynthesis of flavonoids and effects on stress. *Current Opinion in Plant Biology, 5*, 218–223.

# Chapter 10
# Plant Tissue Culture-Based Approaches for the Production of Pharmaceutically Important Bioactive Compounds from Apocynaceae Members

## Introduction

Plants have been the major and economic source of human and animal medicine for decades. Plant-derived metabolites are widely popular due to the wide knowledge of traditional medicine, their cost effectiveness, wide availability, and low toxicity compared to synthetic compounds. It is estimated that over 80% of the world's population relies on traditional and plant-derived medicine.

Plant products and phytochemicals are widely used in pharmaceuticals, cosmetics, nutraceuticals, etc. However, the overexploitation and injudicious use of medicinal plants have resulted in the extinction of thousands of medicinally valuable species across the globe (De Pasquale, 1984). Due to the presence of several biosynthetic pathways, such as photosynthesis, glycolysis, the Krebs cycle, and shikimate pathways, plant cells are able to produce a variety of complex and biologically active secondary metabolites. These metabolites are grouped into several classes depending on their synthesis pathways, structure and functions. It includes alkaloids, tannins, flavonoids and phenolics, terpenes, saponins, enzymes, steroids, etc. Due to their medicinal value and increasing applications, it is necessary to explore alternate approaches to obtain these pharmaceutically/medicinally important secondary metabolites of plant origin. A number of biotechnological routes can be used for commercial and sustainable production of these medicinally important compounds. Plant tissue culture (PTC) is considered an important platform for obtaining medicinally and industrially important secondary metabolites (Table 10.1). Apocynaceae plants have been widely explored for micropropagation, *in vitro* propagation, and hairy root culture using plant biotechnology tools. Other approaches, such as callus and cell suspension culture, have also been widely used to obtain bioactive products similar to those of native plants. Recently, *Agrobacterium rhizogenes*-mediated transformation has been widely explored for the production of enzymes and hormones using hairy root culture.



R. H. Patil et al., *Apocynaceae Plants*,
https://doi.org/10.1007/978-981-99-5406-3_10

**Table 10.1** Various types of secondary metabolites obtained through *in vitro* callus and cell suspension culture

| Sr. no. | Class of Secondary metabolites | Examples | Method(s) of production | References |
|---|---|---|---|---|
| 1. | Alkaloids | Furoquinolines, lobeline, indoles, isoquinolenes, tropane, Piperidine, thebaine, betalaines, catharicine, catharine,, apparicine, β-carboline, iochrovicine, rosicine, serpentine, leurosinone, cathindine, cavincidine | *In vitro* callus and cell suspension cultures | Chandran et al. (2020) |
| 2. | Steroids | Digoxin, digitoxin, helleborin, Brassinolide, glycosides, Ouabain, physodine. Catasterone, steroidal | Callus and cell suspension cultures using synthetic media | Chandran et al. (2020) |
| 3. | Terpenoids | Diterpenes, artimisinin, Taxol, paclitaxel, sesquiterpenes, treterpenes, urosilic acid, monoterpenes | *In vitro* callus and cell suspension cultures | Chandran et al. (2020) |
| 4. | Phenolics and flavonoids | Rutin, quercetin, plumabagin, lupeol, shikimic acid, 2,3-dihydroxybenzoic acid, salicylic acid; salicylic acid glucoside, benzoic acid, gallic acid, glucovanillin, vanillic acid, 2,5-dihydroxybenzoic acid, 2,5-dihydroxybenzoic acid glucoside,<br>Rutin and quercetin | *In vitro* callus and cell suspension cultures | Chandran et al. (2020) |
| 5. | Quinones | Emodin, Rhein, Uglone, shekonin, Isoflavonoids, tannins, Ferulic acid, Caffeic acid | *In vitro* callus and cell suspension cultures | Chandran et al. (2020) |
| 6. | Phenylpropanoids | Lignans, Isoflavonoids, Phenalinones, anthocyanins | *In vitro* callus and cell suspension cultures | Chandran et al. (2020) |

## Large-Scale Production of Plant Secondary Metabolites Using Callus and Cell Suspension Culture

*The in vitro* culture to obtain plant secondary metabolites consists of two stages. (i) Aggregation of biomass and (ii) synthesis of secondary metabolites (Yue et al., 2016). Plant cells develop a disorganized mass of cells known as a callus when grown *in vitro* with the right inducers and hormones. A high auxin concentration or combination of auxins and cytokinins is typically chosen for callus induction. Callus cultures now have the commercial potential to produce medicinally important secondary metabolites. On the other hand, organized tissues such as shoots, roots, calli, and cell suspensions are also popular for obtaining pharmaceutically important secondary metabolites.

Carbon and nitrogen sources, vitamins, minerals, and phytohormones play an important role in the induction of callus tissue in an *in vitro* system (Perez-Jimenez et al., 2013). Calluse cells produce secondary metabolites intracellularly through extracellular synthesis. Therefore, secondary metabolites from callus tissue can be extracted using suitable solvents (Karuppusamy, 2009). Screening of callus tissues that yield maximum bioactive metabolites and are genetically stable is preferred for commercial purposes. Thus, inoculam with better characteristics is preferred for cell suspension culture, which ultimately gives a better yield of the desired compound. Cell suspension culture involves the cultivation of callus tissue in submerged cultures where the callus tissue remans in a free floating medium. Cell suspension culture in a submerged culture improves nutrient utilization in plant cells compared to callus culture in solid medium. Moreover, it also yields maximum cell biomass in less time (Raharjo et al., 2010). In cell suspension culture, the yield of the desired metabolites can be increased by adding the appropriate precursors and optimizing the various physico-chemical parameters of the bioprocess. Elicitation may enhance metabolite production. This is achieved by applying chemical or physical stress in submerged cultures. The use of elicitors has become a popular method to obtain an increased production of metabolites that are normally produced at lower levels.

## Media for Callus and Cell Suspension Culture

Murashige and Skoog's (MS) medium (Murashige & Skoog, 1962) is commonly used for plant tissue culture experiments. For the preparation of media, fresh stock solutions of macrosalts, microsalts, vitamins, iron, potassium iodide, meso-inositol, glycine, and plant growth regulators (PGRs) are prepared separately. The stock solutions of MS medium are prepared in double distilled water and can be stored at low temperature (4 °C) for a month. The compositions of all stock solutions in MS medium and PGRs commonly used are shown in Tables 10.2 and 10.3, respectively. The pH of the medium was adjusted before use. Semisolid media can be prepared by adding agar powder as a gelling agent.

The plant tissue culture room is sterilized before use by fumigation using a mixture of potassium permanganate and formaldehyde (2:3). Fumigation is carried out at least two times with an interval of 24 h before every inoculation.

**Table 10.2** Chemical composition of MS basal medium

| Constituents | Composition (mg/L) |
|---|---|
| *Macrosalts* | |
| $NH_4NO_3$ | 1650 |
| $KNO_3$ | 1900 |
| $CaCl_2 \cdot 2H_2O$ | 440 |
| $KH_2PO_4$ | 170 |
| $MgSO_4 \cdot 7H_2O$ | 370 |
| *Microsalts* | |
| $H_3BO_3$ | 6.2 |
| $Na_2MoO_4 \cdot 2H_2O$ | 0.25 |
| $CoCl_2 \cdot 6H_2O$ | 0.025 |
| $CuSO_4 \cdot 5H_2O$ | 0.025 |
| $ZnSO_4 \cdot 7H_2O$ | 8.6 |
| $MnSO_4 \cdot 4H_2O$ | 22.3 |
| *Source of potassium iodide* | |
| KI | 0.83 |
| *Source of iron* | |
| $Na_2EDTA$ | 37.25 |
| $FeSO_4 \cdot 7H_2O$ | 27.85 |
| *Meso-inositol* | 100 |
| *Source of vitamins* | |
| Thiamine HCl | 0.1 |
| Nicotinic acid | 0.5 |
| Pyridoxine HCl | 0.5 |
| *Source of amino acid* | |
| Glycine | 2 |
| *Gelling agent* | |
| Agar | 0.85% |

**Table 10.3** List of PGRs used in callus culture

| PGRs | Chemical name and formula | Molecular weight |
|---|---|---|
| *Auxins* | | |
| NAA | α- naphthalene acetic acid ($C_{12}H_{10}0_2$) | 186.20 |
| 2,4D | 2,4 Dichlorophenoxy acetic acid ($C_8H_6Cl_2O$) | 221 |
| IAA | Indole 3 acetic acid ($C_{10}H_9$ $NO_2$) | 203.20 |
| IBA | Indole-3-butric acid ($C_{12}H_{13}N0_2$) | 175 |
| *Cytokinins* | | |
| BAP | 6-benzyl-amino purine ($C_{12}H_{11}N_{15}$) | 225 |
| Kinetin | 6-furfurylaminopurine ($C_{10}H_9N_5O_5$) | 215.20 |

## Practical Approaches for Callus Induction in Apocynaceae Plants (*T. divaricata* and Seeds of *C. carandas*)

The explant of the selected plant is first subjected to surface sterilization before inoculation in the media. Different sterilants can be used to achieve effective surface sterilization of explants. The juvenile leaves from the healthy plant of *T. divaricata* and seeds of *C. carandas* can be used as explants. The explant surface sterilization procedure was studied and optimized by Bhadane and Patil (2016). Various sterilants have been tested for surface sterilization, such as 0.1% $HgCl_2$, 70% ethanol, 0.1% benzalkonium chloride (available 4% in commercial bleach Lizol), detergent (5% Teepol), Tween 20, antibiotics (0.1% streptomycin sulfate and 0.2% Bavistin, etc.

### *Sterilization of Explants*

Before surface sterilization, plant material is thoroughly washed with running tap water and air dried. After prewashing, explants are placed into sterile glass bottles, washed with sterile water containing detergent and Tween 20 for 30 min and rinsed with sterile distilled water three times. After washing, the explants were treated with antibiotic solution (0.1% streptomycin sulfate and 0.2% Bavistin) for 15 min. Antibiotic-treated explants are washed three times using sterile double distilled water inside the LAF cabinet.

After surface sterilization using different combinations of sterilants, the explants are finally rinsed with sterile distilled water to wash out the residues from the surface of the explant (Jain et al., 2013) and air dried. Thereafter, the excised leaves of *T. divaricata* can be transferred to MS media fortified with 2.0 mg/L NAA + 0.5 mg/L BAP to check the efficiency of sterilants. Seeds of *C. carandas* are transferred to MS media devoid of PGRs to check the efficiency of sterilants. Data on contamination, explant survival percent and mortality rate, etc., is recorded to evaluate the sterilant efficiency.

### *Cultural Conditions for Callus Induction*

After standardization of the surface sterilization process, callus induction experiments are carried out. Keng et al. (2008) provided the detailed process of callus induction for Apocynaceae plants. The surface-sterilized leaves of *T. divaricata* and leaves of aseptic seedlings of *C. carandas* were cut to obtain approximately 1 $cm^2$ pieces. The aseptically prepared explants are cultured on MS medium sucrose and PGRs (Perez-Jimenez et al., 2013; Yang et al., 2008). Plant growth regulators (PGRs), such as auxin and cytokinins, are used for callus induction in Apocynaceae plants. PGRs are normally insoluble in water; hence, water miscible solvents such as absolute alcohol or 0.1 N HCl can be used to solubilize them prior to use. Stock solutions of PGRs were prepared in double distilled water, filtered to sterilize using a membrane filter and added to MS medium. Some PGRs are sensitive to heat; hence, they are added to the culture medium after autoclaving. MS medium supplemented with various concentrations of auxins (NAA and 2,4 D) either alone or in combination with cytokinins such as BAP and Kinetin can be used for callus induction studies (Rashmi & Trivedi, 2014).

## *Inoculation and Maintenance of in Vitro Cultures*

The maintenance of sterile conditions during inoculation is important for the success of plant tissue culture work. Before inoculation, hands and arms are washed and sanitized with hand sanitizers such as Dettol® and Savlon. Metalware (forceps, scalpels, spatula, etc.) can be sterilized using 95% ethanol followed by flaming over a burner flame. The laminar air flow (LAF) cabinet was swabbed with Dettol and then with rectified spirit prior to inoculation. The autoclaved media containing glassware, spirit lamps, and necessary metalware are kept in the LAF cabinet. Thereafter, sterilization of the LAF cabinet area is achieved by closing the hood and switching on the UV light for an hour. The UV light was switched off for 10 min, and inoculation work was carried out under LAF (De Pasquale, 1984).

Surface sterilized explants are inoculated in culture vessels containing sterile MS media with PGRs. The culture bottles and test tubes are sterilized over a flame before inoculation of explants. The inoculated cultures are placed in a tray and placed onto the center of culture racks under cool white fluorescent light (Bhojwani & Razdan, 1996). All controlled conditions, such as temperature 25 ± 2 °C, relative humidity 55 ± 5% and 16/8 h (dark/light) photoperiod with the intensity of light 2000 lux provided by white fluorescent light during the incubation, were maintained in an incubation room (Patil & Bhalsing, 2015).

## *Subculture*

After 4 weeks of incubation, a well-developed and proliferated callus from explants can be seen, and it is therefore ready to subculture on fresh medium. Subculturing is carried out at 2-week intervals on fresh MS medium containing the same combinations and concentrations of PGRs. Each subculture passage is carried out up to a selective interval to avoid callus browning (Kotomori & Murashige, 1965; Morel, 1972). The developed callus can be measured by measuring the fresh and dry weights. For determination of the fresh weight of callus, all tissues of callus can be picked up from the media surface using flat-ended tweezers and put into a previously weighed sterile glass vial. Glass vials containing callus tissue were further weighed on a digital weighing balance. The moisture content of calli can be calculated during fresh and dry weight measurements (Rashmi & Trivedi, 2014). The moisture content can be calculated by using the following formula:

$$\text{Moisture content } (\%) = (\text{B} - \text{A}) - (\text{C} - \text{A})/(\text{B} - \text{A}) \times 100$$

where A = weight of empty glass vial; B = weight of glass vial with fresh callus tissue, and C = weight of glass vial with dried callus material.

## Extraction of Phytochemicals from Callus Tissue

A well-developed callus mass can be obtained after successful contamination-free incubation (Fig. 10.1).

Callus tissues showing better growth and the highest fresh and dry weights can be selected for the extraction of bioactives. The general procedure for the extraction of metabolites from callus tissue is shown in Fig. 10.2. Dried callus was ground to obtain a fine powder. Powdered calli are usually extracted in methanol using the cold maceration method, and phytoconstituents can be extracted (Rajaram et al., 2013). Generally, the dry callus powder is mixed with 1:10 methanol and kept at 4 °C for 24 h. After cold maceration, the extract is centrifuged and filtered using Whatman filter paper.

## Cell Suspension Culture

The growth of plant cells in a free moving liquid culture is studied by cell suspension culture. A well-developed callus obtained after standardization of callogenesis and used as inoculum for cell suspension culture. A cell suspension culture can be established in liquid MS medium with continuous stirring. The concentrations of PGRs are crucial for callus induction. Mustafa et al. (2011) optimized the

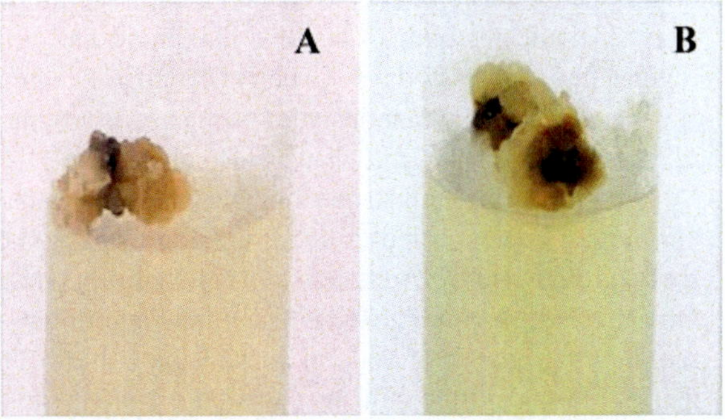


**Fig. 10.1** A well-developed callus of (**a**) *T. divaricata* callus (**b**) *C. carandas*

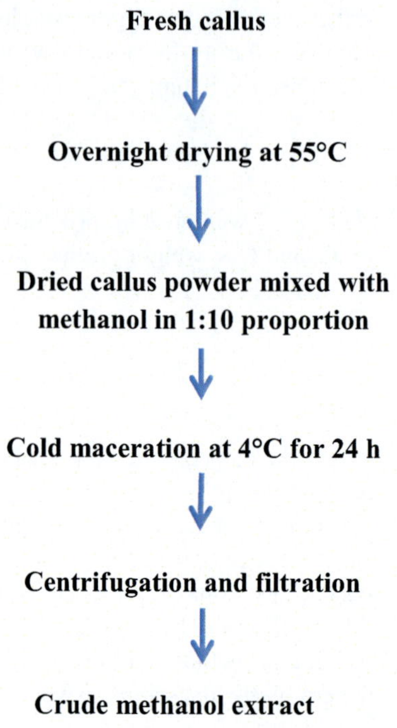


**Fig. 10.2** General procedure for extraction of callus material

experimental procedure for cell suspension culture. One gram of friable callus of the plant was inoculated separately in an Erlenmeyer flask containing sterile liquid MS medium fortified with sucrose and the respective PGRs.

To initiate the cell suspension culture of *T. divaricata*, approximately 1 g of friable callus was placed in a sterile Petri dish and chopped to obtain pieces. These pieces of callus are then inoculated in liquid MS medium containing NAA and BAP. A friable callus from *C. carandas* can also be prepared in the same way as reported for *T. divaricata* and inoculated in MS liquid medium containing NAA and BAP. After inoculation, the culture flasks were incubated at 25 ± 2 °C with continuous shaking in a shaking incubator (Fig. 10.3). Various stages in the cell suspension culture of *C. carandas* and *T. divaricata* are shown in Fig. 10.4.

For maintenance of cell suspension culture, cultures are periodically subcultured to fresh liquid medium to achieve maximum cell growth. Subculturing is done at regular intervals of 5 days. Usually, 10 mL of culture is aseptically transferred to fresh medium. Simultaneously, at each subculture interval, the growth and viability of cells in suspension culture was also checked.

Cell growth in a cell suspension culture is measured by several methods based on various parameters, such as determination of packed cell volume, medium conductivity, protein and DNA content and cell viability (Hahlbrock, 1975; Bradford, 1976; Dolezel et al., 2007). Packed cell volume determination and cell viability is used to measure the growth of cells in suspension culture. The packed cell volume (PCV) using a five-day-old cell suspension culture at every subculture can be taken

**Fig. 10.3** Shaking cooling incubator used for plant cell suspension culture

aseptically into a Falcon tube and centrifuged at 1000 rpm for 5 min. The PCV of the culture is pelleted at the base of the tube and expressed as percent growth, which can be measured by a calibrated arm on the tube.

The *T. divaricata* suspension culture showed a maximum PCV of 14% (v/v) at the second passage (Fig. 10.5). On subsequent passages, it remains static. A decrease in the PCV of cells represents a lower yield of biomass or the presence of nonviable cells in culture. A similar trend was observed in the cell suspension culture of *C. carandas*. The maximum biomass yield with a PCV of 21% (v/v) was found at the third passage (Fig. 10.6). No further increase in biomass was reported in subsequent passages. An overall decrease in PCV was reported as the cells were transferred to a new environment (Bhadane, 2018).

During maintenance of the cell line, the cell biomass becomes the reason for the unsuitability of the medium composition for cell growth (Keng et al., 2008). After every five-day interval, a similar size of inoculum is transferred to fresh medium and observed for cell growth and viability. Approximately 20% inoculum gives

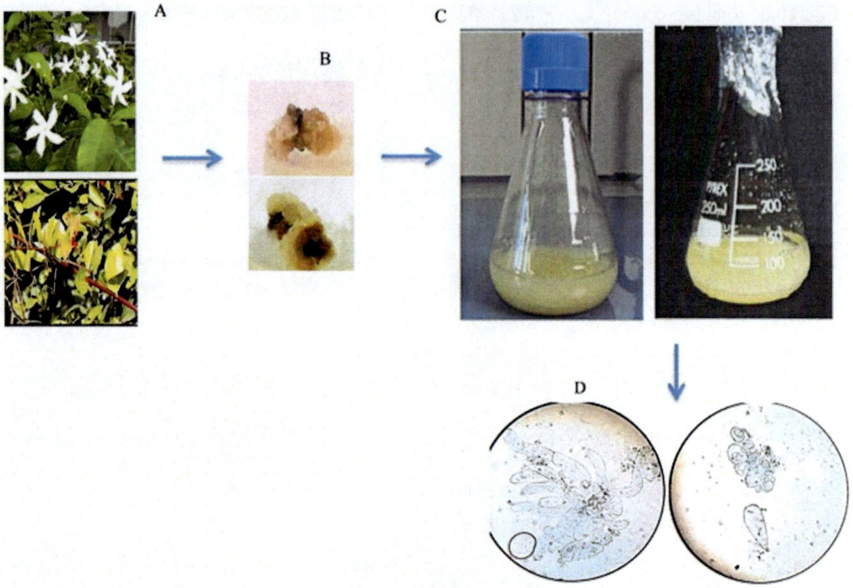


**Fig. 10.4** Various stages of cell suspension culture development of two Apocynaceae plants. (**a**) Healthy young plant of *C. carandas* and *T. divaricata*. (**b**) Well developed callus in MS media. (**c**) 100 mL cell suspension culture. (**d**) Cell viability test using Evans blue staining from suspension culture showing live and dead cells

**Fig. 10.5** Cell growth (PCV) in *T. divaricata* suspension culture

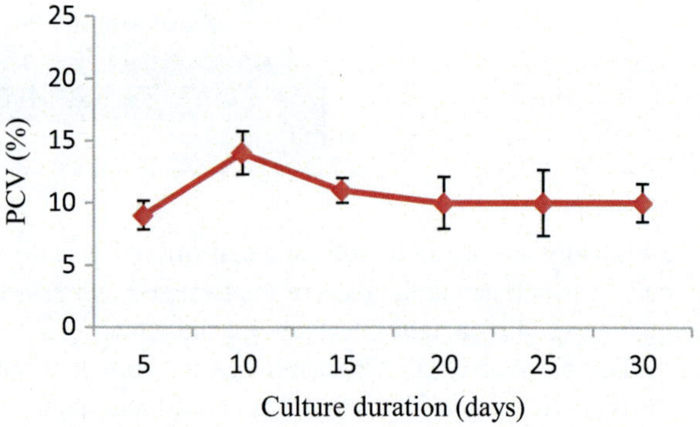


comparatively more growth than 10 and 15% in cell suspension cultures of *T. divaricata* and *C. carandas*. Determining cell viability is also one of the important aspects when studying cell suspension culture (Bhadane, 2018). It is used as an important parameter to determine growth in cultures. The Evans blue (EB) microscopic assay of Baker and Mock (1994) was used to measure the number of living cells.

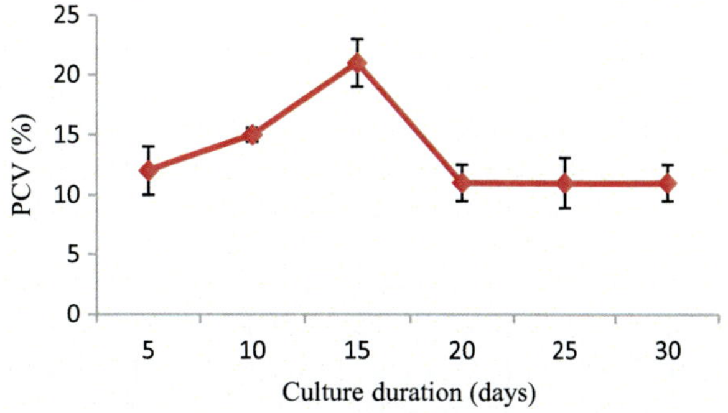


**Fig. 10.6** Cell growth (PCV) in *C. carandas* suspension culture

## Biological Activities of the Callus Extract

Plant secondary metabbolites are known to have diverse biological and medicinal activities. However, the amount of bioactive secondary metabolites produced by plant cells is limited, and the production of these metabolites is affected by several biotic and abiotic factors, seasonality, etc. Moreover, their extraction and purification from plant material is time consuming and involves technical difficulties. In the last few decades, plant cell culture-based technologies have gained importance for the increased production of secondary metabolites (Eibl et al., 2018; Potenza et al., 2020). The pharmacological activities of the callus extracts of several members of the family Apocynaceae are well documented. The callus from *H. antidysenterica* was reported for the presence of various bioactive metabolites. A number of phytoconstituents, such as terpenes, alkaloids and steroids, were isolated from the callus of this plant (Heble et al., 1976). The *in vitro*-grown callus extract showed excellent antimicrobial activity against human pathogens such as *S. aureus*, *S. typhimurium* and *E. coli* (Mahato et al., 2013). Taha et al. (2011) isolated several glycosides, such as thevetin B, peruvoside, digitoxigenin and neriifolin, from the callus of *T. peruviana* using HPLC. Some reports showed an inferior amount of bioactive metabolites in calli; hence, cell suspension culture was preferred for the large-scale production of phytoconstituents. A large number of reports are available on cell suspension cultures of *C. roseus* and *T. divaricata* and other plants of the Apocynaceae family for the production of alkaloids (Stafford et al., 1985; Van der Heijden et al., 1987; Dagnino et al., 1993; Panda et al., 1992; El-Sayed & Verpoorte, 2002; Zabala et al., 2010; Ibrahim et al., 2007, 2009; Arias et al., 2016).

## Hairy Root Culture

Due to their capacity to create secondary metabolites by expressing tDNA from the Ri plasmid of *Agrobacterium rhizogenes*, hairy root cultures (HRCs) have attracted attention from all around the world (Maldonado-Mendoza et al., 1993). Hairy roots have the ability to grow due to the development of lateral roots from the first day to the first week, and this ability can be enhanced by exogenously supplying auxin (Baiza et al., 1999). In contrast to plant roots, hairy roots have long-term genetic stability and distinctive secondary metabolite synthesis skills (Parr & Hamill, 1987). In contrast to cell suspension cultures, they can also develop secondary metabolites that coincide with root growth, making it possible to continuously acquire them throughout hairy root culture (Holmes et al., 1997).

For many plants with therapeutic value, HRC has been effectively established. Only *Catharanthus roseus* and *Rauvolfia serpentina* have been identified *for Agrobacterium rhizogenes*-mediated gene transfer using hairy root culture in the Apocynaceae family. There are numerous reports on hairy root cultures of *Rauvolfia serpentina* and *Catharanthus roseus* for the synthesis of alkaloids. Throgh infection with Agrobaccterium rhizogenes strain 15,834, Bhadra et al. (1993) discovered that seedlings of *Catharanthus roseus* produced hairy roots that produced indole alkaloids. The hairy roots of Rauvolfia serpentina were used to make alkaloids using the same strain and strain A4.

Sim et al. (1994) demonstrated the synthesis of catharanthine and ajmalicine using *A. rhizogenes* (15834)-mediated hairy roots. Catharanthine and ajmalicine were also produced from hairy roots of *Catharanthus roseus.* The production of catharanthine using nitric oxide elicitation (Zhou et al., 2010) and terpenoidal indole alkaloids (ajmalicine and serpentine) was also reported from hairy roots of *Catharanthus roseus* (Batra et al., 2004). New monoterpene alkaloids were obtained from hairy root cultures of *Rauvolfia serpentina* (Sheludko et al., 2002). Recently, Hanafy et al. (2016) reported the synthesis of anticancer compounds such as vinblastine, vincristine and catharanthine in hairy root cultures of *Catharanthus roseus* using MS medium. Table 10.4 provides details of specific metabolites produced using HRC.

In comparison to suspension culture, HRC has a number of benefits, including genotypic and biochemical stability, the ability to differentiate cells, and the ability to grow and multiply cells in hormone-free media. These are significant factors that influence the generation of secondary metabolites in an *in vitro* system. A further benefit is that hairy roots provide an ongoing source for the creation of useful secondary metabolites due to their rapid growth, low doubling time, ease of care, and ability to synthesize a wide variety of chemical compounds (Bourgaud et al., 2001). HRC has been used to create a variety of plant secondary metabolites (Giri & Narasu, 2000).

Hairy root culture offers new opportunities for the *in vitro* production of valuable plant secondary metabolites. It has good prospects for the commercial production of pharmaceutically important secondary metabolites from Apocynaceae and other plants. The applications of HRCs are promising due to their stable, high-level secondary metabolite production, fast growth and suitability for growth and adaptation into bioreactor systems. It also has several challenges, such as the unusual rheological properties of hairy root, which makes the purification of metabolites tedious and time consuming.

**Table 10.4** Some examples of high-value compounds produced using hairy root cultures

| Sr. no. | Secondary metabolite | Plant | Production method | Reference |
|---|---|---|---|---|
| 1. | Anthraquinone | *Rubia tinctoria* | HRC using transformation with *agrobacterium rhizogenes* | Saito et al. (2001) |
| 2. | Atropine and hysocyamine | *A. belladonna* | Ri T-DNA of *agrobacterium rhizogenes* and *in vitro* root growth | Jung and Tepfer (1987) |
| 3. | Hysocyamine and scopolamine | *Hyoscyamus niger* | Adventitious root culture | Shimomura et al. (1991) |
| 4. | Hysocyamine | *Datura innoxia* | Adventitious root culture | Shimomura et al. (1991) |
| 5. | Quinine, quinidine | *Cinchona ledgeriana* | Agrobacterium mediated transformation | Hamill et al. (1989) |
| 6. | Scopolamine | *Atropa belladonna* | HRC using transformation with *agrobacterium rhizogenes* | Bonhomme et al. (2000) |
| 7. | Scopolamine | *Duboisia leichhardtii* | Agrobacterium mediated transformation | Mano et al. (1989) |
| 8. | Scopolamine | *Hyoscyamus niger* | Agrobacterium mediated transformation | Zhang et al. (2004) |
| 9. | Scopolamine and hysocyamine | *D. quercifolia* | Agrobacterium mediated transformation | Dupraz et al. (1994) |
| 10. | Scopolamine and hysocyamine | *D. candida* | Agrobacterium mediated transformation | Christen et al. (1990) |
| 11. | Solasodine | *Solanum khasianum* | Agrobacterium mediated transformation | Jacob and Malpathak (2004) |
| 12. | Thiophene | *Tagetes patula* | Agrobacterium mediated transformation | Croes et al. (1989) |
| 13. | Valpotriates | *Valeriana officinalis* | Agrobacterium mediated transformation | Gränicher et al. (1992) |

# References


Arias, J. P., Zapata, K., Rojano, B., & Arias, M. (2016). Effect of light wavelength on cell growth, content of phenolic compounds and antioxidant activity in cell suspension cultures of *Thevetia peruviana*. *Journal of Photochemistry and Photobiology B: Biology, 163*, 87–91.

Baiza, A. M., Quiroz-Moreno, A., Ruíz, J. A., & Loyola-Vargas, V. M. (1999). Genetic stability of hairy root cultures of *Datura stramonium*. *Plant Cell, Tissue and Organ Culture, 59*(1), 9–17.

Baker, C. J., & Mock, N. M. (1994). An improved method for monitoring cell death in cell suspension and leaf disc assays using Evans blue. *Plant Cell, Tissue and Organ Culture, 39*(1), 7–12.

Batra, J., Dutta, A., Singh, D., Kumar, S., & Sen, J. (2004). Growth and terpenoid indole alkaloid production in *Catharanthus roseus* hairy root clones in relation to left-and right-termini-linked Ri T-DNA gene integration. *Plant Cell Reports, 23*(3), 148–154.

Bhadane, B. S. (2018). *Studies on in vitro propagation and biological activities of some members of family Apocynaceae*. Ph.D Thesis,. KBC North Maharashtra University.

Bhadane, B. S., & Patil, R. H. (2016). Data on the cost effective surface sterilization method for C. carandas (L.) seeds and callus induction from aseptic seedling. *Data in Brief, 7*, 1551–1555.

Bhadra, R., Vani, S., & Shanks, J. V. (1993). Production of indole alkaloids by selected hairy root lines of *Catharanthus roseus*. *Biotechnology and Bioengineering, 41*(5), 581–592.

Bhojwani, S. S., & Razdan, M. K. (1996). *Plant tissue culture: Theory and practice* (Vol. 5). Elsevier.

Bonhomme, V., Laurain-Mattar, D., Lacoux, J., Fliniaux, M. A., & Jacquin-Dubreuil, A. (2000). Tropane alkaloid production by hairy roots of Atropa belladonna obtained after transformation with Agrobacteriumrhizogenes 15834 *and agrobacterium tumefaciens* containing rolA, B, C genes only. *Journal of Biotechnology, 81*(2–3), 151–158.

Bourgaud, F., Gravot, A., Milesi, S., & Gontier, E. (2001). Production of plant secondary metabolites: A historical perspective. *Plant Science, 161*, 839–851.

Bradford, M. M. (1976). A rapid and sensitive method for the quantitation of microgram quantities of protein utilizing the principle of protein-dye binding. *Analytical Biochemistry, 72*(1–2), 248–254.

Chandran, H., Meena, M., Barupal, T., & Sharma, K. (2020). Plant tissue culture as a perpetual source for production of industrially important bioactive compounds. *Biotechnology Reports, 26*, e00450.

Christen, P., Roberts, M. F., David Phillipson, J., & Evans, W. C. (1990). Alkaloids of hairy root cultures of a *Datura candida* hybrid. *Plant Cell Reports, 9*, 101–104.

Croes, A. F., Van den Berg, A. J., Bosveld, M., Breteler, H., & Wullems, G. J. (1989). Thiophene accumulation in relation to morphology in roots of *Tagetes patula*: Effects of auxin and transformation by agrobacterium. *Planta, 179*, 43–50.

Dagnino, D., Schripsema, J., & Verpoorte, R. (1993). Alkaloid metabolism in *Tabernaemontana divaricata* cell suspension cultures. *Phytochemistry, 32*(2), 325–329.

De Pasquale, A. (1984). Pharmacognosy: The oldest modern science. *Journal of Ethnopharmacology, 11*(1), 1–16.

Dolezel, J., Greilhuber, J., & Suda, J. (2007). Estimation of nuclear DNA content in plants using flow cytometry. *Nature Protocols, 2*(9), 2233–2244.

Dupraz, J. M., Christen, P., & Kapetanidis, I. (1994). Tropane alkaloids in transformed roots of *Datura quercifolia*. *Planta Medica, 60*(02), 158–162.

Eibl, R., Meier, P., Stutz, I., Schildberger, D., Hühn, T., & Eibl, D. (2018). Plant cell culture technology in the cosmetics and food industries: Current state and future trends. *Applied Microbiology and Biotechnology, 102*, 8661–8675.

El-Sayed, M., & Verpoorte, R. (2002). Effect of phytohormones on growth and alkaloid accumulation by a *Catharanthus roseus* cell suspension cultures fed with alkaloid precursors tryptamine and loganin. *Plant Cell, Tissue and Organ Culture, 68*(3), 265–270.

Giri, A., & Narasu, M. L. (2000). Transgenic hairy roots: Recent trends and applications. *Biotechnology Advances, 18*, 1–22.

Gränicher, F., Christen, P., & Kapetanidis, I. (1992). High-yield production of valepotriates by hairy root cultures of *Valeriana officnalis* L. var. sambucifolia Mikan. *Plant Cell Reports, 11*, 339–342.

Hahlbrock, K. (1975). Further studies on the relationship between the rates of nitrate uptake, growth and conductivity changes in the medium of plant cell suspension cultures. *Planta, 124*(3), 311–318.

Hamill, J. D., Robins, R. J., & Rhodes, M. J. (1989). Alkaloid production by transformed root cultures of *Cinchona ledgeriana*. *Planta Medica, 55*(04), 354–357.

Hanafy, M. S., Matter, M. A., Asker, M. S., & Rady, M. R. (2016). Production of indole alkaloids in hairy root cultures of *Catharanthus roseus* L. and their antimicrobial activity. *South African Journal of Botany, 105*, 9–18.

Heble, M. R., Narayanaswamy, S., & Chadha, M. S. (1976). Studies on growth and steroid formation in tissue cultures *of Holarrhena antidysenterica. Phytochemistry, 15*(5), 681–682.

Holmes, P., Li, S. L., Green, K. D., Ford-Lloyd, B. V., & Thomas, N. H. (1997). Drip-tube technology for continuous culture of hairy roots with integrated alkaloid extraction. In *Hairy roots: Culture and application* (pp. 201–208). Harwood Academic Publisher.

Ibrahim, A. K., Khalifa, S., Khafagi, I., Youssef, D., Khan, I., & Mesbah, M. (2007). Stimulation of oleandrin production by combined *agrobacterium tumefaciens* mediated transformation and fungal elicitation in Nerium oleander cell cultures. *Enzyme and Microbial Technology, 41*(3), 331–336.

Ibrahim, A., Khalifa, S., Khafagi, I., Youssef, D., Khan, I., & Mesbah, M. (2009). Enhancement of oleandrin production in suspension cultures of *Nerium oleander* by combined optimization of medium composition and substrate feeding. *Plant Biosystems, 143*(1), 97–103.

Jacob, A., & Malpathak, N. (2004). Green hairy root cultures of *Solanum khasianum* Clarke—A new route to *in vitro* solasodine production. *Current Science*, 1442–1447.

Jain, S., Sharma, P., Ghule, S., Jain, A., & Jain, N. (2013). *In vivo* anti-inflammatory activity of Tabernaemontana divaricata leaf extract on male albino mice. *Chinese Journal of Natural Medicines, 11*(5), 472–476.

Jung, G., & Tepfer, D. (1987). Use of genetic transformation by the Ri T-DNA of *agrobacterium rhizogenes* to stimulate biomass and tropane alkaloid production in Atropa belladonna and *Calystegia sepium* roots grown *in vitro*. *Plant Science, 50*(2), 145–151.

Karuppusamy, S. (2009). A review on trends in production of secondary metabolites from higher plants by *in vitro* tissue, organ and cell cultures. *Journal of Medicinal Plants Research, 3*(13), 1222–1239.

Keng, C. L., See, K. S., Hoon, L. P., & Lim, B. P. (2008). Effect of plant growth regulators and subculture frequency on callus culture and the establishment of Melastoma malabatricum cell suspension cultures for the production of pigments. *Biotechnology, 7*(4), 678–685.

Kotomori, S. N., & Murashige, T. (1965). Some aspects of aseptic propagation of orchids. *American Orchid Society Bulletin, 34*, 484–489.

Mahato, S., Mehta, A., & Roy, S. (2013). Studies on antibacterial effects of bark, seed and callus extracts of *Holarrhena antidysenterica* wall. *The Bioscan, 8*(2), 717–721.

Maldonado-Mendoza, I. E., Ayora-Talavera, T., & Loyola-Vargas, V. M. (1993). Establishment of hairy root cultures of *Datura stramonium.* Characterization and stability of tropane alkaloid production during long periods of subculturing. *Plant Cell, Tissue and Organ Culture, 33*(3), 321–329.

Mano, Y., Ohkawa, H., & Yamada, Y. (1989). Production of tropane alkaloids by hairy root cultures of Duboisia leichhardtii transformed by *agrobacterium rhizogenes*. *Plant Science, 59*(2), 191–201.

Morel, G. (1972). Morphogenesis of stem apical meristem cultivated *in vitro*: Application to clonal propagation. *Phytomorphology, 22*, 265–277.

Murashige, T., & Skoog, F. (1962). A revised medium for the rapid growth and bioassay with tobacco tissue cultures. *Physiol Plant, 15*, 473–497.

Mustafa, N. R., De Winter, W., Van Iren, F., & Verpoorte, R. (2011). Initiation, growth and cryopreservation of plant cell suspension cultures. *Nature Protocols, 6*(6), 715–742.

Panda, A. K., Mishra, S., & Bisaria, V. S. (1992). Alkaloid production by plant cell suspension cultures of *Holarrhena antidysenterica*: I. Effect of major nutrients. *Biotechnology and Bioengineering, 39*(10), 1043–1051.

Parr, A. J., & Hamill, J. D. (1987). Relationship between *agrobacterium rhizogenes* transformed hairy roots and intact, uninfected Nicotiana plants. *Phytochemistry, 26*(12), 3241–3245.

Patil, K. S., & Bhalsing, S. R. (2015). Efficient micropropagation and assessment of genetic fidelity of Boerhaavia diffusa L-high trade medicinal plant. *Physiology and Molecular Biology of Plants, 21*(3), 425–432.

Perez-Jimenez, M., López-Soto, M. B., & Cos-Terrer, J. (2013). *In vitro* callus induction from adult tissues of peach (Prunus persica L. Batsch). *In Vitro Cellular and Developmental Biology-Plant, 49*(1), 79–84.

Potenza, L., Minutelli, M., Stocchi, V., & Fraternale, D. (2020). Biological potential of an ethanolic extract from "*Mela Rosa Marchigiana*" pulp callus culture. *Journal of Functional Foods, 75*, 104269.

Raharjo, T. J., Eucharia, O., Chang, W. T., & Verpoorte, R. (2010). Callus induction and phytochemical characterization of Cannabis sativa cell suspension cultures. *Indonesian Journal of Chemistry, 6*(1), 70–74.

Rajaram, K., Moushmi, M., Prakash, M. V. D., Kumpati, P., Ganasaraswathi, M., & Sureshkumar, P. (2013). Comparative bioactive studies between wild plant and callus culture of Tephrosia tinctoria Pers. *Applied Biochemistry and Biotechnology, 171*(8), 2105–2120.

Rashmi, R., & Trivedi, M. P. (2014). Effect of various growth hormone concentration and combination on callus induction, nature of callus and callogenic response of Nerium odorum. *Applied Biochemistry and Biotechnology, 172*(5), 2562–2570.

Saito, K., Sudo, H., Yamazaki, M., Koseki-Nakamura, M., Kitajima, M., Takayama, H., & Aimi, N. (2001). Feasible production of camptothecin by hairy root culture of Ophiorrhiza pumila. *Plant Cell Reports, 20*, 267–271.

Sheludko, Y., Gerasimenko, I., Kolshorn, H., & Stöckigt, J. (2002). New alkaloids of the Sarpagine group from *Rauvolfia serpentina* hairy root culture. *Journal of Natural Products, 65*(7), 1006–1010.

Shimomura, K., Sauerwein, M., & Ishimaru, K. (1991). Tropane alkaloids in the adventitious and hairy root cultures of *solanaceous* plants. *Phytochemistry, 30*(7), 2275–2278.

Sim, S. J., Chang, H. N., Liu, J. R., & Jung, K. H. (1994). Production and secretion of indole alkaloids in hairy root cultures of *Catharanthus roseus*: Effects of in situ adsorption, fungal elicitation and permeabilization. *Journal of Fermentation and Bioengineering, 78*(3), 229–234.

Stafford, A., Smith, L., & Fowler, M. W. (1985). Regulation of product synthesis in cell cultures of *Catharanthus roseus* (L) G. Don. *Plant Cell, Tissue and Organ Culture, 4*(1), 83–94.

Taha, H. S., Farag, S. H., Shams, K. A., Abdel-Azim, N. S., & Seif El-Nasr, M. M. (2011). *In vivo* and *in vitro* studies on *Thevetia* species growing in Egypt II. Establishment of *in vitro* tissue culture system and production of cardiac glycosides. *Journal of American Science, 7*, 1–12.

Van der Heijden, R., Lamping, P. J., Out, P. P., Wijnsma, R., & Verpoorte, R. (1987). High-performance liquid chromatographic determination of indole alkaloids in a suspension culture of *Tabernaemontana div aricata. Journal of Chromatography A, 396*, 287–295.

Yang, J., Gong, Z. C., & Tan, X. (2008). Induction of callus and extraction of alkaloid from Yi mu Cao (Leonurus heterophylus Sw.) culture. *African Journal of Biotechnology, 7*(8).

Yue, W., Ming, Q. L., Lin, B., Rahman, K., Zheng, C. J., Han, T., & Qin, L. P. (2016). Medicinal plant cell suspension cultures: Pharmaceutical applications and high-yielding strategies for the desired secondary metabolites. *Critical Reviews in Biotechnology, 36*(2), 215–232.

Zabala, M. A., Angarita, M., Restrepo, J. M., Caicedo, L. A., & Perea, M. (2010). Elicitation with methyl-jasmonate stimulates peruvoside production in cell suspension cultures of *Thevetia peruviana. In Vitro Cellular and Developmental Biology-Plant, 46*(3), 233–238.

Zhang, L., Ding, R., Chai, Y., Bonfill, M., Moyano, E., Oksman-Caldentey, K. M., Xu, T., Pi, Y., Wang, Z., Zhang, H., & Kai, G. (2004). Engineering tropane biosynthetic pathway in *Hyoscyamus Niger* hairy root cultures. *Proceedings of the National Academy of Sciences, 101*(17), 6786–6791.

Zhou, M. L., Zhu, X. M., Shao, J. R., Wu, Y. M., & Tang, Y. X. (2010). Transcriptional response of the catharanthine biosynthesis pathway to methyl jasmonate/nitric oxide elicitation in *Catharanthus roseus* hairy root culture. *Applied Microbiology and Biotechnology, 88*(3), 737–750.

# Chapter 11
# Enzymes and Enzyme Inhibitors from Apocynaceae Plants

## Introduction

Enzymes have been extensively utilised in numerous sectors from last few decades. Recent years have seen a rise in interest in and demand for enzymes with unique features as a result of the development of biotechnology. For the commercial manufacturing of enzymes, microorganisms mainly isolated from soil are used. Over several decades, these industrially important enzymes have mainly been obtained from plant and animal sources. However, extracting enzymes from plant or animal sources is a difficult task due to their limited availability and high cost associated with downstream processing. Therefore, alternate sources such as microorganisms have been extensively used for enzyme production in recent times. Due to their numerous advantages, microbial enzymes have completely replaced enzyme extractions from plant and animal sources (Underkofler et al., 1957). Microbial enzymes have certain advantages over their plant and animal counterparts. These can be produced in higher quantities within a shorter period, are nonseasonal and are comparatively more stable. The main advantage of using microorganisms instead of plants and animals is that they can be cultured using simple and low-cost media (Wiseman, 1995). Microbial enzymes have wide applications in food, detergent, pharmaceutical, textile, pulp and paper, leather, oleochemical and other industries (Sharma et al., 2001; Hasan et al., 2006). Among industrially important enzymes, hydrolases are widely used. This class of enzymes includes proteases, amylases, and lipases. These enzymes carry out the hydrolysis of their specific substrates, i.e., proteins, starch, and lipids, respectively (Fig. 11.1). Amylases and proteases are used in different industries for various purposes, but it was in the last decade that the use of lipases became more popular. Lipases carry out the hydrolysis of fats and are also involved in esterification and transesterification reactions (Rajendran et al., 2009).

 
R. H. Patil et al., *Apocynaceae Plants*,
https://doi.org/10.1007/978-981-99-5406-3_11

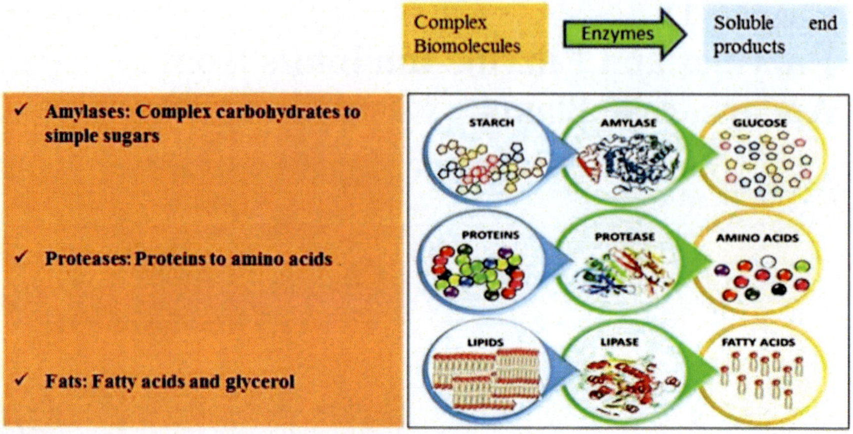


**Fig. 11.1** Biochemical action of digestive enzymes

## *Proteases*

The alternate name for proteases is proteinases or peptidases, which form the major group of enzymes that hydrolyse peptide bonds. Proteolytic enzymes are categorized into endopeptidases and exopeptidases. Peptide bonds present at the center of proteins are cleaved by endopeptidases and are categorized depending on the presence of catalytic groups on the active site, such as serine, cysteine, carboxyl and metallo endopeptidases (Barrett et al., 2012). Precisely, these exopeptidases breakdown the amino (N) terminus (aminopeptidases) or the carboxy (C) terminus (carboxypeptidases). Proteases have vast applications in various sectors, such as the food and detergent ndustry, pharmaceutical and agrochemical industries (Gupta et al., 2002; Bougatef et al., 2007).

## *Amylases*

Amylase is an O-glycosyl hydrolase (EC 3.2.1.) families, which are a ubiquitous set of enzymes that break glycosidic bonds present among two or more sugars or between a sugar and a noncarbohydrate moiety. Amylases are classified into end amylases and exo amoylases. Many bacteria produce amylase, which has wide applications in the field of bioengineering and in the food and beverages industry. Traditionally, amylases were obtained from plant, animals, and microbial (bactrial as well as fungal) sources. However, microorganisms are the major source of amylases to fulfil the growing industrial requirements of amylases (Guzman & Paredes, 1995; Van der Maarel et al., 2002; Reddy et al., 2003).

### *Lipases*

Lipases (triacylglycerol acylhydrolases, E.C. 3.1.1.3) are hydrolases that carry out lipolysis at the lipid-water interface. Lipases are omnipresent in nature and therefore can be isolated from various plants, animals, fungi and bacteria. Lipases catalyze the hydrolysis of triacylglycerol and release glycerol and free fatty acids. Triacylglycerol (TAG) is sparingly soluble in water, and it forms an emulsion, which is the site where lipases perform their action. Triacylglycerol (TAG) has very low solubility in water, and it forms an emulsion, which is the site where lipases perform their action. Lipases are extensively used as biocatalysts in organic chemistry due to their interesting properties, which include easy availability in large quantities, as many of them can be produced in higher amounts from microbes. Microbial lipases are produced intracellularly and secreted extracellularly (Sharma et al., 2012). Commercially, lipases are obtained from *Bacillus, Candida, Rhizopus, Pseudomonas,* etc.

## Latex Bearing Apocynaceae Members as an Important Source of Biomedical Enzymes

Plants have been an important source of enzymes such as amylase, proteases, invertase, lipases, papain, bromelain, ficin and lipoxygenase. These enzymes have a number of applications in the food, pharmaceutical, dairy, bakery and alcoholic beverage industries. Several higher plants from varius families are used as economical factories for enzyme production.

The Apocynaceae family is also known as the family of latex-bearing plants. Common examples of latex-bearing Apocynaceae plants include *Calotropis procera, Calotropis gigantea, Synadenium umbellatum, Nerium oleander, Tarbanamontana divaricate, Tinospora cardifolia* and *Wrightia tinctoria.* Latex components include organic compounds, inorganic compounds, minerals and many hydrolytic enzymes (Silva et al., 2003). Latex hydrolytic enzymes include proteases, esterases and lipases, and amylases, which act as anticoagulant enzymes (Yagami et al., 1998). Hydrolytic enzymes play a significant role in plants' resistance to infectious diseases. Latex contains cysteine or serine proteases that possess diverse biological and pharmacological activities. The anticoagulants of plant latex are used in folk medicine, as well as in coronar heart diseases such as strokes and pulmonary embolism (Battjes et al., 2015).

A number of Apocynaceae plants are known to harbor anticoagulating and other proteases in their latex. Latex of *Calotropis gigantea, Ficus benghalensis, Croton bonplandianum, Ficus carica, Nerium oleander, Synadenium umbellatum, Artocarpus heterophyllus, Manilkara zapota, Ficus religiosa, Carica papaya* and *Plumeria acutifolia* have been widely studied for their pharmacological activities (Table 11.1). On the other hand, little work has been done on the latex of Apocynaceae plants for their anticoagulant enzymes.

**Table 11.1** Some examples of Apocynaceae and non-Apocynaceae plants as a source of enzymes

| Enzymes | Source plant | Application | Reference |
|---|---|---|---|
| Anticoagulating enzymes | *Euphorbia milii* | Surgeries and heart diseases | Khuraishi et al. (2019) |
| | *Jatropha malacophylla* | blood clot treatment | Khuraishi et al. (2019) |
| | *Thevetia peruviana* | Surgeries and blood clot treatment | Khuraishi et al. (2019) |
| | *Euphorbia pulcherimma* | Surgeries and blood clot treatment | Khuraishi et al. (2019) |
| | *Artocarpus altilis* | Surgeries and blood clot treatment | Khuraishi et al. (2019) |
| Anticoagulating and proteolytic enzymes | *Calotropis gigantea* | blood clotting and milk clotting activity | Mishra et al. (2017) |
| | *Nerium oleander* | milk clotting activity | Flemmig and Melzig (2013) |
| | *Carica papaya* | blood clotting | Flemmig and Melzig (2013) |
| | *Artocarpus heterophyllus* | Blood and milk clotting activity | Flemmig and Melzig (2013) |
| | *Plumeria acutifolia* | Blood and milk clotting activity | Flemmig and Melzig (2013) |
| Fibrinolytic enzymes | *Wrightia tinctoria* | Coagulation and fibrinolysis | Flemmig and Melzig (2013) |
| Protease | *Euphorbia nerifolia* | Fibrinolysis | Mahajan et al. (2015) |
| Proetases and coagulating enzymes | *Asclepias curassavica* | Fibrinolysis | Karpagam et al. (2013) |
| coagulating enzymes | *Calotropis gigantea* | Milk clotting activity | Anusha et al. (2013) |
| Protease | *A. cathartica* | Milk clotting activity | Anusha et al. (2013) |
| Bromelain (mixture of proteases) | *Carica papaya* | Anti-inflammatory, meat tenderization | Meshram et al. (2019) |
| Papain (protease) | *Carica papaya* | Anti-inflammatory agent | Meshram et al. (2019) |
| Urease | *Canavalia ensiformis* | Biomedical, blood urea detection | Meshram et al. (2019) |
| Beta amylase | *Hordeum vulgare* | Brewing industry | Meshram et al. (2019) |
| Ficin | *Ficus carica* | Food industry | Meshram et al. (2019) |
| Peroxidase | *Cochlearia armoracia)* | Diagnostic applications | Meshram et al. (2019) |
| Beta gulucanase | *Hordeum vulgare* | Brewing industry | Meshram et al. (2019) |

## Enzyme/Protease Inhibitors from Apocynaceae Plants

Several metabolites of Apocynaceae and other plants function as enzyme/protease inhibitors (EIs/PIs), which modulate various functions such as activity of endogenous proteases and protects against biotic and abiotic stresses. PIs are widely studied dmetabolites involved in plant defense which play a significant role insect attack. Small molecular proteins/peptides act as PIs or involved in cell signaling that leads to synthesis of PIs. Pearce (2011) isolated a signal peptide, Systemin, from tomato leaves which was found to be involved in synthesis of multiple defense components. Depending on the kind of PI, PIs can be found in a variety of cell organelles, including the cytoplasmic matrix, cell wall, and nucleus, in reproductive and vegetative tissues. The quantity and kind of PIs present in a plant vary depending on the plant variety, developmental stage, location, harvesting period, and environmental stressors. PIs from plants have generally been categorised according to how specific they are to different kinds of proteases, specifically cysteine, serine, aspartic, and metallo-proteases (Birk, 2003; Barbole et al., 2022) (Fig. 11.2).

## Therapeutic Applications of PIs

Plants have a number of physiological and molecular mechanisms of protection against biotic stress like phytopathogens plant and insects. By deactivating the proteases of the pathogenic organism during infection or infestation, PIs play a significant part in plant defence. For instance, PIs inhibit the proteolysis of ingested dietary proteins by competitively binding to the catalytic region of insect proteases. Due to a decrease in amino acid absorption in the insect stomach, growth and development are delayed, which decreases fertility and survival of the insect (Mishra et al., 2010). PIs from various plants have been thoroughly investigated at the biochemical and molecular level, in vitro insect feeding experiments etc., to investigate their role in insect control. Enzyme or protease inhibitors of plant origin have attracted the attention of researchers because of their application in treating different metabolic and infectious diseases. Medicinal plants are regarded as vital source of bioactive secondary metabolites which are known for their enzyme inhibitory potential. The applications of EIs as novel drugs or lead compounds are enormous and therefore these compounds have been used for the treatment of various metabolic and physiological disorders. Highly specific regulation of proteases is essential for physiological processes scuhas food digestion, blood clotting, apoptosis, and modulationg blood pressure. Due to their unique structure and high specificity, PIs have been explored as drugs to treat a number of physiological and infectious diseases. Several PIs have been successfully explored in cancer treatment, heart diseases, inflammatory and neurological disorders as well as in the managemen of and obesity. For example, Swedberg et al. (2009) isolated a trypsin inhibitor-1 (SFTI-1), a small peptide from sunflower, which can blocked tumor formation and prevented metastasis through blocking human kallikrein-related peptidase 4. In another study, Pin-II-type inhibitors was isolated from potato and successfully used as anti-obesity molecules (Peters et al., 2011). The ensuing paragraphs describe

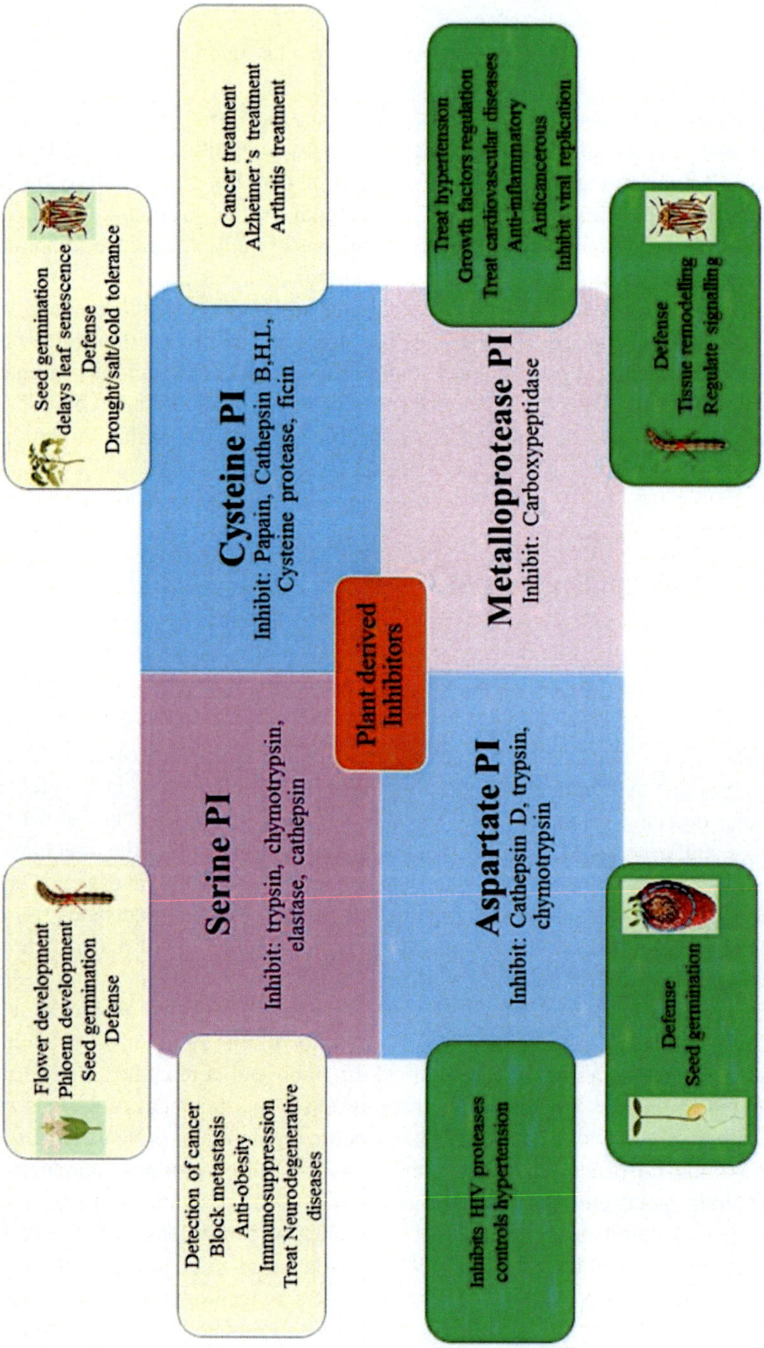


**Fig. 11.2** Classification of protease inhibitors (Source: Barbole et al., 2022)

the structure, function and applications of some important class of PIs of plant origin (Fig. 11.2).

## *α-Glucosidase Inhibitors (AGIs) for Diabetes Control*

Type 2 diabetes affects over 450 million people worldwide, or 6.28% of the world's population. In the entire world, type 2 diabetes is now one of the leading causes of death. It currently reports over 10 lakh deaths annually, with a widespread reporting rate of 6000 cases per 100,000; by the end of 2030, the rate is expected to rise to 7100 patients per 100,000 (Yang et al., 2018). The growing prevalence of diabetes mellitus worldwide, particularly in industrialised nations like America and Europe, calls for rapid attention. A class of oral medications known as alpha-glucosidase inhibitors (AGIs) is used to treat diabetes mellitus type 2. They may be used either alone or in conjunction with other medications. AGIs are popularly used in the patients having low glucose tolerance and delayed-type 2 diabetes. AGIs act by decreasing gastrointestinal assimilation of glucose and thereby prevent sugar absorption.

The presence of new AGIs are being investigated in a variety of plant and fungus species. They lessen postprandial blood glucose and insulin levels, enhancing the ability of the insulin cell to become affected and reducing the stress on the beta-cell (van de Laar, 2008). Nitrogen-containing heterocyclic substances like triazole, quinazoline, imidazole, thiazole, and pyrazole have been reported to possess alpha-glucosidase inhibitors but do not include glycosyl groups. In addition, a number of synthetic and natural flavonoids, including luteolin, anthocyanins, naringenin and baicalein, coumarin, and chromones, as well as their derivatives, have been found as powerful AGIs (Bhatnagar & Mishra, 2022). Triazole and quinazoline compounds have drawn a lot of attention because of their significant pharmacological characteristics. Triazole compounds with a carbazole bond have been found to have inhibitory effect against alpha glucosidase. It has also been demonstrated that substituted quinazolines have antihyperglycemic and anti-glucosidase inhibitory effects. AGIs, in contrast to other oral antidiabetic medications, primarily affect the gut rather than the body's metabolic functions. As a result, a lot of research has been done on the use of AGIs in the management of prediabetic illnesses like impaired glucose tolerance and impaired fasting glucose (Table 11.2).

## *α-Amylase Inhibitors for Pest Control (α-AIs)*

Digestive amylases are enzymes from the endo/exo-enzyme family that hydrolyse the glycosidic bonds of large α-linked polysaccharides like starch and glycogen, to produce smaller oligosaccharides or monosaccharides to meet energy requirements. Thus, α-amylases play a significant role in energy generation (Pelegrini et al., 2006; Kaur et al., 2014).

**Table 11.2** Natural products from Apocynaceae and other plants known as alpha-glucosidase inhibitors (Source: Bhatnagar & Mishra, 2022)

| Sr. no. | alpha-glucosidase inhibitor | Metabolite class | Source | Reference |
|---|---|---|---|---|
| 1. | Acarbose | Pseudotetrasaccharide | *Actinoplanes* | Li Ting (2018) |
| 2. | Akebonoic acid | Triterpenoid | *Akebia trifoliata* | Dirir et al. (2021) |
| 3. | Antroquinonol | Meroterpenoid | *Fungus* | Jhong and Chia (2015) |
| 4. | Berberine | Alkaloid | *Berberis sp.* | Pan et al. (2003) |
| 5. | Catechin | Flavonoid | *Various Plant* | Matsui et al. (2007) |
| 6. | Curcumin | beta-diketone | *Curcuma longa* | Nabavi et al. (2015) |
| 7. | Docosanol | Aliphatic alcohol | *Toona sinensis* | Jhong and Chia (2015) |
| 8. | Eupafolin | Flavonoid | *Artemisia s.* | Jhong and Chia (2015) |
| 9. | Fisetin | Flavonoid | *Fruits and vegetable* | Yue et al. (2018) |
| 10. | Morusin | Flavonoid | *Morus alba* | Jang et al. (2015) |
| 11. | Myricetin | Flavonoid | *Syzygium* | Kang et al. (2015) |
| 12. | Psoralidin | Coumestans | *Psoralea corylifolia* | Wang et al. (2013) |
| 13. | Quercetin | Flavonoid | *Various Plant* | Oboh and Ademosun (2015) |
| 14. | Rhaponticin | Polyphenol | *Rhei Rhizoma* | Choi et al. (2006) |
| 15. | Taxumariene F | Diterpenoid | *Taxus mairei* | Chen et al. (2020) |

Inhibition of $\alpha$-amylase through the binding of α-AIs results in prevention of digestion and absorption of dietary starch (Shamki et al., 2012). The α-AI(s) are the natural productssuch as proteins or synthetic chemical(s) that control the activity of α-amylase (McEwan et al., 2010). According to Kadziola et al. (1998), a potent amylase inhibitor must be effective against the target amylase while remaining non-interfering with the actions of endogenous -amylases (such as those expressed during seed germination) and mammalian enzymes.

In addition to being plentiful in cereals and legumes as part of their natural defence mechanisms, $\alpha$ -amylase inhibitors ($\alpha$ -AIs) are also equally effective candidates for this usage (Guzman-Partida et al., 2007). Due to their function in the metabolic control of starch, -AIs from cereal plants exhibit inhibitory effect against endogenous amylases (Weselake et al., 1985; Giri & Kachole, 1998). It is well known that plant seeds develop a range of enzyme inhibitors that defend them against infections and insects. The most well researched inhibitors in this class are proteinaceous ones (Christou et al., 2006). Pure inhibitor(s) of insect digestive enzymes generated from plants seem to be a secure an appears to be appealing solution for insect cotrol. A earlier review of the potential of $\alpha$ -AIs as bioinsecticidal compounds can be found in Carlini and Grossi-de-Sa (2002). For instance, rye, common bean, wheat, and amaranth -AIs work by delaying the rate at which insects digest the food they consume (Giri & Kachole, 1998; Franco et al., 2002; Dias et al., 2005). According to Bunyatang et al. (2016), the *α*-AIs from *Hevea brasiliensis* regulates the absorption of reserve starch and is in charge of sugar accretion for seed growth. Reports on the use of insect amylase and - *α* -AIs from plant species to regulate postprandial glucose levels predominate over reports on the use of microbial-α-AIs for the same activities (Table 11.3). Large-scale production of α-AIs from plant sources seems to be difficult because it require larger space for growing the plants. Moreover, the extraction procedures for *α* –Ais from plant sources is complex and time consuming as compare to microbial sources (Jayaraj et al., 2013).

## $H^+/K^+$-ATPase Inhibitors

Increased secretion of acid plays significant role in the development of gastric ulcers and elated disorders. The use of plant-derived substances forms the basis of many modern pharmaceuticals designed for the prevention of diseases such as peptic ulcers. H+/K + -ATPase is the proton pump responsible for controlling gastric secretions. Increased secretions of these biomolecules lead to hyperacidity. Plant-derived metabolites have been explored for as inhibitors of H+/K + -ATPase. Compounds from Apocynaceae and several other plants have been explored to alleviate gastric acid secretion-related diseases, and thus, these compounds could be potentially used as new lead compounds for the management of peptic ulcer diseases (Hay et al., 2019). However, there is a massive gap in the discovery of effective inhibitors for proton pump inhibitory effects. (Adinortey & N'guessan, 2022). Earlier reports have shown that natural products prevented the gastric ulcers through inhibition of acid secretion in an *in vivo* animal studies. Medicinal plants and such as *Solanum nigrum, Arctium lappa, Garcinia mangostana, Azadirachta indica* have been explored for controlling the peptic ulcers. The roots of *Hedranthera barteri* (Onasanwo et al., 2011) and the fruits of *Solanum nigrum* (Jainu & Devi, 2006) showed substantial $H^+/K^+$-ATPase inhibitory potential at 71 and 200μg/ml. Some plants showed minimum $H^+/K^+$-ATPase inhibitory potential (with $IC_{50}$ values above 201μg/ml) (Mesía-Vela et al., 2007; Lakshmi et al., 2010). Several other members with substantial $H^+/K^+$ ATPase inhibitory potential are shown in Table 11.4 (Adinortey & N'guessan, 2022).

## Inhibitors of Other Enzymes of Biomedical Importance

Some of the crucial proteins involved in disease and pathogenesis are activated by proteases and other enzymes from pathogens like bacteria, fungi, and viruses. The ability of the majority of viral proteases to cleave host proteins to reduce host immunity and promote viral replication is also well-known. Two proteases are implicated in the activation of the SARS-CoV-2 replication machinery and the

**Table 11.3** Important α-amylase inhibitors of plant origin used for pest control (Kasar et al. 2017, 2021, 2022)

| Sr. no. | Source plant | Plant part | Mol. Wt. | Host amylase inhibited | Reference |
|---|---|---|---|---|---|
| 1. | *Colocasia* | Seeds | 13.9 | *C. chinensis, T. castaneum, C. cephalonica* | Kumari et al. (2012) |
| 2. | *Carica papaya* | Seeds | 4.28 | *C. maculatus* | Farias et al. (2007) |
| 3. | *Capsicum annuum* | Seeds | 9 | Mammalian | Diz et al. (2011) |
| 4. | *Coffea canephora* | Seeds | 9 | Mammalian | Zottich et al. (2011) |
| 5. | *D. esculenta* | Tubers | 16 | *Coffee bean* and *Sweet potato weevil* | Sasikiran et al. (2004) |
| 6. | *Dipteryx alata* | Seeds | 14.8 | *A. grandis, C. maculatus, A. obtectus,* Porcine pancreas | Bonavides et al. (2007) |
| 7. | *Eleusine coracana* | Seeds | 14 | Porcine pancreatic, *C. chinensis* | Saxena et al. (2010) |
| 8. | *Leucas aspera* | Whole plant | 28 | Mammalian, Maize, *B. subtilis, A. oryzae* | Meera et al. (2017) |
| 9. | *Mucuna pruriens* | Seeds | 25.6 | Human salivary α-amylase | Bharadwaj et al. (2018) |
| 10. | *Phaseolus vulgaris* | Seeds | 30 | Mammalian | Ali et al. (2006) |
| 11. | *Streptomyces sp. CC5* | Cells | 8.21 | Human pancreatic amylase, Porcine pancreatic amylase | Sun et al. (2015) |
| 12. | *Triticale seeds* | Seeds | 29.2 | *Eurygaster integriceps* Human salivary α-amylase | Mehrabadi et al. (2011) |
| 13. | *Triticum aestivum* | Seeds | 20.0 | Human salivary, Porcine pancreatic, Chick pea | Saunders and Lang (1973) |
| 14. | *Vigna radiate* | Seeds | 27 | *C. maculatus* | Wisessing et al. (2010) |
| 15. | *Vigna sublobata* | Cotyledon | 14 | Larvae of *C. analis* | Kokiladevi et al. (2005) |
| 16. | *Vigna unguiculata* | Seeds | 25.1 | *C. maculatus, A. obtectus, A. grandis, A. oryzae* | Alves et al. (2009) |

beginning of its pathogenesis: papain-like proteases and chymotrypsin-like cysteine proteases (Vadnere & Patil, 2022). *Staphylococcal* enzymes that degrade host tissue components include nucleases, coagulase, proteases (metalloproteases, serine, and cysteine proteases), hyaluronidase, and lipases. Lipases are known to support the persistence of *S. aureus* in fatty secretions in mammalian skin and therefore directly contribute to their pathogenic potential. Additionally, lipase prevents host granulocytes from phagocytosing pathogenic lipase-producing S. aureus cells, demonstrating a direct role for lipase in pathogenesis (Christensson et al., 1985; Rollof et al., 1988).

Another exoenzyme, hyaluronidase (also called hyaluronate lyase HysA), cleaves the hyaluronic acid (HA) polymer at the β-1,4 glycosidic bond, yielding disaccharide

**Table 11.4** Plants with $H^+/K^+$ ATPase inhibitory activity (Adinortey & N'guessan, 2022)

| Plant name | Family | Part used | Study model | Reference |
|---|---|---|---|---|
| *Acalypha wilkesiana* | Euphorbiaceae | Leaf | In vitro | Gupta and Pradeepa (2013) |
| *Annona squamosa* | Annonaceae | Twigs<br>Twigs<br>Twigs | In vitro<br>In vitro<br>In vitro | Yadav et al. (2012) |
| *Arctium Lappa L.* | Asteraceae | Root | In vitro | Dos Santos et al. (2008) |
| *Carissa carandas* | Apocynaceae | Leaf | In vitro | Shukla et al. (2016) |
| *Cecropia glazioui* | Cecropiaceae | Aerial | In vitro | Souccar et al. (2008) |
| *Cissus quadrangularis L* | Vitaceae | Tubers | In vitro | Yadav et al. (2012) |
| *Decalepis hamiltonii* | Asclepiadaceae | Roots | In vitro | Naik et al. (2007) |
| *Delonix regia* | Fabaceae | Stem bark | In vitro | Sachan et al. (2017) |
| *Garcinia kola* Heckel. | Guttiferae | Seeds | In vitro | Onasanwo et al. (2011) |
| *Garcinia mangostana* | Guttiferae | Fruit peel | In vitro | Nanjarajurs et al. (2014) |
| *Hedranthera barteri* | Apocynaceae | Roots | In vitro | Onasanwo et al. (2011) |
| *Pongamia pinnata (L.)* | Leguminosae | Seeds | In vitro | Belagihally et al. (2011) |
| *Scoparia dulcis L* | Scrophulariaceae | Aerial parts | In vivo | Mesía-Vela et al. (2007) |
| *Solanum nigrum* | Solanaceae | Aerial parts | In vitro | Jainu and Devi (2006) |
| *Tectona grandis* | Verbenaceae | Leaf | In vitro | Lakshmi et al. (2010) |
| *Xylocarpus granatum* | Meliaceae | Fruit | In vitro | Lakshmi et al. (2010) |

units of *N*-acetylglucosamine and d-glucuronic acid. HA is synthesized by plasma membrane of mammalian cells, and serves as a cementing structure of connective tissues (Hynes & Walton, 2000). Tissues with relatively higher HA concentrations are frequently infected with *S. aureus* due to their ability to produce hyaluronidase (Patil et al., 2019). Lipase and hyaluronidase are the major virulence factors during staphylococcal infections. The inhibition of these virulent enzymes can be a simple and effective approach to controlling drug-resistant Staphylococci. A number of medicinal plants have been evaluated for the presence of effective inhibitors targeting several enzymes involved in disease causation and pathogenesis and modulating the host immune response. A number of in vitro and in vivo techniques have been developed for the isolation, screening and evaluation of the inhibitory potential of inhibitors of plant origin. Recently, molecular docking has become an important tool to understand drug-target interactions which is a key step in drug design and discovery (Patil et al., 2022). The docking principles were effectively used to identify the target sites different receptors/targetsof HIV 1 (Schames et al., 2004). Using online tool-AutoDock, this receptor was effectively used to inhibit HIV integrase. The other inhibitors screened using this approach include Type I TGF-beta receptor kinase inhibitor (Singh et al., 2003), Aurora Kinase An inhibitor (Park et al., 2018), Dopamine D3 receptor for anticancer molecules (Varady et al., 2003), etc. Table 11.5 provides the summary of various target proteins/receptors that

**Table 11.5** Protease/enzyme inhibitors from various sources and their biological significance (Source: Patil et al., 2022)

| Sr. no. | Target enzyme/protein | Inhibitor/legend | Source plant | Biological significance | Reference |
|---|---|---|---|---|---|
| 1. | Monoacylglycerol lipase | *Rutin* | *Phaseolus vulgaris* | To attenuate the inflammatory response | Saxena et al. (2010) |
| 2. | Cyclophilin D | *ZINC161011368* | *Secale cereale* | For the treatment of acute pancreatitis. | Cen et al. (2022) |
| 3. | SARS-CoV-2 3-Chymotrypsin-Like Protease. | *Chicoric acid* | *Wrightia religiosa* | preventive nutraceuticals and/or antiviral drugs in COVID-19. | Gyebi et al. (2021) |
| 4. | SARS-CoV-2 3-Chymotrypsin-Like Protease. | *Rosmarinic acid* | *Triticum aestivum,* | preventive nutraceuticals and/or antiviral drugs in COVID-19. | Gyebi et al., 2021 |
| 5. | SARS-CoV-2 3-Chymotrypsin-Like Protease. | *Myricetin* | *Wrightia religiosa* | preventive nutraceuticals and/or antiviral drugs in COVID-19. | Gyebi et al. (2021) |
| 6. | SARS-CoV-2 3-Chymotrypsin-Like Protease. | *Rosmarinic acid* | *Triticum aestivum,* | preventive nutraceuticals and/or antiviral drugs in COVID-19. | Gyebi et al. (2021) |
| 7. | SARS-CoV-2 3-Chymotrypsin-Like Protease. | *Isorhamnetin* | *Wrightia religiosa* | preventive nutraceuticals and/or antiviral drugs in COVID-19. | Gyebi et al. (2021) |
| 8. | SARS-CoV-2 3-Chymotrypsin-Like Protease. | *Luteolin* | *Triticum aestivum,* | preventive nutraceuticals and/or antiviral drugs in COVID-19. | Gyebi et al. (2021) |
| 9. | SARS-CoV-2 3-Chymotrypsin-Like Protease. | *Vernomenin* | *Wrightia religiosa* | preventive nutraceuticals and/or antiviral drugs in COVID-19. | Gyebi et al. (2021) |
| 10. | Stearyl coenzyme A desaturase enzyme 1 | *Stearoyl-Coenzyme A* | *Sorghum bicolar* | To alter the metabolism of tumor cells | Huang et al. (2020) |
| 11. | Stearyl coenzyme A desaturase enzyme 1 | *Aramchol* | *Zea mays* | To alter the metabolism of tumor cells | Huang et al. (2020) |
| 12. | mycobacterial InhA | *Enamine 57,340* | *Oryza sativa* | therapeutic applications against TB. | Jayaraman et al. (2021) |
| 13. | mycobacterial InhA | *Enamine 57,294* | *Vigna unguiculata* | therapeutic applications against TB. | Jayaraman et al. (2021) |
| 14. | Lactate dehydrogenase | *Quercetin,* | *Carisa carandas* | To arrest tumor invasion and metastasis | Li et al. (2021) |
| 15. | Lactate dehydrogenase | *Luteolin,* | *Carisa carandas* | To arrest tumor invasion and metastasis | Li et al. (2021) |

| | | | | | |
|---|---|---|---|---|---|
| 16. | Lactate dehydrogenase | *Ursolic acid* | *Carisa carandas* | To arrest tumor invasion and metastasis | Li et al. (2021) |
| 17. | phosphodiesterase 10A | MA8 | Secale cereale | alleviating the negative and cognitive symptoms of schizophrenia | Mayasah et al. (2018) |
| 18. | phosphodiesterase 10A | MA98 | Eleusine coracana | alleviating the negative and cognitive symptoms of schizophrenia | Mayasah et al. (2018) |
| 19. | Main protease (Mpro) of SARS-CoV-2 | Kazinol T | | lead compound against the main protease of SARS-CoV-2 | Singh et al. (2022) |
| 20. | Aldehyde dehydrogenases (ALDHs) | | Eleusine coracana | In the treatment of cyclophosphamide and cisplatin resistance cancer. | Narendra et al. (2021) |
| 21. | acetylcholinesterase inhibitors (AChEIs) | compound D | *Dipteryx alata* | In the treatment of Alzheimer's disease | Pablo et al. (2015) |
| 22. | butyrylcholinesterase | compound D | *Dipteryx alata* | In the treatment of Alzheimer's disease | Pablo et al. (2015) |
| 23. | Covid 19 main protease. | *Penciclovir* | *Catharanthus roseus* | to develop effective therapeutics to deal with the Covid 19 pandemic | Padhi et al. (2021) |
| 24. | Covid 19 main protease. | *Jasmonic acid,* | *Catharanthus roseus* | to develop effective therapeutics to deal with the Covid 19 pandemic | Padhi et al. (2021) |
| 25. | Covid 19 main protease. | *Jasmonic acid methyl ester,* | *Catharanthus roseus* | to develop effective therapeutics to deal with the Covid 19 pandemic | Padhi et al. (2021) |
| 26. | Covid 19 main protease. | *Putaminoxin B* | *Catharanthus roseus* | to develop effective therapeutics to deal with the Covid 19 pandemic | Padhi et al. (2021) |
| 27. | Covid 19 main protease. | *Putaminoxin D* | *Catharanthus roseus* | to develop effective therapeutics to deal with the Covid 19 pandemic | Padhi et al. (2021) |
| 28. | Haemoglobinase Falcipain-2 | *Nonpeptidic inhibitors* | *Triticum aestivum* | To develop antimalarial drugs to treat resistant strains of *Plasmodium.* | Rajguru et al. (2022) |
| 29. | Covid 19 main protease. | *Bieckol* | *Phaseolus vulgaris* | In the treatment of Covid 19 | Rauf et al. (2021) |
| 30. | Covid 19 main protease. | *7-Hydroxyeckol hepta-acetate* | *Amaranthus hypochodracus* | In the treatment of Covid 19 | Rauf et al. (2021) |

(continued)

**Table 11.5** (continued)

| Sr. no. | Target enzyme/protein | Inhibitor/legend | Source plant | Biological significance | Reference |
|---|---|---|---|---|---|
| 31. | Covid 19 main protease. | *Met165* | *Triticum aestivum* | In the treatment of Covid 19 | Rauf et al. (2021) |
| 32. | Covid 19 main protease. | *5-Hydroxy-cystofurano-quinol* | *Secale cereale* | In the treatment of Covid 19 | Rauf et al. (2021) |
| 33. | Covid 19 main protease. | *Sargaquinoic acid* | *Triticum aestivum* | In the treatment of Covid 19 | Rauf et al. (2021) |
| 34. | Covid 19 main protease. | *Methoxybifurcarenone* | *Secale cereale* | In the treatment of Covid 19 | Rauf and Jehan (2017) |
| 35. | Covid 19 main protease. | *rutin* | *Eleusine coracana* | In the treatment of Covid 19 | Rauf and Jehan (2017) |
| 36. | ERK2 (protein kinase) | Hypo1 | *Calotropis procera* | inhibition cell proliferation and oncogenesis | Sofiene et al. (2014) |
| 37. | *Glycoprotein Receptor-Binding Domain* | *solanine, acetoside, and rutin* | *Solanum melongina* | In the treatment of Covid- 19 | Teli et al. (2021) |
| 38. | *Covid 19 Main protease (Mpro)* | *Solanine, acetoside, and rutin* | *Solanum melongina* | In the treatment of Covid 19 | Teli et al. (2021) |
| 39. | Dipeptidyl-peptidase-4 | *Butirosin* | *Amaranthus hypochodracus* | In the treatment of Type II Diabetes Mellitus. | Xiong et al. (2021) |
| 40. | Dipeptidyl-peptidase-4 | *Alogliptin.* | *Wrightia religiosa* | In the treatment of Type II Diabetes Mellitus | Xiong et al. (2021) |
| 41. | Dipeptidyl-peptidase-4 | *Rutin* | *Triticum aestivum,* | In the treatment of Type II Diabetes Mellitus | Xiong et al. (2021) |
| 42. | Dipeptidyl-peptidase-4 | *Homatropine* | *Hordeum vulgare* | In the treatment of Type II Diabetes Mellitus. | Xiong et al. (2021) |
| 43. | histone acetyltransferase | *Chlorogenic acid (CA)* | *Phaseolus vulgaris* | anti-nonalcoholic fatty liver disease | Xu et al. (2021) |
| 44. | Protein tyrosine phosphatase 1B | *H1 to H2O* | *Triticum aestivum* | In the management of diabetes | Yang et al. (2020) |

are inhibited using a legand with its PDB ID, binding sites, software used in the study, and the biological significance of the study (Patil et al., 2022).

## References


Adinortey, M. B., & N'guessan, B. B. (2022). H+/K+-ATPase inhibitors from plants: a potential source for drug discovery. In *Natural products as enzyme inhibitors: An industrial perspective* (pp. 125–136). Springer Nature Singapore.

Ali, H., Houghton, P. J., & Soumyanath, A. (2006). α-Amylase inhibitory activity of some Malaysian plants used to treat diabetes; with particular reference to Phyllanthus amarus. *Journal of Ethnopharmacology, 107*(3), 449–455.

Alves, D. T., Vasconcelos, I. M., Oliveira, J. T., Farias, L. R., Dias, S. C., Chiarello, M. D., Maria-Neto, S., & Franco, O. L. (2009). Identification of four novel members of Kunitz-like α-amylase inhibitors family from Delonix regia with activity toward Coleopteran insects. *Pesticide Biochemistry and Physiology, 95*(3), 166–172.

Anusha, R., Singh, M. K., & Bindhu, O. S. (2013). Screening of latex producing plants for their milk clotting activity. *Research Journal of Pharmaceutical, Biological and Chemical Sciences., 4*(4), 757–761.

Barbole, R. S., Saikhedkar, N., & Giri, A. (2022). Plant peptides as protease inhibitors for therapeutic and agricultural applications. In *Natural products as enzyme inhibitors: An industrial perspective* (pp. 25–57). Springer Nature Singapore.

Barrett AJ, Woessner JF, Rawlings ND, editors. Handbook of proteolytic enzymes. Vol. 1. Elsevier; 2012.

Battjes, E., Harsh, K., Langton, E., & Huber, B. (2015). What antihypertensive agents are least likely to cause erectile dysfunction? *Evidence-Based Practice, 18*(6), 7–8.

Belagihally, S. M., Rajashekhar, S., Jayaram, V. B., Dharmesh, S. M., & Thirumakudalu, S. K. C. (2011). Gastroprotective properties of karanjin from Karanja (Pongamia pinnata) seeds; role as antioxidant and H+, K+-ATPase inhibitor. *Evidence-Based Complementary and Alternative Medicine, 2011*.

Bharadwaj, R. P., Raju, N. G., & Chandrashekharaiah, K. S. (2018). Purification and characterization of alpha-amylase inhibitor from the seeds of underutilized legume, Mucuna pruriens. *Journal of Food Biochemistry, 42*(6), e12686.

Bhatnagar, A., & Mishra, A. (2022). α-Glucosidase inhibitors for diabetes/blood sugar regulation. In *Natural products as enzyme inhibitors: An industrial perspective* (pp. 269–283). Springer Nature Singapore.

Birk, Y. (2003). *Plant protease inhibitors: Significance in nutrition, plant protection, cancer prevention and genetic engineering*. Springer Science & Business Media. 192p.

Bonavides, K. B., Pelegrini, P. B., Laumann, R. A., Grossi-de-Sá, M. F., JrC, B., Melo, J. A., & Franco, O. L. (2007). Molecular identification of four different α-amylase inhibitors from Baru (Dipteryx alata) seeds with activity toward insect enzymes. *BMB Reports, 40*(4), 494–500.

Bougatef, A., Souissi, N., Fakhfakh, N., Ellouz-Triki, Y., & Nasri, M. (2007). Purification and characterization of trypsin from the viscera of sardine (*Sardina pilchardus*). *Food Chemistry, 102*, 343–350.

Bunyatang, O., Chirapongsatonkul, N., Bangrak, P., Henry, R., & Churngchow, N. (2016). Molecular cloning and characterization of a novel bifunctional α-amylase/subtilisin inhibitor from Hevea brasiliensis. *Plant Physiology and Biochemistry, 101*, 76–87.

Carlini, C. R., & Grossi-de-Sá, M. F. (2002). Plant toxic proteins with insecticidal properties. A review on their potentialities as bioinsecticides. *Toxicon, 40*(11), 1515–1539.

Cen, X. Z., Yu, X. Y., Dai, C., Huang, Q. Y., Shen, Y., Wang, J., Hu, Y., & Lin, Z. H. (2022). Identification of potent CypD inhibitors via pharmacophore based virtual screening, docking and molecular dynamics simulation. *Journal of Molecular Structure, 1247*, 131355.

Chen, K., Liu, X. Q., Wang, W. L., Luo, J. G., & Kong, L. Y. (2020). Taxumarienes A–G, seven new α-glucosidase inhibitory taxane-diterpenoids from the leaves of Taxus mairei. *Bioorganic Chemistry, 94*, 103400. https://doi.org/10.1016/j.bioorg.2019.103400

Choi, S. B., Ko, B. S., Park, S. K., Jang, J. S., & Park, S. (2006). Insulin sensitizing and α-glucoamylase inhibitory action of sennosides, rheins and rhaponticin in Rhei Rhizoma. *Life Sciences, 78*(9), 934–942. https://doi.org/10.1016/j.lfs.2005.05.101

Christensson, B., Fehrenbach, F. J., & Hedstrom, S. A. (1985). A new serological assay for Staphylococcus aureusinfections: Detection of IgG antibodies to S. aureuslipase with an enzyme-linked immunosorbent assay. *The Journal of Infectious Diseases, 152*, 286–292.

Christou, P., Capell, T., Kohli, A., Gatehouse, J. A., & Gatehouse, A. M. (2006). Recent developments and future prospects in insect pest control in transgenic crops. *Trends in Plant Science, 11*(6), 302–308.

Dias, N., Vezin, H., Lansiaux, A., & Bailly, C. (2005). Topoisomerase inhibitors of marine origin and their potential use as anticancer agents. *DNA Binders and Related Subjects*, 89–108.

Dirir, A. M., Daou, M., Yousef, A. F., & Yousef, L. F. (2021). A review of alpha-glucosidase inhibitors from plants as potential candidates for the treatment of type-2 diabetes. In *Phytochemistry reviews* (Vol. 1). Springer Netherlands. https://doi.org/10.1007/s11101-021-09773-1

Diz, M. S., Carvalho, A. O., Ribeiro, S. F., Da Cunha, M., Beltramini, L., Rodrigues, R., & Gomes, V. M. (2011). Characterization, immune localization and antifungal activity of a lipid transfer protein from chili pepper (Capsicum annuum) seeds with novel α-amylase inhibitory properties. *Physiologia Plantarum, 142*(3), 233–246.

Dos Santos, A. C., Baggio, C. H., Freitas, C. S., Lepieszynski, J., Mayer, B., Twardowschy, A., . Marques, M. C. (2008). Gastroprotective activity of the chloroform extract of the roots from *Arctium lappa* L. The Journal of Pharmacy and Pharmacology, 60(6), 795–801.

Farias, L. R., Costa, F. T., Souza, L. A., Pelegrini, P. B., Grossi-de-Sá, M. F., Neto, S. M., Bloch, C., Jr., Laumann, R. A., Noronha, E. F., & Franco, O. L. (2007). Isolation of a novel *Carica papaya* α-amylase inhibitor with deleterious activity toward *Callosobruchus maculatus*. *Pesticide Biochemistry and Physiology, 87*(3), 255–260.

Flemmig, M., & Melzig, M. F. (2013). Plant latex proteases with potential influence on blood coagulation and fibrinolysis. *Planta Medica, 79*(13), PN33.

Franco, O. L., Rigden, D. J., Melo, F. R., & Grossi-de-Sá, M. F. (2002). Plant α-amylase inhibitors and their interaction with insect α-amylases: Structure, function and potential for crop protection. *European Journal of Biochemistry., 269*(2), 397–412.

Giri, A. P., & Kachole, M. S. (1998). Amylase inhibitors of pigeonpea (Cajanus cajan) seeds. *Phytochemistry, 47*(2), 197–202.

Gupta, R., Beg, Q. K., Khan, S., & Chauhan, B. (2002). An overview on fermentation, downstream processing and properties of microbial alkaline proteases. *Applied Microbiology and Biotechnology, 60*, 381–395.

Gupta, R. K. P., & Pradeepa, M. H. (2013). In vitro antioxidant and H+, K+-ATPase inhibition activities of Acalypha wilkesiana foliage extract. *Journal of Pharmacy & Bioallied Sciences, 5*(3), 214.

Guzman, M. H., & Paredes, L. O. (1995). Amylolytic enzymes and products derived from starch: a review. *Critical Reviews of Food Science and Nutrition, 35*, 373–403.

Guzman-Partida, A. M., Jatomea-Fino, O., Robles-Burgueno, M. R., Ortega-Nieblas, M., & Vazquez-Moreno, L. (2007). Characterization of α-amylase inhibitor from Palo Fierro seeds. *International Journal of Pharmacy and Biological Sciences, 45*(9), 711–715.

Gyebi, G. A., Elfiky, A. A., Ogunyemi, O. M., Ibrahim, I. M., Adegunloye, A. P., Adebayo, J. O., Olaiya, C. O., Ocheje, J. O., & Fabusiwa, M. M. (2021). Structure-based virtual screening suggests inhibitors of 3-Chymotrypsin-Like Protease of SARS-CoV-2 from Vernonia amygdalina and Occinum gratissimum. *Computers in Biology and Medicine, 136*, 104671.

Hasan, F., Shah, A., & Hameed, A. (2006). Industrial applications of microbial lipases. *Enzyme and Microbial Technology, 39*, 235–251.

Hay, E., Lucariello, A., Contieri, M., Esposito, T., De Luca, A., Guerra, G., & Perna, A. (2019). Therapeutic effects of turmeric in several diseases: An overview. *Chemico-Biological Interactions, 310*, 108729.

Huang, Y., Wang, H., Wang, H., Wen, R., Geng, X., Huang, T., Shi, J., Wang, X., & Wang, J. (2020). Structure-based virtual screening of natural products as potential stearoyl-coenzyme a desaturase 1 (SCD1) inhibitors. *Computational Biology and Chemistry, 86*, 107263.

Hynes, W. L., & Walton, S. L. (2000). Hyaluronidases of gram-positive bacteria. *FEMS Microbiology Letters, 183*, 201–207.

Jainu, M., & Devi, C. S. S. (2006). Antiulcerogenic and ulcer healing effects of Solanum nigrum (L.) on experimental ulcer models: possible mechanism for the inhibition of acid formation. *Journal of Ethnopharmacology, 104*(1–2), 156–163.

Jang, Y. J., Leem, H. H., Jeon, Y. H., Lee, D. H., & Choi, S. W. (2015). Isolation and Identification of α -glucosidase inhibitors from Morus Root Bark. 44, 1090–1099.

Jayaraj, S., Suresh, S., & Kadeppagari, R. K. (2013). Amylase inhibitors and their biomedical applications. *Starch-Stärke, 65*(7–8), 535–542.

Jayaraman, M., Lakshmanan, L., Karthikeyan, M., & Krishna, R. (2021). Virtual screening assisted discovery of novel natural products to inhibit the catalytic mechanism of Mycobacterium tuberculosis InhA. *Journal of Molecular Liquids, 335*.

Jhong, C., & Chia, Y. (2015). Research communication screening alpha-glucosidase and alpha-amylase inhibitors from natural compounds by molecular docking in silico. *Biofactors (Oxford, England), 41*(4), 242–251. https://doi.org/10.1002/biof.1219

Kadziola, A., Søgaard, M., Svensson, B., & Haser, R. (1998). Molecular structure of a barley α-amylase-inhibitor complex: implications for starch binding and catalysis. *Journal of Molecular Biology, 278*(1), 205–217.

Kang, S., Park, J. Y., Choi, H., & Kim, J. (2015). α-Glucosidase inhibitory activities of myricetin in animal models of diabetes mellitus. *Food Science and Biotechnology, 24*(5), 1897–1900. https://doi.org/10.1007/s10068-015-0249-y

Karpagam, N., Viswanathan, S., Prabhu, S., Sulochana, S., & Sivanandham, M. (2013). Biochemical and in silico clotting activity of latex from *Asclepias curassavica* L. *International Journal of Pharma and Bio Sciences., 4*(4).

Kasar, S. S., Maheshwari, V. L., & Pawar, P. K. (2022). Bioactive α-amylase inhibitors: Sources, mechanism of action, biochemical characterization, and applications. *Natural Products as Enzyme Inhibitors: An Industrial Perspective*, 59–104.

Kasar, S. S., Marathe, K. R., Bhide, A. J., Herwade, A. P., Giri, A. P., Maheshwari, V. L., & Pawar, P. K. (2017). A glycoprotein α-amylase inhibitor from *Withania somnifera* differentially inhibits various α-amylases and affects the growth and development of *Tribolium castaneum. Pest Management Science, 73*(7), 1382–1390.

Kasar, S. S., Sabale, V. B., Shinde, R. A., Maheshwari, V. L., & Pawar, P. K. (2021). Effect of α-amylase inhibitor from Withania somnifera on growth and development of *Callosobruchus chinensis* and in silico studies on its interactions with insect amylase. *Archives of Phytopathology and Plant Protection., 54*(5–6), 231–251.

Kaur, R., Kaur, N., & Gupta, A. K. (2014). Structural features, substrate specificity, kinetic properties of insect α-amylase and specificity of plant α-amylase inhibitors. *Pesticide Biochemistry and Physiology, 116*, 83–93.

Khuraishi, A. A., Jayalakshmi, M., Harini, A. P., Kunder, A. S., & Mariswamy, M. (2019). Isolation and purification of anticoagulant enzymes from plant latex. *Journal of Applied and Natural Science., 11*(1), 217–222.

Kokiladevi, E., Manickam, A., & Thayumanavan, B. (2005). Characterization of alpha-amylase inhibitor in Vigna sublobata. *Botanical Bulletin of Academia Sinica, 46*.

Kumari, B., Sharma, P., & Nath, A. K. (2012). α-Amylase inhibitor in local Himalyan collections of Colocasia: isolation, purification, characterization and selectivity towards α-amylases from various sources. *Pesticide Biochemistry and Physiology, 103*(1), 49–55.

Lakshmi, V., Singh, N., Shrivastva, S., Mishra, S., Dharmani, P., Mishra, V., & Palit, G. (2010). Gedunin and photogedunin of Xylocarpus granatum show significant anti-secretory effects and protect the gastric mucosa of peptic ulcer in rats. *Phytomedicine, 17*(8–9), 569–574.

Li, W., Cui, X., & Chen, Z. (2021). Screening of lactate dehydrogenase inhibitor from bioactive compounds in natural products by electrophoretically mediated microanalysis. *Journal of Chromatography A, 1656*, 462554.

Mahajan, R., Chaudhari, G., & Chopadaa, M. (2015). Report on biotechnological applications of proteolytic enzymes from lattices of euphorbian plants. *Journal of Applied Biotechnology Reports, 2*(4), 33–37.

Matsui, T., Tanaka, T., Tamura, S., Toshima, A., Tamaya, K., Miyata, Y., Tanaka, K., & Matsumoto, K. (2007). α-Glucosidase inhibitory profile of catechins and theaflavins. *Journal of Agricultural and Food Chemistry, 55*(1), 99–105. https://doi.org/10.1021/jf0627672

Mayasah, A., Anand, G., & Gabriel, A. (2018). Discovery of natural product inhibitors of phosphodiesterase 10A as novel therapeutic drug for schizophrenia using a multistep virtual screening. *Computational Biology and Chemistry, 77*, 52–63.

McEwan, R., Madivha, R. P., Djarova, T., Oyedeji, O. A., & Opoku, A. R. (2010). Alpha-amylase inhibitor of amadumbe (Colocasia esculenta): Isolation, purification and selectivity toward-amylases from various sources. *African Journal of Biochemistry Research, 4*(9), 220–224.

Meera, C., Meenakumari, S., Thirumavalavan, M., & Pachaiappan, R. (2017). Isolation and characterization of α-amylase inhibitor from Leucas aspera (Willd) Link: α-amylase assay combined with FPLC chromatography for expedited identification. *Journal of Plant Biochemistry and Biotechnology, 26*(3), 346–355.

Mehrabadi, M., Bandani, A. R., Saadati, F., & Mahmudvand, M. (2011). α-Amylase activity of stored products insects and its inhibition by medicinal plant extracts. *Journal of Agricultural Science and Technology, 13*(7), 1173–1182.

Meshram, A., Singhal, G., Bhagyawant, S. S., & Srivastava, N. (2019). Plant-derived enzymes: a treasure for food biotechnology. In *Enzymes in food biotechnology* (pp. 483–502). Academic Press.

Mesía-Vela, S., Bielavsky, M., Torres, L. M. B., Freire, S. M., Lima-Landman, M. T. R., Souccar, C., & Lapa, A. J. (2007). In vivo inhibition of gastric acid secretion by the aqueous extract of Scoparia dulcis L. in rodents. *Journal of Ethnopharmacology, 111*(2), 403–408.

Mishra, A., Anusha, P., Lakshmikantha, M. R., Nagamani, T. S., & Mahesh, M. (2017). Comparative studies on anticoagulant enzyme from plant latex samples. *World Journal of Pharmaceutical Research., 6*(13), 375–393.

Mishra, M., Tamhane, V. A., Khandelwal, N., et al. (2010). Interaction of recombinant CanPIs with *Helicoverpa armigera* gut proteases reveals their processing patterns, stability and efficiency. *Proteomics, 10*, 2845–2857.

Nabavi, S., Thiagarajan, R., Rastrelli, L., Daglia, M., Sobarzo-Sanchez, E., Alinezhad, H., & Nabavi, S. (2015). Curcumin: A natural product for diabetes and its complications. *Current Topics in Medicinal Chemistry, 15*(23), 2445–2455. https://doi.org/10.2174/1568026615666150619142519

Naik, Y., Jayaram, S., Nayaka, M. H., & Dharmesh, S. M. (2007). Gastroprotective effect of swallow root (Decalepis hamiltonii) extract: Possible involvement of H+–K+–ATPase inhibition and antioxidative mechanism. *Journal of Ethnopharmacology, 112*(1), 173–179.

Nanjarajurs, S. M., Dharmesh, S. M., Bhimangouder, S. V., Eswaraiah, M. S., & Somasundaram, R. (2014). Health and wellness product from mangosteen (Garcinia mangostana l.) rind: bioactive potentials. *International Journal of Biotechnology for Wellness Industries, 3*(4), 111–120.

Narendra, G., Raju, B., Verma, H., Sapra, B., & Silakari, O. (2021). Multiple machine learning models combined with virtual screening and molecular docking to identify selective human ALDH1A1 inhibitors. *Journal of Molecular Graphics and Modelling, 107*, 107950.

Oboh, G., & Ademosun, A. O. (2015). Comparative effect of quercetin and rutin on α-amylase, α-glucosidase, and some pro-oxidant-induced lipid peroxidation in rat pancreas. *Comparative Clinical Pathology, 24*, 1103–1110. https://doi.org/10.1007/s00580-014-2040-5

Onasanwo, S. A., Singh, N., Olaleye, S. B., & Palit, G. (2011). Anti-ulcerogenic and proton pump (H+, K+ ATPase) inhibitory activity of Kolaviron from Garcinia kola Heckel in rodents. *Indian Journal of Experimental Biology, 49*, 461–468.

Pablo, A. N., Rogerio, A. S., Diones, C. B., Lilian, J. L., Cristiane, L. D., Braga, M. M., Denis, B. R., & Joao, B. T. R. (2015). Virtual screening of acetylcholinesterase inhibitors using the lipinski's rule of five and zinc databank. *BioMed Research International, 8*, 20.

Padhi, S., Masi, M., Chourasia, R., Rajashekar, Y., Rai, A., & Evidente, A. (2021). ADMET profile and virtual screening of plant and microbial natural metabolites as SARS-CoV-2 S1 glycoprotein receptor binding domain and main protease inhibitors. *European Journal of Pharmacology, 890*, 173648.

Pan, G. Y., Huang, Z. J., Wang, G. J., Fawcett, J. P., Liu, X. D., Zhao, X. C., Sun, J. G., & Xie, Y. Y. (2003). The antihyperglycaemic activity of berberine arises from a decrease in glucose absorption. *Planta Medica, 69*(7), 632–636. https://doi.org/10.1055/s-2003-41121

Park, H., Jung, H. Y., Mah, S., & Hong, S. (2018). Systematic computational design and identification of low Picomolar inhibitors of Aurora kinase A. *Journal of Chemical Information and Modeling, 58*(3), 700–709.

Patil, M., Patil, S., Maheshwari, V. L., Zawar, L., & Patil, R. H. (2022). Recent updates on in silico screening of natural products as potential inhibitors of enzymes of biomedical and pharmaceutical importance. In *Natural products as enzyme inhibitors: An industrial perspective* (pp. 105–123). Springer Nature Singapore.

Patil, S., Bhadane, B., Shirsath, L., Patil, R., & Chaudhari, B. (2019). Steroidal fraction of Carissa carandas L. inhibits microbial hyaluronidase activity by mixed inhibition mechanism. *Preparative Biochemistry & Biotechnology, 49*(3), 298–306. https://doi.org/10.1080/10826068.2018.1541811

Pearce, G. (2011). Systemin, hydroxyproline-rich systemin and the induction of protease inhibitors. *Current Protein & Peptide Science, 12*, 399–408.

Pelegrini, P. B., Murad, A. M., Grossi-de-Sá, M. F., Mello, L. V., Romeiro, L. A., Noronha, E. F., Caldas, R. A., & Franco, O. L. (2006). Structure and enzyme properties of Zabrotes subfasciatus α-amylase. *Archives of Insect Biochemistry and Physiology, 61*(2), 77–86.

Peters, H. P. F., Foltz, M., Kovacs, E. M. R., et al. (2011). The effect of protease inhibitors derived from potato formulated in a minidrink on appetite, food intake and plasma cholecystokinin levels in humans. *International Journal of Obesity, 2005*(35), 244–250.

Rajendran, A., Palanisamy, A., & Thangavelu, V. (2009). Lipase catalyzed ester synthesis for the food processing Industries. *Brazilian Archives of Biology and Technology, 52*(1), 207–219.

Rajguru, T., Dipshikha, B., & Mahendra, K. M. (2022). Identification of promising inhibitors for Plasmodium haemoglobinase Falcipain-2, using virtual screening, molecular docking, and MD simulation. *Journal of Molecular Structure, 1248*, 131427.

Rauf, A., & Jehan, N. (2017). Natural products as a potential enzyme inhibitors from medicinal plants. In M. Senturk (Ed.), *Enzyme inhibitors and activators*. IntechOpen. https://www.intechopen.com/chapters/54038

Rauf, B., Rashid, U., Khalil, A., Khan, S., Anwar, S., Alafnan, A., Alamri, A., Kannan, R. R., & Rengasamy. (2021). Docking-based virtual screening and identification of potential COVID-19 main protease inhibitors from brown algae. *South African Journal of Botany, 143*, 428–434.

Reddy, N. S., Nimmagada, A., & Sambasiva, R. S. (2003). An over view of the microbial alpha amylase mini review. *African Journal of Biotechnology, 2*, 645–648.

Rollof, J., Braconier, J. H., Soderstrom, C., & Nilsson-Ehle, P. (1988). Interference of Staphylococcus aureuslipase with human granulosyte function. *European Journal of Clinical Microbiology & Infectious Diseases, 7*, 505–510.

Sachan, N., Chandra, P., & Pal, D. (2017). Effect of Delonix regia (Boj. Ex Hook.) Raf. stem bark extract against experimentally induced ulcers in rats. *Indian Journal of Experimental Biology, 55*, 49–54.

Sasikiran, K., Rekha, M. R., & Padmaja, G. (2004). Purification and partial characterization of proteinase and α-amylase inhibitors from lesser yam (Dioscorea esculenta). *International Journal of Food Properties, 7*(2), 185–199.

Saunders, R. M., & Lang, J. A. (1973). α-Amylase inhibitors in Triticum aestivum: purification and physical-chemical properties. *Phytochemistry, 12*(6), 1237–1241.

Saxena, L., Iyer, B. K., & Ananthanarayan, L. (2010). Purification of a bifunctional amylase/protease inhibitor from ragi (Eleusine coracana) by chromatography and its use as an affinity ligand. *Journal of Chromatography B, 878*(19), 1549–1554.

Schames, J. R., Henchman, R. H., Siegel, J. S., Sotriffer, C. A., Ni, H., & McCammon, J. A. (2004). Discovery of a novel binding trench in HIV integrase. *Journal of Medicinal Chemistry, 47*(8), 1879–1881.

Shamki, A. W., Ali, A., & Abdulaziz, R. S. (2012). Purification and characterization of amylase inhibitor extracted from white kidney bean (Phaseolus vulgaris). *Journal of Cell Plant Science, 3*(1), 17–21.

Sharma, A. K., Sharma, V., & Saxena, J. (2012). A review on applications of microbial lipases. *International Journal of Biotech Trends and Technology (IJBTT), 19*, 1–5.

Sharma, R., Chisti, Y., & Banerjee, U. (2001). Production, purification, characterization, and applications of lipases. *Biotechnology Advances, 19*, 627–662.

Shukla, A., Verma, S., Bishnoi, R., & Jain, C. (2016). In vitro Carissa H-K ATPase inhibitory potential of methanolic extract of carandas Linn. leaves. *Asian Journal of Pharmacy and Pharmacology, 2*(5), 132–135.

Silva, J. R., Amaral, A. C., Siani, A. C., Rezende, C. M., Felcman, J., & Pinto, A. C. (2003). Contribution to the study of *Himatanthus sucuuba*: latex macromolecule, microelements and carbohydrates. *Acta Amazonica, 33*, 105–110.

Singh, J., Chuaqui, C. E., Boriack-Sjodin, P. A., Lee, W.-C., Pontz, T., & Corbley, M. J. (2003). Successful shape-based virtual screening: The discovery of a potent inhibitor of the type I TGFβ receptor kinase (TβRI). *Bioorganic & Medicinal Chemistry Letters, 13*(24), 4355–4359.

Singh, M. B., Sharma, R., Kumar, D., Khanna, P., Khanna, L., Kumar, V., Kumari, K., Gupta, A., Chaudhary, P., Kaushik, N., & Choi, E. H. (2022). An understanding of coronavirus and exploring the molecular dynamics simulations to find promising candidates against the Mpro of nCoV to combat the COVID-19: A systematic review. *Journal of Infection and Public Health.*

Sofiene, L., Chaker, B. S., Houssem, H., & Kamel, B. (2014). In silico screening and study of novel ERK2 inhibitors using 3D QSAR, docking and molecular dynamics. *Journal of Molecular Graphics & Modelling, 53*, 1–12.

Souccar, C., Cysneiros, R. M., Tanae, M., Torres, L., Lima-Landman, M., & Lapa, A. (2008). Inhibition of gastric acid secretion by a standardized aqueous extract of Cecropia glaziovii Sneth and underlying mechanism. *Phytomedicine, 15*(6–7), 462–469.

Sun, Z. L. W., Liu, P., Wang, H., Huang, Y., Zhao, Y., Kong, Y., & Cui, Z. (2015). Isolation and characterization of a proteinaceous α-amylase inhibitor AAI-CC5 from Streptomyces sp. CC5, and its gene cloning and expression. *Antonie Leeuwenhoek, 107*(2), 345–356.

Swedberg, J. E., Nigon, L. V., Reid, J. C., et al. (2009). Substrate-guided design of a potent and selective kallikrein-related peptidase inhibitor for kallikrein 4. *Chemistry & Biology, 16*, 633–643.

Teli DM, Shah MB, Chhabria MT (2021) In silico screening of natural compounds as potential inhibitors of SARS-CoV-2 main protease and spike RBD: targets for COVID-19. 429.

Ting, L., Zhang, X. D., Song, Y. W., & Liu, J. W. (2018). A microplate-based screening method for alpha-glucosidase inhibitors. *Chinese Journal of Clinical Pharmacology and Therapeutics, 10*(10), 1128–1134.

Underkofler, L. A., Barton, R. R., & Rennert, S. S. (1957). Production of Microbial Enzymes and their Application. *Microbiological Process Report*, 212–221.

Vadnere, N. C., & Patil, N. P. (2022 May). Natural Protease Inhibitors and Their Therapeutic Potential against SARS-CoV-2. *Natural Products as Enzyme Inhibitors: An Industrial Perspective., 4*, 205–226.

Van de Laar, F. A. (2008). Alpha-glucosidase inhibitors in the early treatment of type 2 diabetes. *Vascular Health and Risk Management, 4*(6), 1189–1195.

Van der Maarel, M. J., Van der Veen, B., Uitdehaag, J. C., Leemhuis, H., & Dijkhuizen, L. (2002). Properties and applications of starch-converting enzymes of the a-amylases family. *Journal of Biotechnology, 94*, 137–155.

Varady, J., Wu, X., Fang, X., Min, J., Hu, Z., & Levant, B. (2003). Molecular modelling of the three-dimensional structure of dopamine 3 (D3) subtype receptor: Discovery of novel and potent D3 ligands through a hybrid pharmacophore-and structure-based database searching approach. *Journal of Medicinal Chemistry, 46*(21), 4377–4392.

Wang, T.-X., Yin, Z.-H., Zhang, W., Peng, T., & Kang, W.-Y. (2013). [Chemical constituents from Psoralea corylifolia and their antioxidant alpha-glucosidase inhibitory and antimicrobial activities]. Zhongguo Zhong yao za zhi = Zhongguo zhongyao zazhi =. *China Journal of Chinese Materia Medica, 38*(14), 2328–2333. http://europepmc.org/abstract/MED/24199566

Weselake, R. J., MacGregor, A. W., & Hill, R. D. (1985). Endogenous alpha-amylase inhibitor in various cereals. *Cereal Chemistry, 62*(2), 120–123.

Wiseman, A. (1995). Introduction to principles. In A. Wiseman (Ed.), *Handbook of enzyme biotechnology* (3rd ed., pp. 3–8). Ellis Horwood Ltd., T. J Press Ltd..

Wisessing, A., Engkagul, A., Wongpiyasatid, A., & Choowongkomon, K. (2010). Biochemical characterization of the α-amylase inhibitor in mungbeans and its application in inhibiting the growth of Callosobruchus maculatus. *Journal of Agricultural and Food Chemistry, 58*(4), 2131–2137.

Xiong, F., Xiaoyu, D., Hao, Z., Xiaomin, L., Kaixian, C., Hualiang, J., Cheng, L., & Xu, H. (2021). Discovery of novel reversible monoacylglycerol lipase inhibitors by docking-based virtual screening. *Bioorganic & Medicinal Chemistry Letters, 41*, 127986.

Xu, J., Liangqin, G., Huiqing, L., Shaoliang, Z., Penghua, L., & Shaodong, C. (2021). Evidence for the anti-NAFLD effectiveness of chlorogenic acid as a HAT inhibitor using in vivo experiments supported by virtual molecular docking. *Phytomed Plus, 1*(4).

Yadav, P., Ganeshpurkar, A., & Rai, G. (2012). In vitro H+-K+ ATPase inhibitory potential of methanolic extract of Cissus quadrangularis Linn. *Pharmacognosy Res., 4*(2), 123.

Yagami, T., Sato, M., Nakamura, A., Komiyama, T., Kitagawa, K., Akasawa, A., & Ikezawa, Z. (1998). Plant defense–related enzymes as latex antigens. *Journal of Allergy and Clinical Immunology, 101*(3), 379–385.

Yang, W., Dall, T. M., Beronjia, K., Lin, J., Semilla, A. P., Chakrabarti, R., Hogan, P. F., & Petersen, M. P. (2018). Economic costs of diabetes in the U.S. in 2017. *Diabetes Care, 41*(5), 917–928. https://doi.org/10.2337/dci18-0007

Yang, Y., Tian, J. Y., Ye, F., & Xiao, Z. (2020). Identification of natural products as selective PTP1B inhibitors via virtual screening. *Bioorganic Chemistry, 86*.

Yue, Y., Chen, Y., Geng, S., Liang, G., & Liu, B. (2018). Antioxidant and α-glucosidase inhibitory activities of fisetin. *Natural Product Communications, 13*(11), 1489–1492. https://doi.org/10.1177/1934578x1801301119

Zottich, U., Da Cunha, M., Carvalho, A. O., Dias, G. B., Silva, N. C., Santos, I. S., do Nacimento, V. V., Miguel, E. C., Machado, O. L. T., & Gomes, V. M. (2011). Purification, biochemical characterization and antifungal activity of a new lipid transfer protein (LTP) from *Coffea canephora* seeds with α-amylase inhibitor properties. *Biochimica et Biophysica Acta: General Subjects, 1810*(4), 375–383.

# Chapter 12
# Endophytes in Apocynaceae Plants

## Introduction

Endophytes are defined as microbes that inhabit the internal parts of plants regardless of the outcome of the association. They gain entry into the seed, leaf, stem, flower and root of a plant, and they show no harm to the host plant (Yadav, 2018). Endophytes are omnipresent alomost in every plant in the internal tissues. They live in intracellular tissues of their plant host host plant, and establish a symbiotic relationship with the host. Sometimes, endophytes may revert to some form of pathogenesis in the host (Christel et al., 2000; Tejesvi et al., 2007). Endophytes have been isolated successfully from terrestrial, aquatic and desert plants. Both, bacterial as well as fungal endophytes are reported. However, fungal endophytes are more ubiquitous, versatile, and widespread microorganisms found in the plants growing in diverse environmental conditions. Plants growing in the Arctic, Antarctic (Suryanarayanan et al., 2005), geothermal soils (Fisher et al., 1995), deserts (Rosa et al., 2009), oceans (Wang et al., 2006), mangroves and coastal (Lin et al., 1996) regions harbors highly diverse bacterial and fungal endophytes. Despite their huge diversity, a very endophytic microorganisms have been isolated and studied systematically. Endophytes have been recognized as chief and sustainable source of biologically active natural products and secondary metabolites. Exploring the plant-based drugs using endophytic microorganisms will help overcome geographical and political restrictons involved in the transportation of plants. The metabolites from endophytic microorganisms weer explored for their applications in agriculture, medicine, bioremediation, biofuel, etc. (Russell et al., 2011; Shaw et al., 2015). The ability of endophytes to produce a variety of novel secondary metabolites with therapeutic efficacy has recently received much interest in the scientific community (Gupta et al., 2020). Despite this, there is yet to be a substantial breakthrough in the utilization and commercialization of endophytic metabolites containing bioactive compounds on a wide scale (Venugopalan & Srivastava, 2015a, 2015b).

 
R. H. Patil et al., *Apocynaceae Plants*,
https://doi.org/10.1007/978-981-99-5406-3_12

## Plant Endophyte Interactions

Endophytes are microorganisms with the Greek prefixes "endo" (within) and "phyton" (plant). The interaction between the plant and endophyte are believed to be over 400 million years old, and it observed that the plats needs association of endophytes for their survival in harsh environments (Rodriguez & Redman, 2008). Plants can "select" as endophytes those bacteria or fungi among the highly diverse microbial community that provide them with some benefit. The makeup of a plant's root exudates plays an important role in defining the type of microbe that can be linked with it (Andreozzi et al., 2019). As a result, the ability of endophytic bacteria to utilize the exudates produced by plant roots as a source of energy is critical to their relationship with plants (Kandel et al., 2017). Endophytes can effectively promote plant growth by utilizing various modes of activity and boosting plant tolerance to severe circumstances (Yadav et al., 2018).

According to a study, plants are connected with microbiota that increase stress resistance systems under stress conditions (Marasco et al., 2012; Santoyo et al., 2016). Endophytes are though to promote plant growth by secreting phytohormones, enehancing nutrient assimilation as well as by improves plant health by protecting them from phytopathogens (Andreozzi et al., 2019). Plant-endophyte interactions cause plants to protect themselves from adverse environmental situations such as heavy metals and drought.

Endophytes posseses a very high number of genes that code for plant-beneficial features and have modulatory effects on the growth and development of host plant (Khare et al., 2018). They play an important role in growth promotion by producing the various phytohormones, nutrient acquisition, promoting biodegradation increasing the stress tolerance in the host plant. These endophytic bacteria have been shown to have a variety of biocontrol mechanisms. They use various strategies to combat phytopathogens, including ecological niche competition, antimicrobial agent production, siderophore generation, and induced systemic resistance (ISR). Endophytic microorganisms help boost plant fitness through several direct and indirect mechanisms, such as plant protection from phytopathogens, secretion of lytic enzymes, and phosphate solubilization. The general mechanisms employed by endophytic microbes for plant protection are represented in Fig. 12.1.

The association between the plant and endophytic fungi has begun from the time of evolution as indicated by the fossil record (Rodriguez & Redman, 2008). Endophytes develop a variety of associations with their host plants during their growth inside the plant tissues. These relashinships may be symbiotic, mutualistic even parasitic. The levels and complexities of these relationships between the endophytic microorganism and its host vary from host to host and microbe to microbe (Verma et al., 2009). The colonization process is the same for endophytes and fungal pathogens infecting plants, which requires host establishment, including recognition, penetration, and germination. During colonization, fungi have to overcome plant defense mechanisms for stable establishment within their plant host. An endophyte enter inside the plant tissue through various mechanisms. Entry and spread through

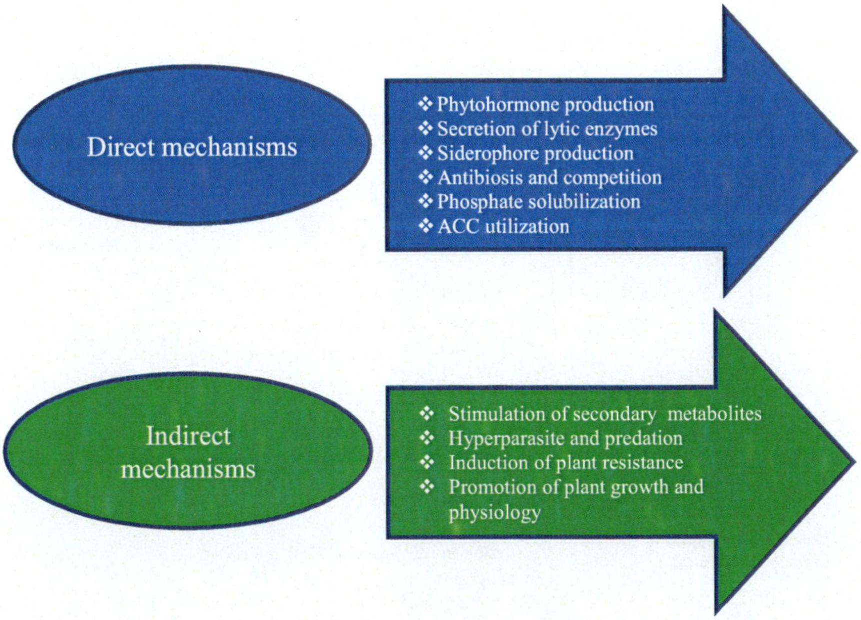


**Fig. 12.1** Mechanisms of plant protection employed by endophytes

**Fig. 12.2** Life cycle of systemic endophytes

the seeds and spreading in different tissues of the plant is well understood (Fig. 12.2).

In the mutualistic relationship, endophytes derive the nutrients from the plant tissues. On the other hand, the endophyte supports the sed germination and produce the phytohormones and other metabolites that favors the growth and the survival

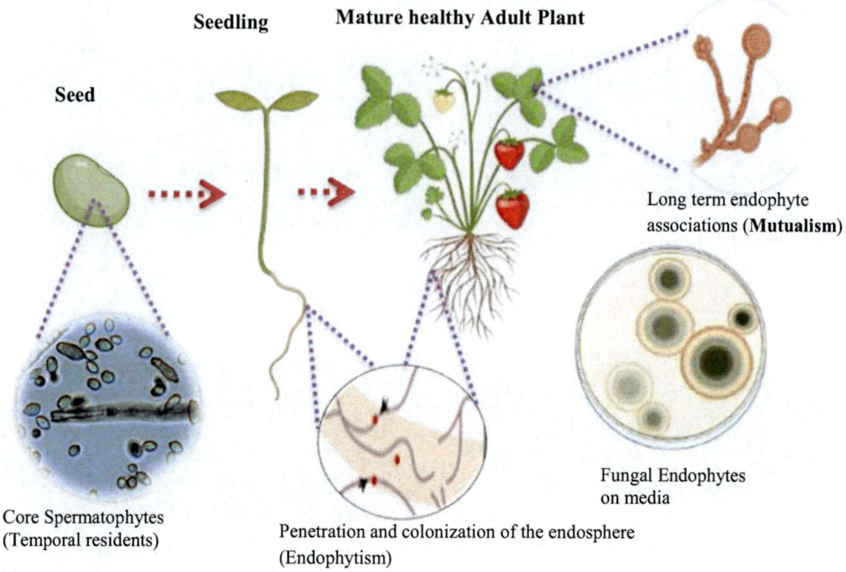


**Fig. 12.3** Mechanism of endophytism

benefits to plant in adverse environmental settings (Verma et al., 2009). The exchange of genes between the pant hhost and its endophyte that may have occurred evolution favored the symbiotic nature of the interaction between the endophytes and their plant host. As a result of it, the bioactive compounds synthesized by endophytes are normally do not show toxicity in the eukaryotic host system. This character of endophyte derived metabolites makes them suitable for their application as a bioactive pharmaceuticals.

The fungal hyphae grow inside the internal tissues of all the aboveground parts of the plant such as stem, twigs, leaves and fruits and seeds. Presence of endophytes in the seeds helps vertical transfer and spread of endophytes fromhe mother plant to the offsprings (Fig. 12.3). According to recent research (Clay and Holah, 1999), horizontal transmission is reported in natural grass populations. Fungal spores frequently recognize the host via lectin-like molecules. After germination, the fungi enter plant tissue by breaking the cuticle mechanically or by weakening the cuticle and the epidermal cell wall.

The fungal hyphae then enter in latent state for lifetime or for an extended period. After establishing a stable association with the host, the endophytes offers several survival benefits to their host like increases the stress tolerance, provides protection against the phytopathogens, increases tolerance for drought, salinity, and variations in temperature (Clay, 1990; Rice et al., 1990; Malinowski, 1997; Saikkonen, 1998).

## Strategies for Metabolite Production Using Endophytic Microorganisms

Different researchers have employed a variety of rational methods to acquire endophytic microorganisms with bioactive potential. A crucial first step in the hunt for potential endophytic microorganisms is choosing the plant. The host plant's location, latitude, longitude, and altitude are used to determine its position. The plant(s) under investigation are located, identified, and external tissues like stem, twig, and leaf are taken. These tissues are then transported to the lab processed for endophyte isolation. The surface of the plant material should be treated with 0.5% sodium hypochlorite or 70% ethanol, completely washed in running tap water, and then dried under the sterile air oy by usinf the laminar flow hood. The plant material is cut into small piees (1.5–2 $cm^2$), the external tissues are carefully removed and the exposed internal tissues are aseptically placed onto water agar plates. Until the hyphae emerge from the plant tissue explant, the plates are incubated a room temperature or at 28–30 °C for a 12-hour dark/light cycle under illumination. To prevent bacterial contamination, antiobiotics are usually added in the isolation medium at the recommended concentrations. During the incubation, the emerging hyphae from the plant explants are transferred to sterile media plates containing Czapek Dox agar or Potato Dextrose agar. The ability of the isolated endophytes to grow in vitro utilizing submerged or solid-state growth methods is examined. The isolated endophytic microorganism (spores in the case of fungal endophytes) is donewas stored at −70 °C using 15% glycerol as a preservative (Fig. 12.4).

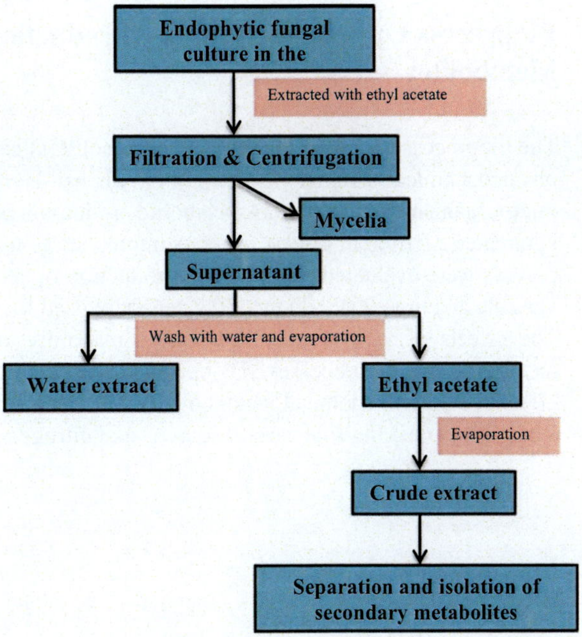


**Fig. 12.4** Extraction scheme for isolation of bioactive compounds

Following isolation, the isolated endophytic microorganism is subjected to identification using morphological and molecular methods. Observations made at the microscopic level are crucial in the tentative identification of endophytic bacteria and fungi. Conidiophore structure and morphology are characterized by macronematous conidiophores taken from the border of conidiogenous pustules during the maturity of conidia for morphological identification of the endophytic fungus (Patil et al., 2015). The most common method for identifying endophytic fungi at the molecular level is ITS sequence analysis. Using 18S rDNA and 5.8S rDNA, several researchers have carried out molecular identification of endophytic fungi (Debbab, 2012).

The isolated endophytic microorganism was cultivated at the shake flask level using synthetic medium at 28 °C for 2–3 weeks on a rotary shaker to provide aeration and agitation. During the incubation period, the medium was monitored for contamination, growth of endophytes, metabolite production and change in *p*H of the medium. After fermentation (at the stationary phase of the culture), the fermented broth is subjected to recovery of metabolites using organic solvents such as ethyl acetate or chloroform. The fermented broth was filtered using a muslin cloth to remove fungal mycelia. The filtered broth was centrifuged to remove other impurities. Extracted cell debris was discarded, and the filtrate containing the organic phase was separated from the medium (water phase) using a separatory funnel. To remove the salts and polar constituents, the ethyl acetate phase was washed with water twice and concentrated under reduced pressure using a rotary vacuum evaporator (Patil, 2016). An overview of metabolite isolation and characterization is summarized in Fig. 12.4.

## Bioprocess Optimization for the Production of Active Metabolites

The bioprocess for maximum yield of metabolite(s) can be optimized using various physicochemical parameters. This not only helps the rapid growth of the producing microorganism but also favors metabolite production *in* in vitro cultivation. Physical parameters*:* Different effects of temperature, aeration and agitation, and incubation periods were evaluated for optimum production of secondary metabolites. The one variable at a time (OVAT) model is generally used for process optimization studies. The effects of various chemical factors markedly affect the yield of secondary metabolites in bioprocesses. To optimize the production of the metabolites, the effect of different chemical factors by OVAT was studied. The chemical parameters studied included the sole carbon source, sole nitrogen source and *p*H.

## Screening of Bioactivities of the Crude Extracts and Isolated Compounds

The main focus of many studies on microbial endophytes is to investigate the potential of endophytic bacteria for industrial microbiology (Aly et al., 2010). Endophytes can produce a wide range of secondary metabolites that are identical to those of their host plant because they replicate their biology. Compounds are useful for developing new drugs, according to numerous studies (Joseph & Priya, 2011). Additionally, it has been discovered that some endophytes can produce bioactive secondary metabolites that their plant hosts typically cannot. It has been discovered that the bioactive substances isolated from endophytic microorganisms such as bacteria and fungi possess a variety of therapeutic activities, including antibacterial, antiparasitic, antiviral, anticancer, insecticidal, cytotoxic, neuroprotective, enzymatic, and antioxidant potential.

### *Antimicrobial Compounds from Endophytic Microorganisms*

Antimicrobial compounds are usually substances synthesized by microorganisms as secondary metabolites that inhibit the growth of other microorganisms even at low concentrations. Several metabolites showing antimicrobial properties have been extracted and purified from endophytic microorganisms. The metabolites of endophytic fungi have a broad range of antimicrobial activities. Endophytic microorganisms synthesize chemically diverse structural groups such as terpenoids, alkaloids, phenolics, steroids, aliphatic compounds and flavonoids (Mousa & Raizada, 2013). These compounds show antibacterial, antifungal and antiviral activities not only in humans but also in plants and animals. Phomol, an excellent antibacterial as well as antifungal compound, was produced by an endophytic fungus, *Phomopsis spp.* E02018, isolated the twig issues of *Erythrina crista-galli* (Weber et al., 2004). The endophytic fungus *Xylaria spp.* Isolated from the *Ginkgo biloba* L. produced the novel e compound 7-amino-4-methylcoumarin, which inhibited the growth of several human pathogens and the compound was successfully used as natural food preservative (Liu et al., 2008). Two new compounds called xanalteric acids I and II were obtained from an endophytic fungus, *Alternaria spp.* Obtained from the Chinese mangrove *Sonneratia alba* (Patil, 2016).

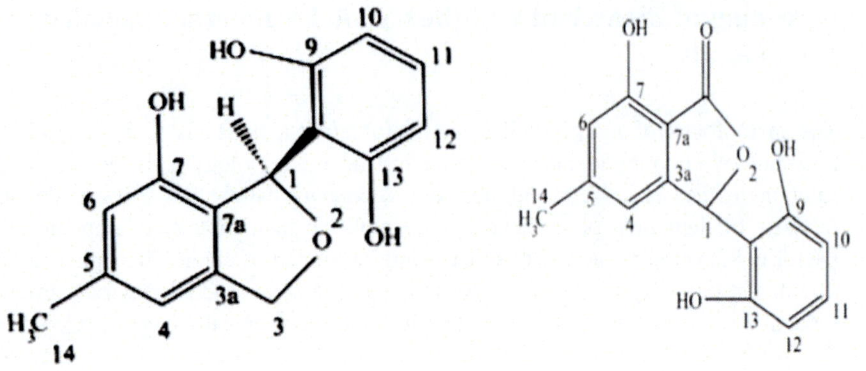


**Fig. 12.5** Antioxidant metabolites obtained from *Pestalotiopsis microspora*

## *Endophytic Metabolites with Antioxidant Potential*

Reactive oxygen species (ROS) play an important role in the development of degenerative conditions such as aging, cancer, inflammation, and atherosclerosis. (Sandesh et al., 2014). ROS such as hydroxyl radicals ($OH^-$), superoxide anions ($O_2^-$), and nitric oxide radicals ($NO^-$) alters the membrane fluidity, causes protein denaturation and leads to lipid peroxidation through generating oxidative stress that leads to injury, induce apoptosis and death (Anderson, 1996).

A number of antioxidant compounds have been successfully isolated from the broth extracts of endophytic microrganisms. Pestacin and isopestacin have been known for their antioxidant potential. These metabolites were obtained from *Pestalotiopsis microspore*, an endophytic fungus isolated from *Terminalia merobensis* (Fig. 12.5).

## Anticancer Compounds

Adter coronary heart diseases (CHDs), cancer is the second largest cause mortality in the United States and Europe. Breast cancer accounts for 31% of total deaths among adult females, while prostate cancer (31%) is found to be the highest in adult males (American Cancer Society, Cancer Facts and Figures, 2007). Numerous anticancer substances have extracted from from endophytic fungi. A few examples of anticancer compounds produced by endophytes are shown in Fig. 12.6. The world's first multibillion-dollar anticancer medicine, paclitaxel (Taxol), was produced from

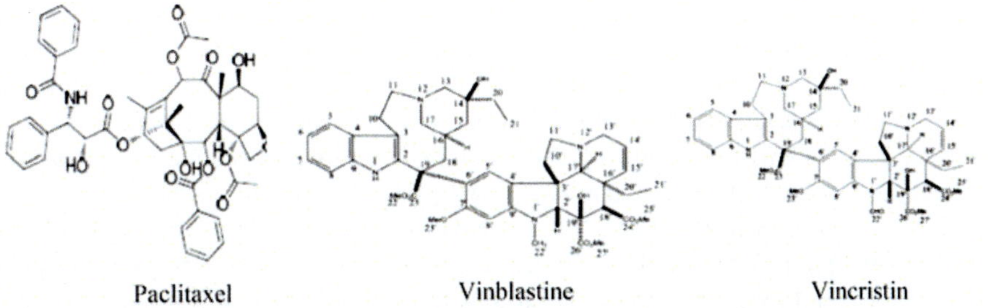


**Fig. 12.6** Some examples of anticancer drugs produced by endophytic microorganisms

*Taxomyces andreanae*, an endophytic fungus isolated from the yew plant *Taxus brevifolia* (Suffness, 1995). The discovery of taxol from endophytic fungi proved the bioactive potential of endophytic fungi. Following that, numerous endophytes with the ability to produce the diterpenoid Taxol were found all over the world. According to Lin et al. (1996), Taxol is a well-known powerful chemotherapeutic drug used to treat a range of malignancies, including ovarian and breast tumors. Taxol specifically suppresses the growth of tumor cells by arresting the depolymerization of tubulin during the cell division (Schiff & Horowitz, 1980). Vinblastine and vincristine are potent anticancer drugs isolated from endophytic *Fusarium oxysporum,* isolated from *Catharanthus roseus.* A strain of *Fusarium oxysporum (AA-CRL-6)* produced vinblastine and vincristine in submerged cultures at appreciable amounts (Kumar et al., 2013).

## *Other Activities of Endophyte-Derived Metabolites*

Hypercholesterolemia is one of the important omtributing factor in the development of CHDs such as atherosclerosis and numerous coronary heart disorders. A significant level of blood cholesterol is obtained through endogenous biosynthesis. HMG-CoA reductase (3-hydroxy-3-methylglutaryl-coenzyme A) is a key enzyme innthe cholesterol bosyntheis pathway which catalyzes the conversion of HMG-CoA to mevalonate. The secondary metabolites of fungi known as statins competitively block HMG-CoA reductase, decreasing blood cholesterol levels. These antihypercholesterolemic medications are known to have an important source in endophytic fungi. When grown in solid-state fermentation, the endophytic fungus *Aspergillus niger*, which was isolated from *Taxus baccata*, was capable of producing lovastatin (Raghunath et al., 2012). Other compounds with bioactive potential such as anti-inflammatory, antidiabetic, antimalarial, immunosuppressive, and antihypercholesterolemic agents, are also produced by endophytic fungi. According to numerous reports, endophytic fungi can produce anti-inflammatory substances such as phomol and mevinic acid, which were isolated from the fermentation extratcs of the endophytic *Phomopsis* sp. (Weber et al., 2004). Strong in vitro anti-inflammatory action was demonstrated by both substances. The primary secondary metabolite of *Claviceps purpurea* was initially identified as ergoflavin, a substance with anticancer and anti-inflammatory properties. Additionally, it has been isolated from an endophytic fungus called PM0651480 that was discovered in the Indian medicinal plant Bakul, *Mimosops elengi*. Human TNF- and IL-6 were also dramatically reduced by ergoflavin at very low IC50 values (Deshmukh et al., 2009).

## Biotechnological Potential of Endophytes from Apocynaceae Plants

Endophytes are becoming more relevant in terms of biotechnology and industry due to huge biological potential of secreting diverse secondary metabolites. These metabolites have been explored as novel biocontrol agents, antimicrobial agents, antitumour agents, and immunosuppressants. Endophytes have benn also reported to secrete antiviral compounds and produce natural antioxidants, antidiabetic agents, antibiotics, and insecticidal products (Yadav et al., 2018). By utilizing a range of processes and enhancing plant resistance to unfavourable environmental factors, endophytes can successfully promote development (Yadav et al., 2018). Moreover, industrially important products such as enzymes, vitamins, organic acids, and antibiotics have also been reported from endophytic microorganisms. Antibiotic resistance, which is on the rise, especially in human and animal pathogens can be is a major threat to public health if proper precautions are not taken. A detailed account of the biodiversity, bioactive metabolites, and biological activities of endophytic microorganisms from Apocynaceae plants is provided in Table 12.1.

**Table 12.1** Biodiversity and biological activities of endophytes in Apocynaceae plants

| Sr. no. | Host plant | Endophytes | Bioactive compounds | Biological activity | Assay method(s) | References |
|---|---|---|---|---|---|---|
| 1. | *Nerium indicum* | *Geomyces* sp. B | Vincamine alkaloids | The acetyl cholin ester ase (AchE) inhibitory activity | Ellman's in vitro assay | Hemmati et al. (2020) |
| 2. | *Catharanthus roseus* | *Stemphylium amaranthi* | Vinblastine | antioxidant activity | DPPH and ABTS in vitro assay | Nascimento (2015) |
| 3. | *Catharanthus roseus* | *Gliomastix masseei* | Ajmalicine | antioxidant activity | DPPH and ABTS in vitro assay | Nascimento (2015) |
| 4. | *Calotropis procera* | *Phaeoramularia calotropidis* | Crude extract | Antimicrobial | Agar well Diffusion method | Parthasarathy et al. (2020) |
| 5. | | *Guignardia bidwellii* | Crude extract | Antimicrobial | | Parthasarathy et al. (2020) |
| 6. | *Catharanthus roseus* | *Curvularia verruculosa* | Vinblastine | Cytotoxic activity | In vitro cytotoxicity against HeLa cell line | Yin and Sun (2011) |
| 7. | *Vinca minor* | Unidentified fungus | Vincamine | Not reported | Not reported | Verma et al. (2014) |
| 8. | *Vinca rosea* | *Nigrospora sphaerica* | NigrosphaerinA, | Antileukemic, antileishmanial, antifungal, antibacterial and antimalarial | In vitro methods | Birat et al. (2022) |
| 9. | *Catharanthus roseus* | *Aspergillus fumigatus* EFBL | Epothilone B analogue | antiproliferative activity | In vitro methods *In silico* | Ayob and Simarani (2016) |

(continued)

**Table 12.1** (continued)

| Sr. no. | Host plant | Endophytes | Bioactive compounds | Biological activity | Assay method(s) | References |
|---|---|---|---|---|---|---|
| 10. | *Catharanthus roseus* | *Colletotrichum* sp., *Macrophomina phaseolina*, *Nigrospora sphaerica* and *Fusarium solani* | Hydrolytic enzymes; Cellulases, Protease, Amylase | Hydrolytic activity | Not reported | Jasmine and Agastian (2013) |
| 11. | *Catharanthus roseus* | Actinomycetes | EEA : extract | free radical scavenging activity and marked alpha glucosidase activity | DPPH | Venugopalan and Srivastava (2015a, 2015b) |
| 12. | *Catharanthus roseus* | Fusarium solani | Camptothecin | Anticancer | Not reported | Ranjan and Jadeja (2017) |
| 13. | *Catharanthus roseus* | *Micrococcus yunnanensis* strain rsk5 | Antibacterial compound | Antibacterial activity | *Against Staphylococcus aureus* | El-Sayed et al. (2020) |
| 14. | *Cynancum acutum* | *Sarocladium strictum* | *Cis*-13-octadecenoic acid, sebacic acid, pentamethoxy flavone and n-hexadecanoic acid | Insecticidal activity | Deformation of *S. littoralis* | Manasa and Nalini (2014) |
| 15. | *Tabernaemontana heyneana* Wall | *Colletotrichum, Curvularia, Fusarium, Phomopsis, Verticillium*, and *Volutella* | L-asparginase | L-Asparaginase Activity | Nesslerization | Rao et al. (2015) |
| 16. | *Cryptolepis buchanani* | *Phomopsis liquidambaris* CBR-15 | Antimicrobial metabolites | Antimicrobial activity | Agar well Diffusion method | Gherbawy and Gashgari (2014) |
| 17. | *Calotropis procera* | *Chaetomium globosum Myrothecium verrucaria* | Antifungal compound | Antifungal activity. | Agar well Diffusion method | Nagda et al. (2017) |
| 18. | *Calotropis procera* | Penicillium spp. and Aspergillus spp., | Phenolics and flavonoids | Antioxidant potential | DPPH | Kudalkar et al. (2012) |

| | | | | | | |
|---|---|---|---|---|---|---|
| 19. | *Prestonia trifidi* | *Muscodor sutura* | Thujopsene, chamigrene, isocaryophyllene, and butanoic acid, 2-methyl. | antifungal activity | Agar well Diffusion method | Akpotu et al. (2017) |
| 20. | *Catharanthus roseus* | MR 1B MRB2 | Citreoisocoummarin, Paxillin, nigricinol, fatty acid, cladosporin | Antibacterial and antifungal activity | Agar well Diffusion method | Ramesha and Srinivas (2014) |
| 21. | Plumeria acuminata L | *Colletotrichum sp* | Crude extract | Antibacterial and antifungal activity | Agar well Diffusion method | Akshatha et al. (2014) |
| 22. | Plumeria btusifolia L. | *Fusarium oxysporum* | Crude extract | Antibacterial and antifungal activity | Agar well Diffusion method | Akshatha et al. (2014) |
| 23. | Rauwolfia densiflora | *Unidentified fungi* | Crude extract | Antidiabetic activity | a-amylase inhibitor | Huang et al. (2007) |
| 24. | Tabernaemontana divaricata (L), | *Strain AcapF3* | Phenolic compound | Antioxidant activity | *In vitro* | Khiralla et al. (2015) |
| 25. | Rauwolfia verticillata (Lour.) | *Strain AcapF3* | Phenolic compound | Antioxidant activity | *In vitro* | Khiralla et al. (2015) |
| 26. | Calotropis procera, Catharanthus roseus, | *Aspergillus sp.* | Gallic acid Phenol | Antioxidant activity | *In vitro* | Palem et al. (2015) |
| 27. | Catharanthus roseus | *Talaromyces radicus* | Vinblastine and vincristine | Anticancer activity | HeLa cells | Ramesha (2014) |
| 28. | Nerium oleander | *Cochliobolus Kusanoi* | Oosporein | Antimicrobial and Anticancer activity | Cytotoxicity | Li et al. (2007) |
| 29. | *Trachelospermum jasminoides* | *Verticillium sp. Tj33* | Verticase | fibrinolytic activity | *In vitro* | Zhang et al. (2008) |
| 30. | *Trachelospermum jasminoides* | *Cephalosporium acremonium* IFB-E007 | Cephalosol | Antimicrobial | *In vitro* | Ayob et al. (2017) |
| 31. | *Catharanthus roseus* | *Nigrospora sphaerica* | Vinblastine | Cytotoxic activity | breast cell line cancer (MDA-MB 231) | Corrado and Rodrigues (2004) |

(continued)

**Table 12.1** (continued)

| Sr. no. | Host plant | Endophytes | Bioactive compounds | Biological activity | Assay method(s) | References |
|---|---|---|---|---|---|---|
| 32. | *Aspidosperma tomentosum* | *Phomopsis* sp. | Crude extract | Antibacterial and antifungal activities. | bioautographic TLC agar-overlay technique | Bhagat et al. (2016) |
| 33. | *Vinca rosea* | *Alternaria alternate* | cholinesterase inhibitor 'altenuene | Antioxidant and insecticidal activity | Dot blot assay | Akshatha et al. (2016) |
| 34. | *Rauwolfia densiflora* | *Streptomyces globosus* | Crude extract | Antioxidant activity | DPPH activity | Song et al. (2005) |
| 35. | Trachelospermum jasminoides | Cephalosporium sp. IFB-E001 | Graphislactone A | Antioxidant activity | DPPH activity | Gao et al. (2018) |
| 36. | *Cynanchum auriculatum* | *Stemphylium lycopersici* BG01 | Physcione | Antioxidant activity | DPPH and ABTS scavenging activity | Gao et al. (2018) |

## References


Akpotu, M., Eze, P., Ajaghaku, D., Okoye, F., & Esimone, C. (2017). Antimicrobial activity of endophytic fungi isolated from Catharanthus roseus and Euphorbia hirta. *Chemical Science Review and Letters, 6*, 12–19.

Akshatha, J. V., Prakash, H. S., & Nalini, M. S. (2016). Actinomycete endophytes from the ethno medicinal plants of Southern India: Antioxidant activity and characterization studies. *Journal of Biologically Active Products from Nature, 6*(2), 166–172.

Akshatha, V., Nalini, M., D'souza, C., & Prakash, H. (2014). Streptomycete endophytes from anti-diabetic medicinal plants of the Western Ghats inhibit alpha-amylase and promote glucose uptake. *Letters in Applied Microbiology, 58*, 433–439.

Aly, A. H., Debbab, A., Kjer, J., & Proksch, P. (2010). Fungal endophytes from higher plants: A prolific source of phytochemicals and other bioactive natural products. *Fungal Diversity, 41*, 1–16.

American Cancer Society. (2007). *Cancer facts and figures.*

Anderson, D. (1996). Antioxidant defenses against reactive oxygen species causing genetic and other damage. *Mutation Research, 350*, 103–108.

Andreozzi, A., Prieto, P., Mercado-Blanco, J., Monaco, S., Zampieri, E., Romano, S., et al. (2019). Efficient colonization of the endophytes *Herbaspirillum huttiense* RCA24 and Enterobacter cloacae RCA25 influences the physiological parameters of *Oryza sativa* L. cv. Baldo rice. *Environmental Microbiology, 21*, 3489–3504. https://doi.org/10.1111/1462-2920.14688

Ayob, F. W., & Simarani, K. (2016). Endophytic filamentous fungi from a Catharanthus roseus: Identification and its hydrolytic enzymes. *Saudi Pharmaceutical Journal, 24*(3), 273–278.

Ayob, F. W., Simarani, K., Zainal Abidin, N., & Mohamad, J. (2017). First report on a novel Nigrospora sphaerica isolated from Catharanthus roseus plant with anticarcinogenic properties. *Microbial Biotechnology, 10*(4), 926–932.

Bhagat, J., Kaur, A., Kaur, R., Yadav, A. K., Sharma, V., & Chadha, B. S. (2016). Cholinesterase inhibitor (Altenuene) from an endophytic fungus Alternaria alternata: Optimization, purification and characterization. *Journal of Applied Microbiology, 121*(4), 1015–1025.

Birat, K., Binsuwaidan, R., Siddiqi TO, Mir, S. R., Alshammari, N., Adnan, M., Nazir, R., Ejaz, B., Malik, M. Q., Dewangan, R. P., & Ashraf, S. A. (2022). Report on Vincristine-Producing Endophytic Fungus Nigrospora zimmermanii from Leaves of Catharanthus roseus. *Metabolites, 12*(11), 1119.

Christel, Q. D., Gressier, B., Vasseur, J., Dine, T., Brunet, C., Luyckx, M., Cazin, M., Cazin, J. C., Bailleul, F., & Trotin, F. (2000). Phenolic compounds and antioxidant activities of buckweat hulls and flour. *Journal of Ethnopharmacology, 72*, 35–42.

Clay, K. (1990). Comparative demography of three graminoids infected by systemic, clavicipitaceous fungi. *Ecology, 71*, 558–570.

Clay, K., & Holah, J. (1999). Fungal endophytes symbiosis and plant diversity in successional fields. *Science, 285*, 1742–1744.

Corrado, M., & Rodrigues, K. F. (2004). Antimicrobial evaluation of fungal extracts produced by endophytic strains of Phomopsis sp. *Journal of Basic Microbiology, 44*(2), 157–160.

Debbab, A. (2012). Endophytes and associated marine derived fungi—ecological and chemical perspectives. *Fungal Diversity, 57*, 45–83.

Deshmukh, S. K., Mishra, P. D., Kulkarni-Almeida, A., Verekar, S., Sahoo, M. R., Periyasamy, G., Goswami, H., Khanna, A., Balakrishnan, A., & Vishwakarma, R. (2009). Anti-inflammatory and anticancer activity of ergoflavin isolated from an endophytic fungus. *Chemistry and Biodiversity, 6*, 784–789.

El-Sayed, A. S., Moustafa, A. H., Hussein, H. A., El-Sheikh, A. A., El-Shafey, S. N., Fathy, N. A., & Enan, G. A. (2020). Potential insecticidal activity of Sarocladium strictum, an endophyte of Cynanchum acutum, against Spodoptera littoralis, a polyphagous insect pest. *Biocatalysis and Agricultural Biotechnology, 24*, 101524.

Fisher, P. J., Graf, F., Petrini, L. E., Sutton, B. C., & Wookey, P. A. (1995). Fungal endophytes of Dryas octopetala from a high arctic polar semi desert and from the Swiss Alps. *Mycologia, 87*, 319–323.

Gao, H., Li, G., & Lou, H. X. (2018). Structural diversity and biological activities of novel secondary metabolites from endophytes. *Molecules., 23*(3), 646.

Gherbawy, Y. A., & Gashgari, R. M. (2014). Molecular characterization of fungal endophytes from Calotropis procera plants in Taif region (Saudi Arabia) and their antifungal activities. *Plant Biosystems, 148*(6), 1085–1092.

Gupta, S., Chaturvedi, P., Kulkarni, M. G., & Van Staden, J. (2020). A critical review on exploiting the pharmaceutical potential of plant endophytic fungi. *Biotechnology Advances, 39*, 107462. https://doi.org/10.1016/j.biotechadv.2019.107462. In press.

Hemmati, N., Azizi, M., Spina, R., Dupire, F., Arouei, H., & Saeedi, M. (2020). Dominique Laurain-Mattar, Accumulation of ajmalicine and vinblastine in cell cultures is enhanced by endophytic fungi of Catharanthus roseus cv. Icy Pink. *Industrial Crops and Products, 158*.

Huang, W.-Y., Cai, Y.-Z., Xing, J., Corke, H., & Sun, M. (2007). A potential antioxidant resource: endophytic fungi from medicinal plants. *Economic Botany, 61*, 14–30. https://doi.org/10.1663/0013-0001(2007)61[14:APAREF]2.0.CO;2

Jasmine, D. J., & Agastian, P. (2013). In vitro antioxidant activity and in vivo alpha glucosidase activity of endophytic actinomycetes isolated from Catharanthus roseus (l.) G. Don. *Journal of Pharmacy Research, 6*(6), 674–678.

Joseph, B., & Priya, R. M. (2011). Bioactive compounds from endophytes and their potential in pharmaceutical effect: A review. *American Journal of Biochemistry and Molecular Biology, 1*, 291–309.

Kandel, S., Joubert, P., & Doty, S. (2017). Bacterial endophyte colonization and distribution within plants. *Microorganisms, 5*, 77. https://doi.org/10.3390/microorganisms5040077

Khare, E., Mishra, J., & Arora, N. K. (2018). Multifaceted interactions between endophytes and plant: Developments and prospects. *Frontiers in Microbiology, 9*, 1–12. https://doi.org/10.3389/fmicb.2018.02732

Khiralla, A., Mohamed, I., Thomas, J., Mignard, B., Spina, R., Yagi, S., et al. (2015). A pilot study of antioxidant potential of endophytic fungi from some Sudanese medicinal plants. *Asian Pacific Journal of Tropical Medicine, 8*, 701–704.

Kudalkar, P., Strobel, G., Riyaz-Ul-Hassan, S., et al. (2012). Muscodor sutura, a novel endophytic fungus with volatile antibiotic activities. *Mycoscience, 53*, 319–325.

Kumar, A., Patil, D., Rajamohanan, P. R., & Ahmad, A. (2013). Isolation, purification and characterization of vinblastine and vincristine from endophytic Fungus *Fusarium oxysporum* Isolated from *Catharanthus roseus*. *PLoS One, 8*(9).

Li, Y., Shuang, J. L., Yuan, W. W., Huang, W. Y., & Tan, R. X. (2007). Verticase: A fibrinolytic enzyme produced by Verticillium sp. Tj33, an Endophyte of Trachelospermum jasminoides. *Journal of Integrative Plant Biology, 49*(11), 1548–1554.

Lin, X., Hezari, M., Koepp, A. E., Floss, H. G., & Croteau, R. (1996). Mechanism of taxadiene synthase, a diterpene cyclase that catalyzes the first step of taxol biosynthesis in Pacific Yew. *Biochemistry, 35*, 2968–2977.

Liu, X., Dong, M., Chen, X., Jiang, M., Liv, X., & Zhou, J. (2008). Antimicrobial activity of an endophytic *Xylaria sp.YX-28* and identification of its antimicrobial compound 7-amino-4-methylcoumarin. *Applied Microbiology and Biotechnology, 78*, 241–247.

Malinowski, D., Leuchtmann, A., Schmidt, D., & Neosberger, J. (1997). Growth and water status in meadow fescue is affected by *Neotyphodium* and *Phialophora species* endophytes. *Agronomy Journal, 89*, 673–678.

Manasa, C., & Nalini, MS (2014). L-Asparaginase activity of fungal endophytes from Tabernaemontana heyneana Wall. (Apocynaceae), Endemic to the Western Ghats (India). *Hindawi Publishing Corporation International scholarly research notices volume* (Article ID 925131, 7 pages).

Marasco, R., Rolli, E., Ettoumi, B., Vigani, G., Mapelli, F., Borin, S., Abou-Hadid, A. F., El-Behairy, U. A., Sorlini, C., Cherif, A., Zocchi, G., & Daffonchio, D. (2012). A drought resistance-promoting microbiome is selected by root system under desert farming. *PLoS One, 7*, e48479. https://doi.org/10.1371/journal.pone.0048479

Mousa, W. K., & Raizada, M. N. (2013). The diversity of antimicrobial secondary metabolites produced by fungal endophytes: an interdisciplinary perspective. *Frontiers in Microbiology, 4*, 65.

Nagda, V., Gajbhiye, A., & Chhatwani, D. K. (2017). Isolation and characterization of endophytic fungi from calotropis procera for their antioxidant activity. *Asian Journal of Pharmaceutical and Clinical Research, 10*(3), 254–258.

Nascimento, T. L., Oki, Y., Lima, D. M. M., Almeida-Cortez, J. S., Wilson Fernandes, G., & Souza-Motta, C. M. (2015). Biodiversity of endophytic fungi in different leaf ages of Calotropis procera and their antimicrobial activity. *Fungal Ecology, 14*, 79–86.

Palem, P. P., Kuriakose, G. C., & Jayabaskaran, C. (2015). An endophytic fungus, talaromyces radicus isolated from catharanthus roseus, produces vincristine and vinblastine, which induce apoptotic cell death. *PLoS One, 10*(12), e0144476.

Parthasarathy, R., Shanmuganathan, R., & Pugazhendhi, A. (2020). Vinblastine production by the endophytic fungus Curvularia verruculosa from the leaves of Catharanthus roseus and its in vitro cytotoxicity against HeLa cell line. *Analytical Biochemistry, 593*, 113530.

Patil, M. P. (2016). *Probing the microbial endophytes in indigenous medicinal plant(s) for production of biologically active compounds*., PhD thesis. KBC North Maharashtra University.

Patil, M. P., Patil, R. H., & Maheshwari, V. L. (2015). Biological activities and identification of bioactive metabolite from Endophytic *Aspergillus flavus* L7 isolated from *Aegle marmelos. Current Microbiology, 71*, 39–48.

Raghunath, R., Radhakrishna, A., Angayarkanni, J., & Palaniswamy, M. (2012). Production and Cytotoxicity studies of Lovastatin from *Aspergillus niger* Pn2 an endophytic fungi isolated from *Taxus baccata. International Journal of Applied Biology and Pharmaceutical Technology, 3.*

Ramesha, A. (2014). *Extraction and characterization of secondary metabolites of endophytic fungi isolated from medicinal plants of Apocynaceae*. Ph. D thesis,. Bangalore University.

Ramesha, A., & Srinivas, C. (2014). Antimicrobial activity and phytochemical analysis of crude extractsof endophytic fungi isolated from Plumeria acuminata L. and Plumeria obtusifolia L. European. *Journal of Experimental Biology, 4*.

Ranjan, R., & Jadeja, V. (2017). Isolation, characterization and chromatography based purification of antibacterial compound isolated from rare endophytic actinomycetes Micrococcus yunnanensis. *Journal of Pharmaceutical Analysis, 7*(5), 343–347.

Rice, J. S., Pinkerton, B. W., Stringer, W. C., & Undersander, D. J. (1990). Seed production in tall fescue as affected by fungal endophytes. *Crop Science, 30*, 1303–1305.

Rodriguez, R., & Redman, R. (2008). More than 400 million years of evolution and some plants still can't make it on their own: plant stress tolerance via fungal symbiosis. *Journal of Experimental Botany, 59*(5), 1109–1114. https://doi.org/10.1093/jxb/erm342

Rosa, L. H., Vaz, A. B. M., Caligiorne, R. B., Campolina, S., & Rosa, C. A. (2009). Endophytic fungi associated with the antarctic grass Deschampsia Antarctica Desv. *Polar Biology, 32*, 161–167.

Russell, J. R., Huang, J., Anand, P., Kucera, K., Sandoval, A. G., Dantzler, K. W., Hickman, D. S., Jee, J., Kimovec, F. M., Koppstein, D., Marks, D. H., Mittermiller, P. A., Núñez, S. J., Santiago, M., Townes, M. A., Vishnevetsky, M., Williams, N. E., Vargas, M. P. N., Boulanger, L. A., Bascom-Slack, C., & Strobel, S. A. (2011). Biodegradation of polyester polyurethane by endophytic fungi. *Applied and Environmental Microbiology, 77*, 6076–6084.

Saikkonen, K., Faeth, S. H., Helander, M., & Sullivan, T. J. (1998). Fungal Endophytes: A Continuum of Interactions with Host Plants. *Annual Review of Ecology and Systematics, 29*, 319–343.

Sandesh, P., Velu, V., & Singh, R. P. (2014). Antioxidant activities of tamarind (*Tamarindus Indica*) seed coat extract using in vitro and in vivo models. *Journal of Food Science and Technology, 51*(9), 1965–1973.

Santoyo, S., Moreno-Hagelsieb, G., del Carmen Orozco-Mosqueda, M., & Glick, B. R. (2016). Plant growth-promoting bacterial endophytes. *Microbiological Research, 183*, 92–99. https://doi.org/10.1016/j.micres.2015.11.008

Schiff, P. B., & Horowitz, S. B. (1980). Taxol stabilizes microtubules in mouse fibroblast cells. *Proceedings of National Academy of Science U S A, 77*, 1561–1565.

Shaw, J. J., Spakowicz, D. J., Dalal, R. S., Davis, J. H., Lehr, N. A., Dunican, B. F., Orellana, E. A., Narváez-Trujillo, A., & Strobel, S. A. (2015). Biosynthesis and genomic analysis of medium-chain hydrocarbon production by the endophytic fungal isolate *Nigrograna mackinnonii* E5202H. *Applied Microbiology and Biotechnology, 99*, 3715–3728.

Song, Y.-C., Huang, W.-Y., Sun, C., Wang, F., & Tan, R. X. (2005). Characterization of graphislactone A as the antioxidant and free radical-scavenging substance from the culture of Cephalosporium sp. IFB-E001, an endophytic fungus in Trachelospermum jasminoides. *Biological & Pharmaceutical Bulletin, 28*, 506–509. https://doi.org/10.1248/bpb.28.506

Suffness, M. (1995). *Taxol, science and applications*. CRC Press.

Suryanarayanan, T. S., Wittlinger, S. K., & Stanley, H. F. (2005). Endophytic fungi associated with cacti in Arizona. *Mycological Research., 109*, 635–639.

Tejesvi, M. V., Kini, K. R., Prakash, H. S., Subbiah, V., & Shetty, H. S. (2007). Genetic diversity and antifungal activity of species of Pestalotiopsis isolated as endophytes from medicinal plants. *Fungal Diversity, 24*, 37–54.

Venugopalan, A., & Srivastava, S. (2015a). Endophytes as in vitro production platforms of high value plant secondary metabolites. *Biotechnology Advances, 33*, 873–887.

Venugopalan, A., & Srivastava, S. (2015b). Enhanced camptothecin production by ethanol addition in the suspension culture of the endophyte, Fusarium solani. *Bioresource Technology, 188*, 251–257.

Verma, V. C., Kharwar, R. N., & Strobel, G. (2009). Chemical and functional diversity of natural products of plant associated endophytic fungi. *Natural Products Communication, 4*, 1511–1532.

Verma, P., Khan, S. A., Mathur, A. K., Shanker, K., & Kalra, A. (2014). Fungal endophytes enhanced the growth and production kinetics of Vinca minor hairy roots and cell suspensions grown in bioreactor. Plant Cell. *Tissue and Organ Culture (PCTOC), 118*, 257–268.

Wang, S., Li, X. M., Teuscher, F., Li, D., Diesel, A., Ebel, R., Proksch, P., & Wang, B. G. (2006). Chaetopyranin, a benzaldehyde derivative, and other related metabolites from *Chaetomium globosum*, an endophytic fungus derived from the marine red alga *Polysiphonia urceolata*. *Journal of Natural Products, 69*, 1622–1625.

Weber, D., Sterner, O., Anke, T., Gorzalczancy, S., Martino, V., & Acevedo, C. (2004). Phomol, a new anti-inflammatory metabolite from endophytes of the medicinal plant Erythrina crista-galli. *The Journal of Antibiotics, 57*, 559–563.

Yadav, A. N., Kumar, V., Dhaliwal, H. S., Prasad, R., & Saxena, A. K. (2018). Microbiome in crops: diversity, distribution, and potential role in crop improvement. In R. Prasad, S. S. Gill, & N. Tuteja (Eds.), *Crop improvement through microbial biotechnology* (pp. 305–332). Elsevier.

Yashavantha Rao, H. C., Santosh, P., Rakshith, D., & Satish, S. (2015). Molecular characterization of an endophytic Phomopsisliquidambaris CBR-15 from Cryptolepis buchanani Roem. and impact of culture media on biosynthesis of antimicrobial metabolites. *3 Biotech, 5*(2), 165–173. https://doi.org/10.1007/s13205-014-0204-2

Yin, H., & Sun, Y.-H. (2011). Vincamine-producing endophytic fungus isolated from Vinca minor. *Phytomedicine, 18*(8–9), 802–805.

Zhang, H. W., Huang, W. Y., Chen, J. R., Yan, W. Z., Xie, D. Q., & Tan, R. X. (2008). Cephalosol: An Antimicrobial metabolite with an unprecedented skeleton from endophytic Cephalosporium acremonium IFB-E007. *Chemistry A European Journal, 14*(34), 10670–10674.

# Chapter 13
# Micropropagation-Based Conservation Strategies for Apocynaceae Plants

## Introduction

Being the largest family in the plant kingdom, Apocynaceae consists of over 400 genera with more than 4600 species distributed in five subfamilies: Rauvolfioideae, Apocynoideae, Periplocoideae, Secamonoideae, and Asclepiadoideae. A number of plants from this family are cultivated as ornamentals for their attractive flowers and foliage. The members are most abundant in the tropics and subtropics. Plants of the family Apocynaceae are native in Indian Subcontinent. The countries such as Pakistan, China, Bangladesh and Sri Lanka are known for dominant presence of Apocynaceae plants (Mahmood et al., 2011). In the Indian subcontinent, the Himalayan ranges of the southernmost mountains at altitudes of 600–1200 m above sea level are richly colonized by Apocynaceae plants. The western states of India are geographically and culturally rich and diverse and are widely exploring plant diversity for various ethnomedicinal purposes (Vineeta et al., 2022).

Applications of medicinal plants and their products for the primary health needs is an integral part of day to day life of tribal people. Indian tribal people rich knowledge of medicinal plants and therefore medicinal plants are widely used for health benefits. Historically, tribal tribes in the area have investigated the therapeutic potential of many of the local plants that are a part of their ecology. Alkaloids abundant in this family of plants provide them with great therapeutic value. Additionally, many species are planted for decorative purposes. People in rural regions utilize some Apocynaceae species as food and some of them as poisons (Aiyambo, 2010). Apocynaceae plants have been used in different traditional systems of medicine, such as Indian, Chinese, and Thai (Van Beek et al., 1984; Bhat et al., 2012). The Indian subcontinent has a dominant presence of Apocynaceae plants. Approximately 84 species or 30 genera are present in India. Some important members of this family, viz. *C. roseus*, *A. scholaris*, *R. serpentina*, *H. antidysentriaca*, *N. indicum* and *T. peruviana* have been documented in

 
R. H. Patil et al., *Apocynaceae Plants*,
https://doi.org/10.1007/978-981-99-5406-3_13

Ayurveda, Siddha, Unani and Homeopathy for the treatment of various diseases and disorders.

In African countries such as Ghana, Uganda, Nigeria and Senegal, Apcynaceae members such as *T. crassa, V. africana* and *Rauvolfia vomitoria* are commonly used for similar ethnomedicinal uses. The leaf, stem, root and flower extracts of these plants have been widely used for anti-arthritic activity in the treatment of rheumatism and parasitic skin infections, including leprosy, used as diuretic, insecticidal as well as in the treatment of fever, skin diseases as well as in wound healing (Bhat et al. 2012).

The Western Ghats of India are a biodiversity hotspot and rich in medicinal plants that harbor approximately 3000 plants of medicinal importance (Vineeta et al., 2022). The tribals of Western Ghats are using these plants for their primary health care. The high biodiversity of medicinal plants in the Western part of India is attributed to the varying ecosystems in this region. *Caralluma europaea* is another plant of this family native to Morocco, Algeria, Tunisia, Libya, Egypt, Jordan, Spain, and Italy. Various plants are used traditionally to treat diabetes mellitus, flu, caught, kidney stones, cysts, respiratory infections, cancer, digestive disorders, urogenital infections, metabolic disorders, and cardiovascular problems (Ouassou et al., 2021).

## Micropropagation Based Conservation of Apocynaceae Plants

Injudicious use and ovberharvesting of medicinal plants from the Apocynaceae family, many species of the family are facing the danger of on extinction. Therefore, there is a need of immediate efforts to desigh the conservation stratigies for this family. Plant cells have the totipotent capacity to regenerate the complete plant from explant tissue using the tools and techniques of plant tissue culture technology. Currently, secondary metabolites with applications in medicine, food, and cosmetics are mostly produced on a big scale from medicinal and aromatic plants (Karuppusamy, 2009). Many species of medicinal and aromatic plants have already gone extinct as a result of excessive commercial use, and the ones that survive fall into the category of endangered. The preservation of superior genotypes of plants is thought to be best accomplished through micropropagation. Plant genotypes can be preserved through effective shoot tip/meristem culture, lateral bud culture, and somatic embryogenesis. The development of buds from tobacco explants resulted in the identification of kinetin and other useful cytokinins (Miller & Skoog, 1953). It has been documented that coconut milk can induce somatic embryogenesis in carrot explants (Steward, 1958). Due to the totipotent nature of the cell, single-cell somatic embryogenesis has the capacity to regenerate entire plants. Forster et al. (2007) investigated embryogenesis and organogenesis from pollen culture of datura anthers.

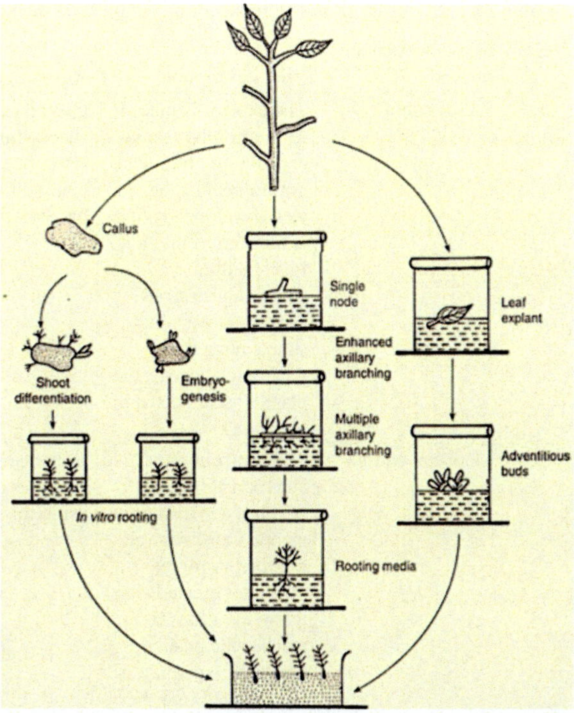


**Fig. 13.1** Various approaches to micropropagation (Verma, 2017)

Micropropagation is widely used for raising genetically identical plants, pathogen-free plants, large-scale plant production, gene conservation, etc. The plant can be multiplied asexually (by vegetative propagation) or sexually (via seed). Gametes are fused during sexual reproduction, which produces seeds with a high degree of variability. Cloning takes place during micropropagation. They are spread either by vegetative propagation or asexual modes of reproduction. In both cases, a single individual is multiplied *in vitro* to create genetically identical plants. Several methods are practiced for the micropropagation of medicinal and other economically important plants. These methods are summarized in Fig. 13.1, Table 13.1.

## Advantages of Micropropagation

1. A large number of clones can be created in a given time using the plant tissue in limited quantities.
2. Micropropagation is used as an effective alternative for the large-scale production of plant species that are difficult to grow using traditional practices.
3. A large number of plantlets can be kept up in small spaces.
4. Clonal expansion is possible throughout the year.

**Table 13.1** Summary of micropropagation methods and their advantages and drawbacks

| Sr. no. | Micropropagation method | Brief description | Advantages | Drawbacks |
|---|---|---|---|---|
| 1. | Callus culture | The explant is placed in an artificial media and cultured for callus development. The well-developed callus is transferred into a a fresh medium containing **desired hormones** for roots, shoots induction. After a few weeks, the plantlet is subjected to hardening | Can be used to successfully accelerate the growth of many different herbaceous plants | High biochemical variability and slow growth rate |
| 2. | Meristem culture | The leaf primordial and meristem tissue are placed in a nutrient medium and allowed to grow. A rooted plantlet is created after a few weeks. When it reaches the desired height or growth, it is transplanted into the soil | A disease-free plant can be produced using this method | |
| 3. | Suspension culture | Individual cells or callus cells are suspended and allowed to grow in liquid media and it is continuously aerated on a rotary shaker at a specific temperature | Suitable for secondary metabolites production | More susceptible to contamination. Less yield of product. The formation of clumps and cells does not separate after division |
| 4. | Protoplast culture | To repair the cell wall and the callus, the isolated explant/plant cell is grown in the appropriate medium. In the right circumstances, the cell forms a cell wall, which is followed by a stimulation of cell division that results in differentiation into a new plant | | It is difficult to strip the cell wall of root tips and embryos and therefore generally not used as an explant for the protoplast culture |

(continued)

**Table 13.1** (continued)

| Sr. no. | Micropropagation method | Brief description | Advantages | Drawbacks |
|---|---|---|---|---|
| 5. | Embryo culture | In this method, the embryos are extracted and placed into a medium with desired plant growth regulators with proper nutrients in aseptic conditions | Has wide application in crop improvement | |

5. The global trade and exchange of medicinal and horticultural plant material without the danger is possible.
6. Micropropagation is an important method for seed creation.
7. Because micropropagation has distinct favorable conditions than conventional engendering techniques, this method has a better chance of producing major plant-based phytopharmaceuticals.

## Applications and Merits of Micropropagation Over Conventional Plant Breeding

1. Small amounts of explant tissue are sufficient for micropropagation to produce thousands of clones in a single year. The same number of plants could not be produced using conventional methods in the same amount of time.
2. Plant species that are either slow-growing or difficult to cultivate using conventional bulk propagation can be effectively propagated using micropropagation.
3. Vegetative propagation of valuable plants for large-scale production is possible through micropropagation. Large-scale production of plant types that are genetically pure and free of diseases is possible, and the creation of new kinds takes much less time.
4. A small amount of physical space is needed for a lot of plants. This aids in the storage of genetic material and the preservation of endangered species.
5. Using meristem tip culture, disease-free varieties are obtained through micropropagation. Expansion of *in vitro* stocks can be performed throughout the year.
6. Increased yields and vigor of plants in floriculture and horticulture species can be achieved.
7. International transfer of plants can be made easier without any risk of spreading of phytopathogens and plant diseases.
8. Because the requirement for genetic conservation is crucial for seed production, micropropagation is especially helpful for the production of seeds in some crops.

Micropropagation, due to its several advantages over conventional propagation methods, has become popular and a prospect for the production of important plant-based bioactive natural products. Irrespective of the availability of plants, micropropagation is considered an alternative approach to conventional methods for the controlled production of medicinally important biochemicals.

## *In Vitro* Propagation of Apocynaceae Members

The shoot apex of the native disease-free plant or aseptically germinated seedling is used for *in vitro* propagation of *C. carandas*. Surface sterilization of explants is not needed when aseptically germinated seedlings are used for multiple shoot induction, which ultimately reduces the chances of contamination. Aseptic seedlings were obtained after seed germination on MS medium. Before inoculation, seeds are surface sterilized and inoculated on MS medium, and the shoot apex of aseptic seedlings can be used for multiple shoot induction in *C. carandas*. The role of cytokinins is critical for shoot multiplication on the *C. carandas* shoot apex. In addition, cytokinins and many other metabolites may have a synergistic or antagonistic effect on shoot induction. For example, when quercetin was used in combination with cytokinins, it favored shoot induction, and the results obtained were comparatively better than those of other treatments (Bhadane et al., 2018).

## Multiple Shoot Induction

The shoot apex of *in vitro* raised aseptic seedlings or the apex tissue from disease-free plants was excised and cultured on MS medium supplemented with different combinations and concentrations of 6-benzylaminopurine (BAP) alone or in combination with kinetin. The effect of several organic and inorganic molecules can also be evaluated. Bhadane et al. (2018) evaluated the effect of quercetin and $AgNO_3$ on multiple shoot induction. Quercetin acts as an auxin transporter inhibitor (Brown et al., 2001), while $AgNO_3$ acts as an inhibitor of ethylene action (Beyer, 1976). In these experiments, shoot apices from aseptic seedlings were cultured on MS medium fortified with the abovementioned combinations of PGRs combined with quercetin at 0.75 and 1.5 mg/L. Similarly, in another set, MS medium augmented combinations of BAP and kinetin combined with $AgNO_3$ at 0.5 and 1.0 mg/L. Explants cultured on MS medium devoid of PGRs served as controls.

Cytokinins possess the ability to break buds and are involved in cell division, shoot initiation, and multiplication from explants (George et al., 2008; Mazri, 2015). BAP and kinetin are commonly employed for multiple shoot induction in the *C. carandas* shoot apex. Several studies reported lower efficiency of cytokinins when they were used alone for shoot multiplication. The combinations of two or more cytokinins were found to be effective for shoot induction in *Curcuma longa*

**Fig. 13.2** (**a**) Shoot apex of *C. carandas* explant on MS medium. (**b**) Multiple shoot induction on MS medium after 4 weeks

(Thingbaijam et al., 2012), *Pyrus pyrifolia* (Kadota & Niimi, 2003), and *Phoenix dactylifera* (Al-Khateeb, 2006). Studies on shoot multiplication showed the effectiveness of BAP over other cytokinins in *Cucumis anguria* (Jeyakumar et al., 2014). A higher BAP to lower kinetin ratio for shoot enhancement in *C. carandas* resulted in effective shoot regeneration (Bhadane et al., 2018). This combination of cytokinins showed a positive effect on shoot induction. The newly developed shoots were initiated from the cut edges of the shoot apex after 13 days of incubation. The use of BAP alone at 1.0 mg/L was unable to show shoot initiation from the explant. However, increasing concentrations of BAP had a poor impact on multiple shoot induction (Fig. 13.2a, b). The maximum length (1.73 ± 0.16 cm) of shoots was obtained in MS media supplemented with 4.0 mg/L BAP and 0.5 mg/L kinetin (Bhadane, 2017). The mean number of shoots per explant and the average length of shoots per explant in all treatments were found to be comparable with each other (Table 13.2).

This indicates that different concentrations of BAP used in combination with kinetin at 0.5 mg/L increase the rate of shoot induction in *C. carandas* (Bhadane et al., 2018). Development hormones, specifically auxins and cytokinins, are the most important components of tissue culture media because they have a higher impact on cell growth and differentiation than other components. It is known that cytokinins may distinguish between shoot buds, shoot tips, and nodal explants. For the initiation of shoot buds from nodal explants, BAP is the most often employed cytokinin (Ram & Shekhawat, 2011).

The shoot regeneration frequency is positively modified by the synergistic impact of cytokinin and auxin. Other growth additives, such as adenine sulfate, polyamines (low molecular weight compounds), diamines, triamine, and casein hydrolysate and tetramine spermine, positively affect cell division and differentiation (Gopal et al., 2020).

Secondary metabolites such as phenolics and flavonoids are known for their role as the inhibitors of auxin transport, *in vitro* as well as *in vivo* (Brown et al., 2001).

**Table 13.2** Effect of cytokinins with quercetin on shoot multiplication (Bhadane, 2017)

| BAP (mg/L) | Kinetin (mg/L) | Quercetin (mg/L) | Mean number of shoot per explant | Mean length (cm) of shootlet per explant |
|---|---|---|---|---|
| 1.0 | – | – | $0 \pm 0.00^{o}$ | $0.00 \pm 0.00^{m}$ |
| 2.0 | – | – | $2 \pm 1.51^{n}$ | $1.20 \pm 1.05^{k,l}$ |
| 3.0 | – | – | $3 \pm 1.35^{m,n}$ | $1.39 \pm 0.53^{j,k,l}$ |
| 4.0 | – | – | $3 \pm 0.67^{l,m}$ | $1.64 \pm 0.21^{i,j}$ |
| 1.0 | 0.5 | – | $2 \pm 1.54^{m,n}$ | $1.08 \pm 0.74^{l}$ |
| 2.0 | 0.5 | – | $4 \pm 1.14^{k,l}$ | $1.45 \pm 0.24^{i,j,k}$ |
| 3.0 | 0.5 | – | $5 \pm 1.30^{j,k}$ | $1.58 \pm 0.19^{i,j}$ |
| 4.0 | 0.5 | – | $5 \pm 1.47^{j,k}$ | $1.73 \pm 0.16^{h,i,j}$ |
| 1.0 | – | 0.75 | $6 \pm 1.08^{i,j}$ | $1.81 \pm 0.30^{h,i}$ |
| 2.0 | – | 0.75 | $7 \pm 1.87^{h,i}$ | $2.18 \pm 0.21^{f,g}$ |
| 3.0 | – | 0.75 | $7 \pm 1.14^{h,i}$ | $2.27 \pm 0.21^{f,g}$ |
| 4.0 | – | 0.75 | $8 \pm 1.04^{g,h}$ | $2.40 \pm 0.28^{e,f}$ |
| 1.0 | – | 1.5 | $9 \pm 1.03^{f,g}$ | $2.03 \pm 0.17^{g,h}$ |
| 2.0 | – | 1.5 | $10 \pm 1.70^{c,d,e}$ | $2.45 \pm 0.23^{d,e,f}$ |
| 3.0 | – | 1.5 | $10 \pm 1.37^{d,e,f}$ | $2.18 \pm 0.27^{f,g}$ |
| 4.0 | – | 1.5 | $9 \pm 1.40^{e,f,g}$ | $2.54 \pm 0.28^{c,d,e,f}$ |
| 1.0 | 0.5 | 0.75 | $11 \pm 2.31^{c,d}$ | $2.81 \pm 0.22^{b,c,d}$ |
| 2.0 | 0.5 | 0.75 | $15 \pm 1.62^{a}$ | $3.24 \pm 0.28^{a}$ |
| 3.0 | 0.5 | 0.75 | $11 \pm 2.37^{c,d}$ | $2.76 \pm 0.30^{b,c,d}$ |
| 4.0 | 0.5 | 0.75 | $12 \pm 1.96^{c}$ | $2.90 \pm 0.28^{b,c}$ |
| 1.0 | 0.5 | 1.5 | $11 \pm 1.79^{c,d}$ | $2.74 \pm 0.25^{b,c,d,e}$ |
| 2.0 | 0.5 | 1.5 | $13 \pm 1.56^{b}$ | $2.92 \pm 0.21^{a,b}$ |
| 3.0 | 0.5 | 1.5 | $11 \pm 1.79^{c,d}$ | $2.77 \pm 0.13^{b,c,d}$ |
| 4.0 | 0.5 | 1.5 | $11 \pm 1.75^{c,d}$ | $2.82 \pm 0.30^{b,c,d}$ |

The values correspond to means ± standard deviation of 10 replicates. Means in each column followed by the same letter (a to o) were not statistically different according to DMRT (Duncan's Multiple Range Test) at $p < 0.05$

Quercetin may have a synergistic effect with cytokinins on shoot multiplication. During plant growth, auxins are synthesized in meristematic tissues and are responsible for apical dominance with reduced growth of axillary buds. Cytokines induce axillary bud formation and stimulate multiple shoot initiation. Therefore, if the medium is supplemented with auxin transport inhibitors such as flavonoids, it can improve the shoot multiplication ability of *in vitro* cultured plants. Bhadane et al. (2018) demonstrated the effect of different concentrations and combinations of PGRs and quercetin on shoot multiplication in a few members of the Apocynaceae family. To induce multiple shoot induction, the synergistic effect of quercetin and kinetin was studied. It was observed that the use of quercetin with kinetin and BAP favored multiple shoot induction. The maximum numbers of shoots were obtained in MS media supplemented with combinations of BAP:kinetin:quercetin at 2.0:0.5: 0.75 mg/L after 4 weeks (Fig. 13.3a, b) This combination of PGRs and quercetin not

**Fig. 13.3** (**a**) Shoot apex explant of *C. carandas* inoculated on MS medium with BAP, kinetin and quercetin combination on day one. (**b**) Multiple shoot induction on MS medium with BAP, kinetin and quercetin after four (Bhadane, 2017)

only increased multiple shoot induction but also showed a significant increase in mean shoot length.

The combination of Quercetin, kinetin and benzyl adenine was used for shoot multiplication studies on *Oldenlandia umbellata* (Krishnan & Siril, 2015). They reported maximum shoot induction in MS medium supplemented with 5 μM BA and 5 μM quercetin. According to them, quercetin has endogenous auxin transport inhibitory potential, which ultimately enhances cytokinin action with an increase in shoot number. On the other hand, synthetic auxin transport inhibitors such as 2,3,5-tri iodobenzoic acid (TIBA) and 1-naphthylphthalamic acid (NPA) showed a progressive effect on shoot multiplication in *Rosa hybrida* (Singh & Symal, 2000) and *Alnus glutinosa* (Lall et al., 2005). Jacobs and Rubery (1988) demonstrated that quercetin acts as a competitive inhibitor of NPA.

Recently, there has been renewed interest and progress in the use of $AgNO_3$ in multiple-shoot induction. It is used as the source of nitrate, which favours the shoot induction process. In addition, it has been reported to inhibit ethylene action in plants. Plants generally evolve or produce ethylene during injury by various mediators. In plant tissue culture, ethylene is produced in aseptically cut explants placed on medium. This ultimately inhibits *the de novo* regeneration of shoots and retards growth. Due to its potency, easy availability, stability, and solubility in water, it has been widely used in plant tissue culture for promoting growth and shoot induction (Kumar et al., 2009).

Sridevi and Giridhar (2014) reported an increase in shoot growth in *Coffea dewevrei* using $AgNO_3$ at 40 μM + BA at 8.88 μM + IAA at 2.85 μM. They also reported that a reduced concentration of $AgNO_3$ in the medium promotes somatic embryogenesis of *Coffea dewevrei* calli. According to studies by Bayer (1976), silver ions are effective in blocking exogenously applied ethylene response towards senescence and growth retardation. These features of $AgNO_3$ were studied for the regeneration of monocot and dicot plants using *in vitro* tissue culture techniques (Bais et al., 2000; Giridhar et al., 2003).

## *In Vitro* Rooting of Shoots

Once the microshoots are obtained from the multiplication of shoot induction with a length of >2 cm, they can be used in *in vitro* rooting to regenerate plantlets. Before cutting the shoots from the clump, they are subjected for washing using sterile deionized water. The dried explants are cultured MS medium fortified with various concentrations of IAA and IBA alone. After 30 days of culture, developed roots can be observed in root lengths, thickness, etc. Studies on C. carandas and *T. divaricata* showed that the quarter strength MS medium favors rooting at lower levels of IAA and IBA. Tiny roots can be seen during the second and third weeks of incubation (Fig. 13.4a, b).

For successful rooting, *in vitro*-raised plantlets are transferred to auxin-containing medium. Previous studies on other members of the Apocynaceae family, Goel et al. (2009) and Nagesh and Shanthamma (2011), showed that IAA, IBA, and NAA were effective for *in vitro* rooting. In addition, the concentration of auxin and major stocks in the basal medium also affects *in vitro* rooting. According to Purohit and Kukda (2004), increased concentrations of medium constituents as well as auxins are the reasons for calluses to shootlets instead of root formation. In their study on *Wrightia tinctoria,* they used reduced concentrations of major salts of MS medium to ¼ strength for *in vitro* rooting. This modified medium favored maximum rooting without callusing. These facts were advantageous for designing the rooting experiment in our study; therefore, we used two different auxins (IAA and IBA) alone at 0.25 mg/L – 2.0 mg/L in ½ strength and ¼ strength MS medium.

**Fig. 13.4** *In vitro* rooting (**a**) Rooting to the shoots cultured on MS medium with IAA. (**b**) *In vitro* rooted plant with multiple roots after 4 weeks (Bhadane, 2017)

## Stepwise Acclimatization (hardening) of *In Vitro* Raised Plantlets

*In vitro* developed plants well developed roots are removed from the culture vessel and the roots of each plant are washed with sterile distilled water. These plantlets are further acclimatized in glass/plastic pots containing sterile garden soil and vermicompost, and irrigated with diluted MS salt solution. The pots can be covered with polybags to maintain proper moisture inside the container. Acclimatization of these plantlets is done in the culture room under controlled conditions at 25 ± 2 °C and 55 ± 5% relative humidity with 16/8 h photoperiod for 4–6 weeks. After acclimatization, these plantlets are transported in a plastic pot containing natural soil. For proper hardening, all pots were placed in a case covered with a polythene sheet and maintained in a shed net house for 4 weeks before subsequent transfer to the field (Bhadane, 2017).

During hardening in a shed net house, plantlets show varying levels of survival. Upon completion of acclimatization and hardening, plantlets can transfer to the field. The entire process takes 8 weeks for the complete regeneration of *C. carandas* (Fig. 13.5).

Upon completion of acclimatization and hardening, plantlets are transferred to the field. The entire process takes 8 weeks (Fig. 13.6).

It can be concluded that *in vitro* propagation techniques developed to conserve *C. carandas* and the other members of this family can be useful for other plants under the endangered category.

Plants such as *Holarrhena antidysenterica* and *Carissa carandas* are pharmacologically very significant, as they possess several bioactivities. Additionally, these plants grow slowly, and their seeds have poor viability and germination. The delicate *Carissa carandas* seeds must be sown right away after being removed from the fruit because they cannot be kept for a longer period of time. Any attempt at preservation causes them to lose their ability to germinate. Therefore, to address these challenges,

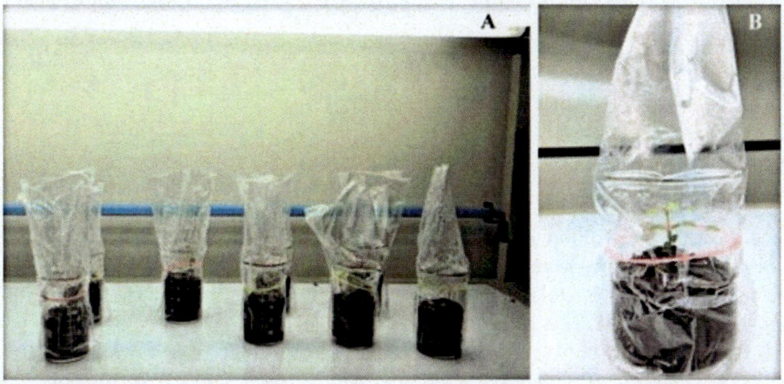


**Fig. 13.5** (**a**) Acclimatization of *in vitro* rooted plantlets in containers covered with polythene bags; (**b**) magnified view (Bhadane, 2017)

**Fig. 13.6** Acclimatized and established plantlets (after 8 weeks) in plastic pots containing natural soil (Bhadane, 2017)

*in vitro*-raised medicinally important, recalcitrant, and fruiting plants of this family can also be effectively conserved by *the ex situ* method. There are reports on phenotypic and genetic variability among wild plants grown in different regions. Hence, there is the possibility of variability *in in vitro* raised and *ex situ* conserved plants. It is possible to determine the genetic variability/stability of *in vitro* raised and *ex situ* conserved plants by molecular markers. Further transcriptome analysis tools can be used to obtain gene sequences to develop molecular markers for the plants under study.

## References

Aiyambo, D. (2010). Traditional uses of selected members of the Apocynaceae family in Namibia. Ministry of agriculture, water and forestry, Windhock.

Al-Khateeb, A. A. (2006). Role of cytokinin and auxin on the multiplication stage of date palm (*Phoenix dactylifera* L.) cv. Sukry. *Biotechnology, 5*(3), 349–352.

Bais, H. P., Sudha, G., Suresh, B., & Ravishankar, G. A. (2000). $AgNO_3$ influences *in vitro* root formation in *Decalepis hamiltonii* Wight and Arn. *Current Science, 79*(6), 894.

Beyer, E. M. (1976). A potent inhibitor of ethylene action in plants. *Plant Physiology, 58*(3), 268–271.

Bhadane, B. S. (2017). *Studies on biological activities and in vitro propagation of some members of family Apocynaceae*. PhD thesis. KBC North Maharashtra University.

Bhadane, B. S., Maheshwari, V. L., & Patil, R. H. (2018). Quercetin and silver nitrate modulate organogenesis in Carissa carandas (L.). *In Vitro Cellular & Developmental Biology-Plant, 54*, 600–605.

Bhadane, B., Maheshwari, V., & Patil, R. (2018). Quercetin and silver nitrate modulates organogenesis in Carissa carandas (L.) by inhibiting auxin transport and ethylene action. In vitro Cellular and Developmental Biology-Plants.

Bhat, P., Hegde, G., & Hegde, G. R. (2012). Ethnomedicinal practices in different communities of Uttara Kannada district of Karnataka for treatment of wounds. *Journal of Ethnopharmacology, 143*(2), 501–514.

Brown, D. E., Rashotte, A. M., Murphy, A. S., Normanly, J., Tague, B. W., Peer, W. A., & Muday, G. K. (2001). Flavonoids act as negative regulators of auxin transport *in vivo* in *Arabidopsis*. *Plant Physiology, 126*(2), 524–535.

Forster, B. P., Heberle-Bors, E., Kasha, K. J., & Touraev, A. (2007). The resurgence of haploids in higher plants. *Trends in Plant Science, 12*(8), 368–375.

George, E. F., Hall, M. A., & De Klerk, G. J. (2008). Plant growth regulators II: Cytokinins, their analogues and antagonists. In *Plant propagation by tissue culture* (pp. 205–226). Springer Netherlands.

Giridhar, P., Indu, E. P., Vijaya Ramu, D., & Ravishankar, G. A. (2003). Effect of silver nitrate on *in vitro* shoot growth of coffee. *Tropical Science, 43*(3), 144–146.

Goel, M. K., Kukreja, A. K., & Bisht, N. S. (2009). *In vitro* manipulations in St. John's wort (Hypericum perforatum L.) for incessant and scale up micropropagation using adventitious roots in liquid medium and assessment of clonal fidelity using RAPD analysis. *Plant Cell, Tissue and Organ Culture, 96*(1), 1.

Gopal, Y., Birla, D., & Lalwani, M. (2020). Reduced switches multilevel inverter integration with boost converters in photovoltaic system. *SN Applied Sciences, 2*, 1–5.

Jacobs, M., & Rubery, P. H. (1988). Naturally occurring auxin transport regulators. *Science, 241*(4863), 346.

Jeyakumar, J. J., Kamaraj, M., & Thiruvengadam, M. (2014). Efficient plant regeneration from petiole explants of West Indian gherkin (*Cucumis anguria* L.) via indirect organogenesis. *Journal of Plant Biochemistry and Biotechnology, 23*(3), 307–315.

Kadota, M., & Niimi, Y. (2003). Effects of cytokinin types and their concentrations on shoot proliferation and hyperhydricity in *in vitro* pear cultivar shoots. *Plant Cell, Tissue and Organ Culture, 72*(3), 261–265.

Karuppusamy, S. (2009). A review on trends in production of secondary metabolites from higher plants by *in vitro* tissue, organ and cell cultures. *Journal of Medicinal Plants Research, 3*(13), 1222–1239.

Krishnan, S. S., & Siril, E. A. (2015). Enhanced *in vitro* shoot regeneration in *Oldenlandia umbellata* L. using quercetin: A naturally occurring auxin-transport inhibitor. *Proceedings of the National Academy of Sciences, India Section B: Biological Sciences*, 1–6.

Kumar, V., Parvatam, G., & Ravishankar, G. (2009). $AgNO_3$: A potential regulator of ethylene activity and plant growth modulator. *Electronic Journal of Biotechnology, 12*(2), 1–15.

Lall, S., Mandegaran, Z., & Roberts, A. V. (2005). Shoot multiplication in cultures of mature *Alnus glutinosa*. *Plant Cell, Tissue and Organ Culture, 83*(3), 347–350.

Mahmood, T., Meer, F., Munir, F., Nazar, N., & Naveed, I. (2011). Genetic diversity of selected Apocynaceae species based on chloroplast gene rps11. *Journal of Medicinal Plants Research, 5*, 4382–4387.

Mazri, M. A. (2015). Role of cytokinins and physical state of the culture medium to improve *in vitro* shoot multiplication, rooting and acclimatization of date palm (*Phoenix dactylifera*). *Journal of Plant Biochemistry and Biotechnology, 24*(3), 268–275.

Miller, C., & Skoog, F. (1953). Chemical control of bud formation in tobacco stems segments. *American Journal of Botany*, 768–773.

Nagesh, K. S., & Shanthamma, C. (2011). Micropropagation and antioxidant activity of *Mollugo nudicaulis* Lam. *Journal of Medicinal Plants Research, 5*(6), 895–902.

Ouassou, H., Bouhrim, M., Daoudi, N. E., Mekhfi, H., Ziyyat, A., Legssyer, A., Aziz, M., & Bnouham, M. (2021). Evaluation of Hepatoprotective Activity of Caralluma europaea stem extract against CCl 4-induced hepatic damage in wistar rats. Advances in Pharmacological and Pharmaceutical Sciences 2021.

Purohit, S. D., & Kukda, G. (2004). Micropropagation of an adult tree-*Wrightia tinctoria. Indian Journal of Biotechnology, 3*(2), 216–220.

Ram, K., & Shekhawat, N. S. (2011). Micropropagation of commercially cultivated Henna (*Lawsonia inermis*) using nodal explants. *Physiology and Molecular Biology of Plants, 17*, 281–289.

Singh, S. K., & Syamal, M. M. (2000). Anti-auxin enhance *Rosa hybrida* L. micropropagation. *Biologia Plantarum, 43*(2), 279–281.

Sridevi, V., & Giridhar, P. (2014). *In vitro* shoot growth, direct organogenesis and somatic embryogenesis promoted by silver nitrate in *Coffea dewevrei. Journal of Plant Biochemistry and Biotechnology, 23*(1), 112–118.

Steward, F. C. (1958). Growth and organized development of cultured cells. III. Interpretations of the growth from free cell to carrot plant. *American Journal of Botany*, 709–713.

Thingbaijam, D. S., Khumallambam, D. D., Kshetrimayum, P., Chongtham, H. S., Shagolsem, B. S., Chingakham, B. S., & Huidrom, S. D. (2012). Silver nitrate and different culture vessels influence high frequency microrhizome induction *in vitro* and enhancement growth of turmeric plantlet during ex vitro acclimatization. *Notulae Scientia Biologicae, 4*(4), 67.

Van Beek, T. A., Verpoorte, R., Svendsen, A. B., Leeuwenberg, A. J. M., & Bisset, N. G. (1984). Tabernaemontana L. (Apocynaceae): a review of its taxonomy, phytochemistry, ethnobotany and pharmacology. *Journal of Ethnopharmacology, 10*(1), 1–156.

Verma R. (2017). *Micropropagation plant tissue culture technique*. Biotech articles.

Vineeta, S. B. C., Tamang, M., Shukla, G., Debnath, M. K., Nath, A. J., & Chakravarty, S. (2022). Floristic diversity, and conservation status of large cardamom based traditional agroforestry system along an altitudinal gradient in the Darjeeling Himalaya, India. *Agroforestry Systems, 96* (8), 1199–1210.

# Chapter 14
# Practical Applications of Apocynaceae Plants in Nanotechnology

## Introduction

Nanotechnology is a branch of applied science and technology whose basis includes the control and synthesis of matter on a molecular level in the range of 1–100 nm. The concepts and ideas of nanoscience and nanotechnology were first used by the physicist Richard Feynman in 1959 (Feynman, 1959). The term "nanotechnology" was defined by Tokyo Science University Professor Norio Taniguchi in 1974. Nanotechnology is the multidisciplinary integration of biotechnology, nanotechnology, chemical processing, material science, and system engineering in biochips, molecular motors, nanocrystals and nanobiomaterials (Huang et al., 2007). Nanoparticles (NPs) are particles less than 100 nm in diameter that have a very high surface area to volume ratio, which leads to both physical and chemical differences in their properties compared to bulk material (Rai et al., 2008; Baker et al., 2013; Iravani, 2011).

A nanometer (nm) is one-billionth of a meter, which is a microscopic particle with at least one dimension less than 100 nm. A DNA molecule is 2.5 nm wide, a protein approximately 50 nm, and a flu virus approximately 100 nm. Human hair is approximately 10,000 nm thick. Nanoparticles are of great scientific interest because they link the gap between bulk materials and atomic or molecular structures. A bulk material has constant physical properties irrespective of its size, but at the nanoscale, it is different. Several well-characterized bulk materials have been found to possess the most interesting properties when studied at the nanoscale (Thakkar et al., 2010).

Noble metal nanoparticle research has great scientific interest due to its potential applications in biomedical, optical, magnetic, electronic fields and information storage. Metal nanoparticles have superior properties to their bulk counterparts, i.e., They have powerful mechanical strength, thermal stability, catalytic activity, electrical conductivity, magnetic properties (Murray et al., 2001), optical properties, etc. Nanoparticles have size-dependent properties on the nanometer scale, while bulk materials have more constant physical properties than their size (Fendler & Tian,



R. H. Patil et al., *Apocynaceae Plants*,
https://doi.org/10.1007/978-981-99-5406-3_14

1998). Due to their smaller size, nanoparticles have optical properties that produce quantum effects (Dai & Bruening, 2002; Klabunde & Richards, 2009).

Research has reported the applications of metal nanoparticles (MNPs) in several fields. MNPs have attractive spectral, optical and electronic properties. MNPs also have wide applications in catalysis and biosensing applications. Biosensing platforms are selective and specific bioanalyte detectors (Doria et al., 2012). The fields of electronics, magnetism and optoelectronics, and information storage are some of the applications of MNPs. Silver nanoparticles (AgNPs) are used in photonics, microelectronics, and medicine, and they are commonly used for their antimicrobial activity against a broad variety of microorganisms (Nangare & Patil, 2020). Copper nanoparticles (CuNPs) are reported as a disinfectant for wastewater, catalysts, conducting applications, antibacterial activity, etc. The CuNPs may cause fluorescence quenching, dye degradation, dye aggregation, dye deaggregation, and fluorescence enhancement for biosensing and biolabelling applications (Din & Rehan, 2017). Ferrous nanoparticles (FeNPs) are widely used in remediation, in the treatment of organic or inorganic pollutants in water, in dye removal and as antibacterial agents (Fahmy et al., 2018).

## Methods of Nanoparticle Synthesis

Nanoparticles are synthesized via physical, chemical and biological methods. NP synthesis methods are broadly classified as top-down and bottom-up approaches (Fig. 14.1). The top-down approach refers to the carving or successive cutting of suitable bulk material into particles at the nanoscale. Bottom-up or self-assembly refers to the buildup of material at the nanoscale from the bottom: atom-by-atom, molecule-by-molecule, or cluster-by-cluster. Lithographic techniques and grinding and milling are typical examples of top-down methods, whereas colloidal dispersion is an example of the bottom-up method (Ranjit & Baquee, 2013).

### *Physical Methods*

#### Mechanical Attrition/Ball Milling

Milling is a characteristic "top-down" approach that belongs to the mechanical alloying process. The purpose of milling is to reduce the particle size and blend particles into new phases, which can be carried out at room temperature. It can be performed on high-energy mills, centrifugal-type mills, vibratory-type mills, and low-energy tumbling mills. Various types of high-energy milling techniques are available to prepare nanoparticles, such as the Attrition Ball Mill, Planetary Ball Mill, Vibrating Ball Mill, Low Energy Tumbling Mill and High Energy Ball Mill (Rajput, 2015; Ghorbani, 2014).

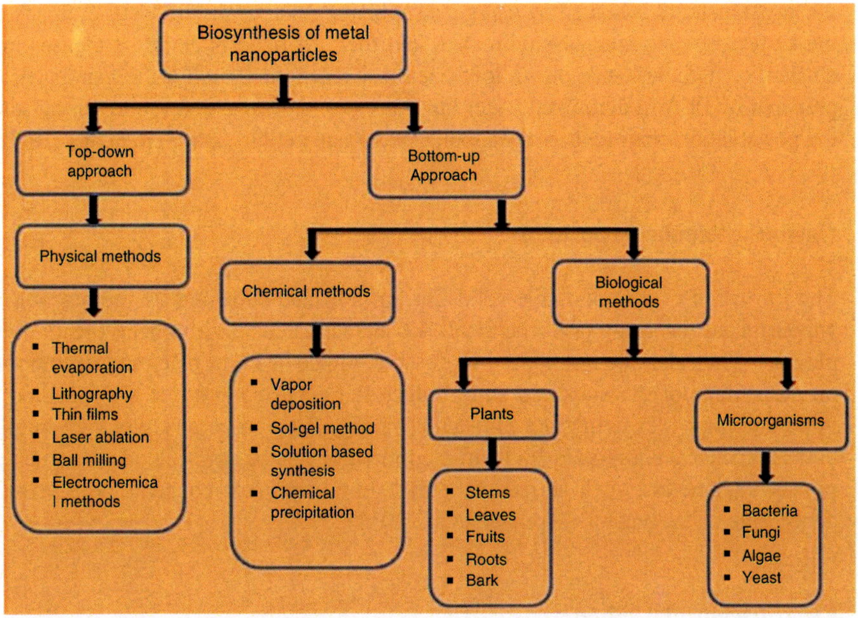


**Fig. 14.1** An overiview of different methods of nanoparticle synthesis

Milling is performed by the stirring action of an agitator. It has a vertical rotator central shaft with horizontal arms, i.e., impellers. The rotation speed and the milling temperature must be in greater control. The common drawbacks of this technique were the low surface area, highly polydispersed size distribution and partially amorphous state of the powder. The powders were highly reactive with oxygen, hydrogen and nitrogen (Rajput, 2015; Ghorbani, 2014).

## Pulsed Laser Ablation (PLA)

Pulsed laser ablation in liquid media (PLAL) is a flexible technique used for the synthesis of several kinds of NPs and metal-oxide NPs, such as noble metals, alloys, oxides and semiconductors. The controlled synthesis of nanomaterials was carried out through rapid reactive quenching of ablated species at the plasma and liquid interface. Using a single-step method, well-crystallized NPs can easily be achieved due to the high energetic state of ablated species. Using different combinations of ablated inorganic species and surfactant molecules in liquids, organic and inorganic nanocomposites can be synthesized (Sasaki et al., 2006; He et al., 2008).

The PLA technique possesses the advantages of simplicity of the procedure, as it requires a small amount of chemical substances and inexpensive equipment for controlling the ablation atmosphere, which cannot be achieved by vapour-based techniques or chemical solution route processes. During this process, the surface of the target material is heated by using a laser beam. As a result, the metal atoms from the target are vaporized or sublimated, and the vaporized material condenses into clusters or particles with or without chemical reaction. Usually, the nanoparticles prepared by PLA in deionized water have a rather large size distribution because of the postablation coalescence of nanoclusters (Sasaki et al., 2006; He et al., 2008).

#### Chemical Vapour Deposition

The formation of a nonvolatile solid film on a heated substrate by the reaction of vapour-phase or gas-phase chemicals in an activated environment (heat, light, plasma). The substrate is exposed to gaseous molecules that react and decompose on the surface of the substrate and produce the expected product. The method is important for synthesizing thin films and coatings. This process is relatively simple, uses inexpensive equipment, and particle formation can be controlled by a variety of process parameters, such as reactor temperature, precursor concentration, or residence time in the reactor (Moravec et al., 2011).

#### Electrochemical Method

Electrochemical deposition is a technique that has been widely used for the synthesis of metal nanoparticles. Electrochemical deposition occurs at the interface of an electrolyte solution containing the metal to be deposited and an electrically conductive metal substrate.

Mostly in the electrochemical method of nanoparticle synthesis, a two-electrode system is used. The method is inexpensive and yields high-purity nanoparticles compared to other bottom-up approaches. By monitoring the electrolytic concentration and the applied current density, the size of the particles can be manipulated. AgNPs are usually synthesized by this method, which does not require any expensive chemicals and materials or special reagents such as surfactants and reducing agents. AgNPs are synthesized electrochemically with 0.01 M $AgNO_3$ as an electrolyte. Glassy carbon electrodes and silver metal were used as working electrodes and counter electrodes, respectively (Singaravelan & Alwar, 2015).

Ultrasonication, photo irradiation, radiolysis, spray pyrolysis, vaporization, solvothermal, chemical reduction and reverse micelles have been successfully used for the synthesis of NPs (Anamika Mubayi et al., 2012).

## *Chemical Methods*

### Sol–Gel Method

The sol–gel method is a wet-chemical technique widely used for nanoparticle synthesis. A two-step sol–gel method to prepare a nanostructured $Pt/CeO_2–ZrO_2–Al_2O_3$ catalyst with high thermal stability was reported. The first step includes the preparation of a dispersion colloid of a noble metal particle suspended with aluminum isopropoxide $Ce(NO_3)_4$ and $Zr(NO_3)_4$ and condensed into a gel, whereas the second step involves the hydrolysis of metal oxide precursors (Long et al., 2007). It is possible to modify the surface of synthesized zinc oxide nanoparticles by the sol–gel method with selected organic compounds. The method is inexpensive, requires a low processing temperature and reliable, repeatable, controllable reaction conditions, and narrow nanoparticles can be produced (Kołodziejczak-Radzimska & Jesionowski, 2014).

### Chemical Precipitation

The chemical precipitation method is attractive due to its low cost, high purity, short preparation time, high homogeneity, well-crystallized product, and relatively low reaction temperature (Lassoued et al., 2017). A wide-ranging literature review concludes that the chemical precipitation method is a notable method of ZnO nanoparticle synthesis (Jay Chithra et al., 2015). Chemical precipitation is used widely in microelectronics and plastic electronics because it is an inexpensive, low-temperature process with controllable process parameters for particle size (Jay Chithra et al., 2015). The method yields the smallest ZnO nanoparticles with the best optical and photocatalytic activity compared with physical and chemical methods (Liu et al., 2012). The synthesis involved a reaction between constituent materials in a suitable solvent. The precipitate was separated by centrifugation. The Cd NPs synthesized by chemical precipitation were heated and dried at 50 °C using double-distilled water as the solvent and thioglycerol as the capping agent (Singh & Chauhan, 2009; Rajput, 2015).

In top-down methods such as size reduction from a suitable starting material, imperfections were observed in the surface structure of the products (Rai et al., 2008). The surface chemistry and other physical properties of nanoparticles are highly dependent on the surface structure, which limits the use of these methods. Other physical methods, such as UV irradiation, aerosol and lithography, are also not considered environmentally friendly. In chemical methods (traditional wet methods), hazardous substances such as sodium borohydride, tetrakishhydrooxy methyl phosphonium chloride (THPLC), poly-n-vinyl pyrrolidone (PVC), and hydroxylamine have been used for the synthesis of nanoparticles. The factors limiting the use of synthesized nanoparticles by these methods for biomedical applications include contamination from precursor chemicals, use of toxic solvents, hazardous byproducts generated during synthesis, etc. (Nath & Banerjee, 2013; Karazhanov & Ravindran, 2010; Narayanan & Sakthivel, 2010; Sharma et al., 2019).

Physical and chemical methods were preferred for narrow particle sizes and uniform distributions. However, specific chemicals, such as reducing and capping agents, which may be environmentally destructive and costly, involve tedious and severe experimental conditions with high energy intake (Ahmed et al., 2016).

Additionally, the use of conventional reducing agents such as sodium borohydride, hydrazine hydrate and polyvinyl pyrrolidone (PVP), sodium polyacrylate and cetyltrimethylammonium bromide (CTAB), etc., as capping agents may adsorb on the surface of the nanoparticle and may be liable for various biological risks due to their general toxicity. Thus, it is important to develop reproducible "green" processes for the biogenic synthesis of nanomaterials (Luque & Varma, 2012; Ahmed et al., 2016).

In the last few years, chemical methods have been used for the production of large quantities of nanoparticles that are complicated, outdated and expensive and produce hazardous toxic wastes that are harmful to the environment as well as human health (Din & Rehan, 2017). Biological entities offer a nontoxic, clean and eco-friendly approach to synthesize a varied range of nanoparticles, physicochemical properties, shapes and composites (Husen & Siddiqi, 2014). Plants have been used for the synthesis of metallic nanoparticles due to their availability, cost-effectiveness, environmentally friendly nature and nonhazardous byproducts (Din & Rehan, 2017).

## Biological Methods

In bottom-up synthesis, the nanostructured building blocks of the nanoparticles are formed first and then assembled to produce the final particle (Thakkar et al., 2010; Mukherjee et al., 2001). Biosynthesis of a nanoparticle is a type of bottom-up approach that employs a biological system or its components for the formation of nanoparticles (Baker et al., 2013). The bioreduction of raw metallic salt into metal nanoparticles is the basis of biogenic nanoparticle synthesis.

The 'biogenic approach' involves nanoparticle synthesis via the biological approach without adding any reducing agent or stabilizer, which are replaced by molecules produced by living organisms, i.e., bacteria, fungi, yeast, algae, higher plants, etc. Biological methods involve the use of plants, animals and microorganisms for the synthesis of nanoparticles.

### *Microorganism-Mediated Nanoparticle Synthesis*

Microorganisms (such as algae, fungi, yeast, bacteria, and actinomycetes) have been widely used for the synthesis of various nanoparticles. Studies on the biosynthesis of silver (Ag), gold (Au), cadmium sulfide (CdS), zinc sulfide (ZnS), platinum (Pt), and palladium (Pd) nanoparticles using microorganisms are well documented (Sastry et al., 2003; Gahlawat & Choudhury, 2019; Talebi et al., 2010). Biosynthesis of nanoparticles using microorganisms is widely used but has limitations such as maintenance of aseptic conditions, culture maintenance, and time-consuming processes (rate of synthesis is slow). Moreover, it is expensive compared to plant-mediated biosynthesis of nanoparticles (Rai et al., 2008). Other disadvantages related to microbes include spontaneous mutations that lead to loss of their biosynthesis potential due to long-term preservation and maintenance of microbial cultures (Baker et al., 2013). Furthermore, nanoparticles with limited size and shapes are synthesized by microbial sources compared to plant-derived materials (Kharissova et al., 2013).

## *Plant-Mediated Synthesis Of Nanoparticles*

Phytomediated nanoparticle synthesis is a simple, eco-friendly and cost-effective method for the synthesis of nanoparticles (Rai et al., 2008). It can be easily scaled up, and plant metabolites are generally free from toxic contaminants (Mittal et al., 2013). Medicinal and aromatic plant usage for various diseases was suggested by the traditional Indian system. This opens up new avenues for the green chemical synthesis of biocompatible nanomaterials using plant parts (Alam et al., 2013). The use of plant extracts for making nanoparticles is inexpensive, easily scaled up and environmentally benign (Mittal et al., 2013). Plant tuber extracts such as *curcuma longa*, *ginger officinale*, and *Dioscorea bulbifera*. were enriched with different reducing and capping agents, which help in the rapid and stable formation of different metal nanoparticles (Ghosh et al., 2012; Singh et al., 2011; Shameli et al., 2012; Sathishkumar et al., 2009).

The synthesis mediated by plant extracts is based on the reduction/oxidation reaction. In this reaction, the phytochemicals and enzymes present in biological materials favor the conversion of metal compounds into specific nanoparticles (Bhuyan et al., 2015). The plant parts used are leaves, roots, latex, seeds, stems and tubers (Thakkar et al., 2010; Kharissova et al., 2013). The metal accumulation potential of plants is a fundamental basis explored for the biosynthesis of nanoparticles (Baker et al., 2013; Iravani, 2011). Plant-mediated nanoparticle synthesis is an easy one-step protocol. In phytomediated nanoparticle synthesis, the rate of reduction of the metal ion is comparatively faster compared to the microbial source. Plant-derived nanoparticles are relatively stable, and the process is reproducible compared to the microbial process (Kharissova et al., 2013). Moreover, plants are a more suitable option than microbes at a large scale due to their easy availability, lack of requirement for aseptic conditions, rapid synthesis, and easy handling (Sathishkumar et al., 2018). The biosynthesis processes predominantly use water extracts that require normal temperature and air pressure, making the process energy efficient (Liu et al., 2012).

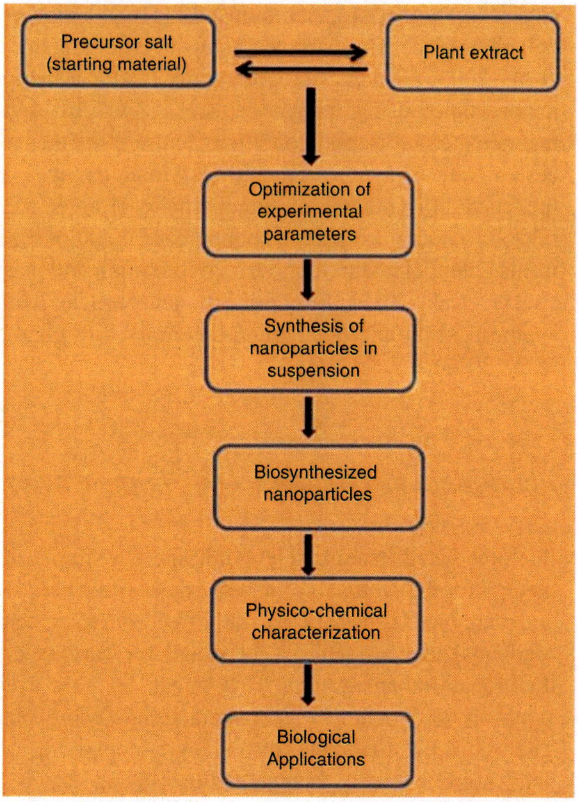


**Fig. 14.2** Flow chart for the biosynthesis of metal nanoparticles

Plant extracts contain various biomolecules that act as a reducing agent (reduction of metallic salt to zero-valent metal nanoparticle) and capping agent/stabilizing agents (provides nanoparticle surface capping) (Mittal et al., 2013; Baker et al., 2013). This is because plant extracts contain different concentrations and combinations of organic reducing agents such as amino acids, vitamins, polysaccharides, polyphenols, organic acids, citric acid, dehydrogenase, flavonoids, functional groups (alcohol, aldehyde, amines, carboxylic acids, ketones and sulfahydryl), heterocyclic compounds hydrogenase, intracellular $CO_2$, membrane proteins, NADP reductase, peptide, phenolics, polyphenol, polyols, proteins, quercetin, reducing sugars, saponins, secondary metabolites, sesquiterpenes, tannic acid, tartaric acid and terpenoids (Akhtar et al., 2013; Baker et al., 2013). The general protocol for the biosynthesis of metal nanoparticles using plant extracts is shown in Fig. 14.2.

In the "Reduction" cum "Stabilization" approach, the plant extract containing biomolecules was simply mixed with the corresponding aqueous metal salt solution at room temperature. As a result, bioreduction occurs within a few minutes. The mono- or divalent states of metal ions were converted to a zero-valent state, which was visually observed by the color change of the solution (Malik et al., 2014; Mittal et al., 2013). In the "chelation" cum "stabilization" approach, polyphenols and flavonoids can act as chelating agents. Due to chelation, the hydroxyl functional group of polyphenols coordinates with the metal ions forming metal-phenolate complexes. Additionally, the complexes undergo thermal decomposition, leading to the formation of metal oxide nanoparticles. In both approaches, the biosynthesized nanoparticles might be capped by the plant extract containing biomolecules, which may further prevent aggregation of nanoparticles. The secondary metabolites of plants that act as capping agents will stabilize the growth of biosynthesized nanoparticles. Amino acids or proteins containing hydroxyl and carbonyl functional groups usually act as capping agents and strongly adhere to the surface of nanoparticles to avoid aggregation with nearby NPs (Imran Din & Rani, 2016; Makarov et al., 2014; Mallikarjuna et al., 2011).

The reduction process of metal ions to metal nanoparticles is affected by different factors, such as the concentration of metallic salt, the concentration of plant extracts, pH, biosynthesis temperature, and reaction time (Baker et al., 2013). Phyotosynthesis of nanoparticles mainly depends on the choice of solvents, reducing agent, capping agent, etc. (Alam et al., 2013). The concentration of extract affects the shape of nanoparticles such as triangular, spherical, hexagonal, etc. (Mittal et al., 2013).

The mechanism of nanoparticle synthesis by plant extract was explained on the basis of the positive functional groups present in the plant extract, such as positive amino groups, sulfahydryl groups and carboxylic groups, which allowed the metal ions to become more closed to binding sites and approved the reduction of metal ions to their zero-state metal ions, i.e., metal nanoparticle formation at selective pH (Iravani, 2011). The oxidation of the polysaccharide hydroxyl group in the carbonyl group plays an important role in the reduction of metal salt into nanoparticles (Akhtar et al., 2013).

## Applications of Nanoparticles

There is a wide range of medical applications that lead to the development of diagnostic, therapeutic and pharmaceutical applications. Due to the similar size of nanoparticles and biological molecules, nanoparticles can be useful for both in vivo and in vitro biomedical research and applications.

### *Medical Applications*

Drug delivery is the most promising application of colloidal nanoparticles, which are used as carriers of a drug for site-specific delivery. For these applications, the size, charge and surface chemistry of magnetic nanoparticles are important and strongly affect both blood circulation time and bioavailability of the particles within the body.

A glucose-sensing platform using gold and silver nanoparticles was also reported (Aslan et al., 2004).

## *Antioxidant Activity*

Nanomaterials represent one of the most promising frontiers in research for improved antioxidants. Inorganic metal nanoparticles (NPs) possess intrinsic antioxidant properties due to the surface properties of the material itself (Valgimigli et al., 2018). Nanoparticles have a higher ability to scavenge free radicals due to their characteristic features of small size and high surface area to volume ratio. The antioxidant activity of many metal nanoparticles, such as silver (Sharma et al., 2012; Bunghez et al., 2012) silver, gold (Veeraapandian et al., 2012), silver, copper, iron and zinc (El-Refai et al., 2018), copper (Das et al., 2013), iron and nickel (Saikia et al., 2010), and zinc (Muthuraman et al., 2014), has been reported.

## *Anticancer Activity*

New tools for cancer imaging and treatment are available to researchers with recent developments in nanotechnology. Nanotechnology has enabled the development of nanoscale devices that can be conjugated with numerous functional molecules simultaneously, comprising anticancer drugs, antibodies, tumor-specific ligands, and imaging probes (Wang et al., 2008). Therapeutic nanoparticles can enhance efficacy and reduce side effects compared with conventional cancer therapeutic drugs (Wang et al., 2009). The high cytotoxic effect against human liver cancer cells (HepG2) was shown by AgNPs biosynthesized using the medicinal plant dandelion *Taraxacum officinale* (Saratale et al., 2018). The reduced cell viability and increased cytotoxicity in HCT-116 colon cancer cells have been shown by biosynthesized gold and silver nanoparticles (Kuppusamy et al., 2016).

The in vitro anticancer efficacy of *guava* and *clove* extracts synthesized from silver and gold NPs against four dissimilar cancer cell lines has been reported (Raghunandan et al., 2011).

A mini-review "On the anticancer activities of silver nanoparticles" summarizing the utilization of AgNPs in cancer treatment is also available in the literature (Abdel-Fattah & Ali, 2018).

## *Antimicrobial Activity*

The antimicrobial activity of nanoparticles has mostly been considered against human pathogenic bacteria and fungi. The bactericidal activity of nanoparticles in part depends on their size, shape and stability. The antimicrobial and medicinal properties of silver are well recognized (Prabhu & Poulose, 2012), whereas silver and its salts have short-term and restricted antimicrobial activity. This problem can be resolved with nanoforms of silver, which can act as suitable antimicrobial agents with inertness and stability (Huang et al., 2011; Ghosh et al., 2012).

Several mechanisms for the microbicidal activity of silver nanoparticles (AgNPs) have been proposed (Dakal et al., 2016): (a) cell membrane structural changes due to deposition of AgNPs on cells (Sondi & Salopek-Sondi, 2004); (b) AgNPs form free radicals (Kim et al., 2007). Ag ions from AgNPs (Feng et al., 2000) deactivating several active enzymes (the thiol group of enzyme interaction with Ag leads to inactivation of enzyme) (Matsumura et al., 2003). Soft acids (Ag) react with soft bases (cells) (Morones et al., 2005). Bacterial signal transduction was altered by AgNPs by destroying cell DNA. Here, soft acids (Ag) and DNA contain sulfur and phosphorus as soft bases (Hatchett & White, 1996).

AgNPs are a real bactericide as AgNP deposition on the bacterial cell wall leads to structural changes that ultimately cause cell death (Sondi & Salopek-Sondi, 2004). Gram-negative bacteria contain less peptidoglycan than Gram-positive bacteria cell walls (Ankanna et al., 2010; Maliszewska & Sadowski, 2009). AgNPs inhibit DNA replication, as fungal DNA forms a complex with AgNPs after penetration (Kolya et al., 2015; Nasrollahi et al., 2011).

Silver and gold nanoparticles have potential applications in antibacterial and antifungal activities against even multidrug registrant pathogenic microbes (Vijayakumar et al., 2017). Furthermore, AgNPs in combination with antibiotics have improved bactericidal activity (Fayaz et al., 2010; Ghosh et al., 2012).

## *Applications in Biosensing*

The current demand for more specific and highly sensitive biomolecular diagnostics can be fulfilled with the application of functionalized MNPs. The type of conjugation between biomolecules and MNPs for biofunctionalization may be electrostatic interactions, chemisorption, and affinity-based (Doria et al., 2012). Nanomaterial optical sensors are a type of chemical sensor classified as luminescence-based and absorbance-based depending on the origin of the optical signal. Luminescence-based sensors are based on the principle of detection by photoluminescence and chemiluminescence, whereas absorbance sensors are based on spectral absorbance changes (Shi et al., 2004). Optical sensors have resistivity to electromagnetic noise, fire resistance and remote control capability and information transfer through an optical fibre network compared to conventional electrochemical sensors (Shi et al., 2004). The optical properties of MNPs vary depending on the spacing between polymeric aggregates, which is explored for colorimetric detection of various kinds of analytes, such as metal ions, organics or biomolecules, within a simple and low-cost way (Shi et al., 2004). The MNP optical properties were also explored to improve the sensitivity of conventional SPR systems. Gold nanoparticle aggregates appear red when interparticle distances are noticeably greater than the average particle diameter, whereas when interparticle distances are less than average, they appear blue (Kreibig & Genzel, 1985; Yang et al., 1995). A highly selective heterogeneous colorimetric polynucleotide detection method based on mercaptoalkyloligonucleotide-modified gold nanoparticle probes was reported in which the color change is due to a redshift in the surface plasmon resonance of the Au nanoparticles (Elghanian et al., 1997; Reynolds et al., 2000; Storhoff et al., 2000). A label-free optical biosensor (Colorimetric Gold Nanoparticle Sensor) to quantify biomolecular interactions in real time on a surface in a commercially available UV–visible spectrophotometer and of a colorimetric end-point assay using an optical scanner was reported. The spectrophotometric sensor shows concentration-dependent binding and a detection limit of 16 nM for streptavidin (Nath & Chilkoti, 2002). An aggregation-based immunoassay for anti-Protein A using gold nanoparticles was reported. The aggregation of gold nanoparticles coated with protein antigens in the presence of their corresponding antibodies results in an absorption change at 620 nm, which was the basis of the assay monitored with an absorption plate reader. A two orders of magnitude dynamic range and a 1 μg/ml limit of detection for anti-protein A were reported. The new technique could be used for fast, high-throughput screening of antibodies in clinical diagnostic applications (Thanh & Rosenzweig, 2002).

### *Applications in Catalysis*

Green catalysis involves the use of environmentally benign catalysts for easy removal from the reaction media and is recyclable with high efficiency. The reactivity of nanoparticles depends on their size, shape, composition and surface atomic arrangement. One of the major reasons is the high surface-to-volume ratio of metal nanoparticles, which makes them promising materials for catalysis. Nanoparticle catalysis has been investigated for both homogeneous and heterogeneous systems. A shape-dependent catalytic activity of platinum nanoparticles in colloidal solution was reported (Narayanan & El-Sayed, 2004). The textile, paper, paint, plastic and pharma industries have extensive usage of organic synthetic dyes, which are pollutants (environmentally unsafe) (Bhakya et al., 2015). Organic dyes can be efficiently degraded using biosynthesized nanoparticles (Indana et al., 2016; Ganapuram et al., 2015; Seralathan et al., 2014).

## Characterization of Nanoparticles

The reduction of a particular metal nanoparticle can be monitored using the UV–vis visible spectrum, which is a simple tool for the detection of the surface plasma resonance (SPR) property of the metal nanoparticle. XRD analysis confirms the metallic state of a particular metal nanoparticle. The diffraction pattern provides information on translational symmetry—size and shape of the unit cell from peak positions and information on electron density inside the unit cell. It also provides information on deviations from a perfect particle if the size is less than 100 or 200 nm, extended defects and microstrain from peak shapes and widths. The surface morphology and size of bioreduced metal nanoparticles were characterized by transmission electron microscopy (TEM) (Ghosh et al., 2011). The FT-IR analysis of metal nanoparticles along with plant extract was performed before and after bioreduction of nanoparticles to identify the probable biomolecules responsible for the reduction and capping of nanoparticles (Shameli et al., 2012).

## *Silver Nanoparticles*

Silver nanoparticles (Ag NPs) are one of the most commercialized nanomaterials and have unique optical, electrical and thermal properties. They are widely applied as biocides for their strong antimicrobial activity. Due to its strong antimicrobial activity, silver is used in both medical treatment and industrial processes for its inhibitory effects. Therefore, silver nanoparticles have wide applications, such as diagnostic applications, antibacterial applications, fabrics, manufacturing materials, electronic devices, and solving environmental problems.

The application of nanoparticles depends upon their size, shape, surface morphology, composition and fine structure (Alam et al., 2013). Silver nanoparticles have numerous applications in the fields of dentistry, clothing, catalysis, mirrors, optics, photography, electronics and the food industry (Rai et al., 2009).

The antimicrobial and medicinal properties of silver are recognized (Panáček et al., 2006; Prabhu & Poulose, 2012). The Ag ion and its salt antimicrobial activity is short duration and restricted, which is resolved by using inert and stable nanoforms of silver (Huang et al., 2011; Ghosh et al., 2012). Additionally, AgNPs in combination with antibiotics have improved bactericidal activity (Fayaz et al., 2010; Ghosh et al., 2012). The green synthesis of silver nanoparticles with antibacterial effects has been reported. (Zhang et al., 2020; Gomathi & Suhana, 2021; Nakhjiri et al., 2017; Pethakamsetty et al., 2017; Rautela & Rani, 2019; Premkumar et al., 2018).

AgNP biosynthesis have reported been using plant part extracts as reducing agents. An overview of the synthesis of silver nanoparticles by physical, chemical, and biological synthesis is reviewed by Iravani et al. (2014). AgNPs synthesis, effects of various reaction parameters, characterization techniques, properties and their applications were summarized by Pethakamsetty et al. (2017). Plant extract-mediated AgNP synthesis for antimicrobial applications was reported by Ahemed et al. (2016). A green synthesis of silver nanoparticles using the different plant products of different taxonomic groups with a single step protocol was reported (Gupta & Chauhan, 2017). A summary of the green phytochemical methods used for AgNP biosynthesis is presented in Table 14.1, highlighting different precursors,

**Table 14.1** Different precursors, plants, plant parts, and the ratio of extract and metal salts used for the biosynthesis of AgNPs (Pawar, 2021)

| Sr. no. | Precursor salt | Precursor concentration (mM) | Plant source | Plant part used | The ratio of extract: metal salt (ml) | Nanoparticle size (nm) | Applications | Reference |
|---|---|---|---|---|---|---|---|---|
| 1 | Silver nitrate | 1 | *Carica papaya* | Green unripe fruits | 10:90 | 10–50 | Antibacterial activity | Jain et al. (2009) |
| 2 | Silver nitrate | 1 | *Ocimum sanctum* | Leaves | 0.5:10 | 3–20 | Not reported | Mallikarjuna et al. (2011) |
| 3 | Silver nitrate | 1 | *Ipomoea batatas L.* (Sweet potato) | Tubers | 10:90 | 1–10 | Antibacterial activity | Sivakumar et al. (2015) |
| 4 | Silver nitrate | 1 | *Dioscorea oppositifolia* | Tuber | 05:95 | 17–25 | Catalytic activity towards reduction of 4-nitrophenol | Ghosh et al. (2016) |
| 5 | Silver nitrate | 1 | *Curcuma longa* | Rhizomes | 02:08 | 18 ± 0.5 | Antibacterial activity | Alsammarraie et al. (2018) |
| 6 | Silver nitrate | 1 | *Momordica cymbalaria* | Fruit and tuber | 10:90 | 10–50 | Antibacterial and cytotoxic potential | Hemalatha et al. (2019) |
| 7 | Silver nitrate | 1 | *Nervalia zeylanica* (Vatha Kodi) | Leaf | 10:90 | 10–50 | Catalytic activity by rreduction of an azo dye methyl orange | Vijayan et al. (2019) |
| 8 | Silver nitrate | 25 | *Ocimum sanctum* | Stem and root | 40:10 | 10 ± 2 | Not reported | Ahmad et al. (2010) |
| 9 | Silver nitrate | 5 | *Jatropha curcas* | Latex | 20:20 | 10–20 | Not reported | Bar et al. (2009) |
| 10 | Silver nitrate | 1 | *Chaetomorpha linum* | Seaweed | 100:20 | 3–44 | Not reported | Kannan et al. (2013) |
| 11 | Silver nitrate | 1 | *Solanum xanthocarpum* L. | Berries | 20:10 | 10–18 | Antimicrobial and Urease Inhibitory Activity | Amin et al. (2012) |

| | | | | | | | | |
|---|---|---|---|---|---|---|---|---|
| 12 | Silver nitrate | 3 | *Theobroma cacao*: | Husk, pulp, seed | 20:40 | 6–18 | Antibacterial and photocatalytic activity | Thatikayala et al. (2019) |
| 13 | Silver nitrate | 1.75 | Banana (*Musa paradisiaca*) | Peel | 1:50 | 23.7 | Antimicrobial activity | Ibrahim (2015) |
| 14 | Silver nitrate | 1.75 | *Premna herbacea* | Leaves | 0.5:10 | 10–30 | Antibacterial activity | Kumar et al. (2013) |
| 15 | Silver nitrate | 1 | *Tinospora cordifolia* | Stem | 15:200 | 9–12 | Antibacterial activity | Singh et al. (2014) |

plant parts used for extracts, their concentrations, and applications of b iosynthesized AgNPs.

A green AgNP biosynthesis method including $AgNO_3$ reduction through *papaya* fruit extract was reported. AgNPs are highly toxic against different multidrug-resistant human pathogens (Jain et al., 2009). An eco-friendly simple route for the reduction of silver salt using the leaf broth of *Ocimum sanctum* was reported (Mallikarjuna et al., 2011). Tuber extract of *sweet potato* was used for the biosynthesis of AgNPs under different reaction times. The biosynthesized AgNPs demonstrated strong antioxidant and effective antibacterial activity against human pathogenic bacteria (Sivakumar et al., 2015).

Biosynthesis of gold (AuNPs) and silver (AgNPs) nanoparticles using *Dioscorea oppositifolia* tuber extract is also reported. The biosynthesized AuNPs and AgNPs demonstrated effective catalytic action in 4-nitrophenol reduction by $NaBH_4$ to 4-aminophenol by pseudofirst-order rate kinetics (Ghosh et al., 2016). Green synthesis of AgNPs was reported using the aqueous extract of turmeric powder, where plant biomaterials were used as reducing agents as well as capping agents. Green synthesized AgNPs show high and efficient antimicrobial activities against two food-borne pathogens (*Escherichia coli* O157:H7 and *Listeria monocytogenes*) (Alsammarraie et al., 2018). Green synthesis of AgNPs by using aqueous fruit and tuber extracts of *Momordica cymbalaria* was reported along with their antibacterial and cytotoxic potential (Hemalatha et al., 2019). A simple, one-pot, inexpensive, and eco-friendly method for the biosynthesis of AgNPs using *Nervalia zeylanica* leaf extract as a reducing and stabilizing agent by a microwave-assisted strategy was reported. The AgNPs showed better scavenging activities than the leaf extract. AgNPs show excellent catalytic activities and antimicrobial activities (Vijayan et al., 2019).

The room-dried stem and root of *Ocimum sanctum* were explored in the biosynthesis of AgNPs at room temperature (Ahmad et al., 2010). *Jatropha curcas* latex was demonstrated as a reducing as well as a capping agent in the biosynthesis of AgNPs (Bar et al., 2009). Seaweed (*Chaetomorpha linum*) extract was explored for the biosynthesis of AgNPs (Kannan et al., 2013). The methanol extract from *Solanum xanthocarpum* berry as a green route for the synthesis of AgNPs was reported. In vitro, the anti-*Helicobacter pylori* activity of synthesized AgNps demonstrated a stronger anti-*H. pylori* activity compared to $AgNO_3$ or metronidazole and equally potent to tertracycline (Amin et al., 2012).

Extracts of different parts of *Theobroma cacao*, such as the husk, pulp, and seed, were reported for the biosynthesis of AgNPs. The biosynthesized AgNPs demonstrated nearly 30% methylene blue degradation within 60 min and antibacterial activity against *Bacillus subtilis* and *Escherichia coli* (Thatikayala et al., 2019). The banana peel extract as a reducing and capping agent was demonstrated in the biosynthesis of AgNPs. The biosynthesized AgNPs had effective antibacterial activity against representative pathogens of bacteria and yeast (Ibrahim, 2015). The aqueous leaf extracts of *Premna herbacea*, a local herb of Bodoland, Assam, India, were reported as reducing as well as capping agents in the biosynthesis of

AgNPs. Significant antibacterial activity was found against *Shigella dysentrieae* and *E. coli* (Kumar et al., 2013).

The biosynthesis of AgNPs from the stem of *Tinospora cordifolia* was performed. The biosynthesized AgNPs were evaluated for their antibacterial activity against MDR strains of *P. aeruginosa* (Singh et al., 2014). Marathe et al. (2021) studied the AgNPs biosynthesis using haloalkaliphilic *Streptomyces* spp. The AgNPs were found significantly active against phytopathogenic fungi, *Fusarium verticillioides* and *Ustilago maydis* (Marathe et al., 2021).

## *Copper Nanoparticles (CuNPs)*

Copper nanoparticles have wide applications as heat transfer systems, antimicrobial materials, superstrong materials, sensors and catalysts. Copper nanoparticles are very reactive because of their high surface-to-volume ratio and can easily interact with other particles (Subhankari & Nayak, 2013; Yallappa et al., 2013). Researchers have reported green biosynthesis of CuNPs with pharmacological activities such as antibacterial, antioxidant and catalytic activities (Thiruvengadam et al., 2019; Gondwal & Joshi nee Pant, 2018; Aher et al., 2017; Tahvilian et al., 2019; Harikumar & Aravind, 2016). One-step biofabrication of copper nanoparticles from *Aegle marmelos* aqueous leaf extract and evaluation of its anti-inflammatory and mosquito larvicidal efficacy was reported (Angajala et al., 2014). Antidiabetic and antioxidant properties of copper nanoparticles synthesized by the medicinal plant *Dioscorea bulbifera* were reported (Ghosh et al., 2015). The green synthesis of copper oxide nanoparticles using *Caesalpinia bonducella* seed extract and evaluation of electrocatalytic properties for electrochemical detection of riboflavin (Vitamin B2) was reported (Sukumar et al., 2020).

### Biosynthesis of CuNPs Using Plant Extracts

A summary of the green phytochemical methods used for CuNP biosynthesis is using extracts of several Apocynaceae and other plants is presented in Table 14.2, showing different precursors, plant parts, combinations of extracts, their concentrations, and applications of CuNPs.

A study indicated the use of an aqueous extract of *Abutilon indicum* for the synthesis of CuO NPs by a superficial method (solution combustion method), which had considerable antioxidant activity and antibacterial activity against *Klebsiella* and *Bacillus subtilis* (Ijaz et al., 2017). CuNP biosynthesis was carried out after the addition of *Alchornea laxiflora* aqueous leaf extract to the copper sulfate solution. The superior catalytic activity of the biosynthesized CuNPs was observed compared to conventional acetic acid in the oxidation of dibenzothiophene (DBT). Biosynthesized CuNPs may be an economical substitute material for oxidative desulfurization in the oil industry (Olajire et al., 2018). *Asparagus adscendens* leaf

**Table 14.2** Different precursors, plants, plant parts, and the ratio of extract and metal salts used for the biosynthesis of CuNPs (Pawar, 2021)

| Sr. no | Precursor concentration (mM) | Precursor | Plant source | Plant part used | Extract conc (ml) | Metal salt (ml) | Nanopaarticle size (nm) | Application | Reference |
|---|---|---|---|---|---|---|---|---|---|
| 1 | – | Cu $(NO_3)_2{\cdot}3H_2O$ | *Abutilon Indicum* | Leaf extract | 0.3 | – | 16.78 | Antimicrobial, antioxidant and photocatalytic dye degradation activities | Ijaz et al. (2017) |
| 2 | 1 | $CuSO_4{\cdot}5H_2O$ | *Alchornea laxiflora* | Leaf extract | 20 | 80 | 3.29 | Catalytic activity | Olajire et al. (2018) |
| 3 | 1 | $CuSO_4{\cdot}5H_2O$ | *Asparagus adscendens Roxb.* | Root and Leaf extract | 10 | 100 | 50–60 | Antimicrobial activities | Thakur et al. (2018) |
| 4 | 10 | Cu $(CH_3COO)_2{\cdot}H_2O$/ methanol | *Borreria hispida* | Linn. extract | 50 | 100 | 84–158 | Antioxidant activity | Venugopalan et al. (2020) |
| 5 | 100 | $CuSO_4{\cdot}5H_2O$ | *Citrus medica* | Linn. (Idilimbu) juice | – | – | 10–60 | Antimicrobial activity | Shende et al. (2015) |
| 6 | 10 | Copper sulfate | *Cissus arnotiana* | Plant extract | 10 | 90 | 60–90 | Antibacterial and antioxidant activity | Rajeshkumar et al. (2019) |
| 7 | 1 | Copper (II) sulfate pentahydrate | *Citrus sinensis* | Aqueous fruit extract | 100 | – | 10.2 | Antibacterial potentials | Jahan et al. (2021) |
| 8 | 20 | $Cu(NO_3)_2$ | *Cocoa pod* | Cocoa pod extract | 1.5 gm | – | 20–50 | Catalytic activity | Nguyen et al. (2020) |
| 9 | 5 | $CuSO_4{\cdot}5H_2O$ | *Ripened Duranta erecta* | Aqueous fruit extract | 15 | 85 | 70 | Catalytic reduction of carcinogenic azo dyes Methyl orange and Congo red | Ismail et al. (2019) |

| | | | | | | | | | |
|---|---|---|---|---|---|---|---|---|---|
| 10 | 40 | $CuSO_4{\cdot}5H_2O$ | *Cordia myxa* | L. Leaves | 100 | – | 20–35 | Reducing, stabilizing and capping agent | Thamer et al. (2018) |
| 11 | 100 | Copper acetate dihydrate solution | *Curcuma longa* | Curcuma longa extract | 50 | 100 | 5–20 | Antimicrobial activity | Jayarambabu et al. (2020) |
| 12 | 5 | Copper nitrate | *Eryngium campestre* | Leaf extract | 10 | 90 | 50–60 | Not reported | Khodaie and Ghasemi (2018) |
| 13 | 40 | Copper (II) ion | *Moringa oleifera* | Leaves extract | 80 | 20 | 35.8–49.2 | Antioxidant and Anti-microbial Activities | Das et al. (2020) |
| 14 | – | Copper sulfate pentahydrate | *Seedless dates* | Seedless dates extract | 100 | 10 | 78 | Not reported | Mohamed (2020) |
| 15 | 25 | $CuCl_2{\cdot}2H_2O$, $Cu(CH_3COO)_2{\cdot}H_2O$ and $Cu(NO_3)_2{\cdot}3H_2O$ | *Vaccinium species* | Bilberry fruits | 9 | 1 | 2–10 | Antimicrobial activity | Benassai et al. (2021) |
| 16 | 10 | $CuSO_4{\cdot}5H_2O$ | *Eucalyptus globulus L.* | Leaves extract | 10 | 100 | 10–130 | Against fruit rot path-ogen of chilli crop | Iliger et al. (2021) |
| 17 | 10 | $(Cu(NO_3)_3{\cdot}3H_2O)$ | *Eryngium caucasicum Trautv* | Leaves extract | 5 | 50 | 40 | Antioxidant and anti-microbial properties | Hasheminya and Dehghannya (2020) |
| 18 | 100 | $CuSO_4{\cdot}5H_2O$ | *Ginger Lily* | Leaves extract | 100 | – | 40 | Antibacterial activities | Nagore et al. (2020) |
| 19 | – | $CuSO_4$ | *Hawthorn berries* | Hawthorn berries extract | 10 | 30 | 200 | The antimicrobial activities | Długosz et al. (2020) |
| 20 | 40 | Copper chloride | *Cochlospermum gossypium* | Bark extract | – | – | 14 | Antimicrobial prop-erty, biomedical application | Seku et al. (2018) |

(continued)

**Table 14.2** (continued)

| Sr. no | Precursor concentration (mM) | Precursor | Plant source | Plant part used | Extract conc (ml) | Metal salt (ml) | Nanopaarticle size (nm) | Application | Reference |
|---|---|---|---|---|---|---|---|---|---|
| 21 | 3 | Copper chloride | *Jatropha curcas* | Leaves extract | 20 | 80 | 10 | CT-DNA binding and photocatalytic activity | Ghosh et al. (2020) |
| 22 | 3 | Cu $(CH_3COO)_2{\cdot}H_2O$ | *Magnolia Champaca* | Floral Extract | 10 | – | 20–40 | Antioxidant & Toxicity Assay using Danio Rerio | Santhosh Kumar and Shanmugam (2020) |
| 23 | – | Copper sulfate | *Mitragyna parvifolia* | Plant Bark extract | 10 | 90 | 12–23 | Antimicrobial study | Kotval et al. (2018) |
| 24 | 3 | $CuSO_4{\cdot}5H_2O$ | *Thymus vulgaris L* | Leaf extract | 50 | 50 | 23–94 | Catalytic activity | Issaabadi et al. (2017) |
| 25 | 1 | $CuSO_4{\cdot}5H_2O$ | *Phoenix Dactylifera L* | Leaves extract | 1 | 100 | 22–28 | Medical application | Berra et al. (2018) |
| 26 | 1 | $CuSO_4$ | *Plectranthus Amboinicus* | Leaves Extract | 2 | 10 | 21.74 | – | Parthasarathy et al. (2020) |
| 27 | – | $CuCl_2$ | *Rheum palmatum L* | Root extract | – | 30 | 10–20 | Catalytic activity | Bordbar et al. (2017) |
| 28 | – | Copper sulfate pentahydrate | *Citrus limon* | Fruit extract | 100 | – | 5–28 | Antibacterial activity | Amer and Awwad (2020) |
| 29 | 5 | $CuSO_4{\cdot}5H_2O$ | *Syzygium alternifolium* | Stem bark | 10 | 100 | 5–13 | Antimicrobial and anticancer activity | Yugandhar et al. (2017) |
| 30 | 1 | Cu $(CH_3COO)_2{\cdot}H_2O$ | *Syzygium aromaticum* | Bud extract | 10 | 50 | 20 | Physical, optical, antimicrobial properties, Fungicidal activity | Rajesh et al. (2018) |
| 31 | 25 | Copper chloride | *Tinospora cardifolia* | Leaves extract | 2 | 10 | 63.3 | Antimicrobial evaluation | Sharma et al. (2019) |

| | | | | | | | | | |
|---|---|---|---|---|---|---|---|---|---|
| 32 | $7.5 \times 10^{-3}$ M | $CuCl_2 \cdot 2H_2O$ | *Azadirachta indica* | Leaves extract | – | – | 48 | – | Nagar and Devra (2018) |
| 33 | 3 | Cu $(CH_3COO)_2 \cdot H_2O$ | *Eclipta prostrata* | Leaves extract | 20 | 80 | 23–57 | Antioxidant and cytotoxic activities | Chung et al. (2017) |
| 34 | 1 | Copper sulfate | *Nerium oleander* | Leaves extract | 100 | 100 | 21 | – | Sebeia et al. (2019) |
| 35 | 5 | $CuCl_2 \cdot 2H_2O$ | *Plantago asiatica* | Leaf extract | 100 | 20 | 7–35 | Cyanation of aldehydes | Nasrollahzadeh et al. (2017) |
| 36 | 20 | Copper sulfate | *Ziziphus spina-christi (L.)Willd* | Fruit extract | – | – | 5–20 | Adsorption of triphenylmethane dye and antibacterial assay | Khani et al. (2018) |

and root extracts were used for the biosynthesis of CuNPs. The antimicrobial efficacy of CuNPs was established against Gram-negative and Gram-positive pathogenic microbes (Thakur et al., 2018).

The crude methanol extract of the whole dried plant *Borreria hispida* (Linn) was used for the synthesis of CuNPs. The antioxidant efficacy of the CuNPs was found to be higher than that of the plant crude extract (Venugopalan et al., 2020). An innovative biogenic synthesis of CuNPs was studied using *citron* juice. Its efficacy against a wide variety of microorganisms, such as *Escherichia coli,* followed by *Klebsiella pneumoniae, Pseudomonas aeruginosa, Propionibacterium acnes*, and *Salmonella typhi,* was studied (Shende et al., 2015). *Cissus arnotiana* extract was reported for the biosynthesis of CuNPs. The biosynthesized CuNPs demonstrated better antibacterial activity against Gram-negative bacteria such as *E. coli* and antioxidant properties than the standard ascorbic acid (Rajeshkumar et al., 2019). The use of orange *Citrus sinensis* fruit extract for the biosynthesis of CuNPs through microwaves was reported. These biosynthesized CuNPs demonstrated their effectiveness in the management of human pathogens, i.e., *Escherichia coli* and *Staphylococcus aureus* (Jahan et al., 2021). Biosynthesis of CuNPs using *cocoa pod* extract as a reducing agent, an efficient and green technique, was reported. CuNPs loaded on different inert supports were tested for catalytic deep oxidation of aromatic hydrocarbons (AHs) (Nguyen et al., 2020).

The fruit extract of *Duranta erecta* was reported for the green synthesis of CuNPs. In the presence of $NaBH_4$ as a reducing agent, CuNPs exhibit outstanding catalytic reduction for methyl orange and Congo red. They can be used to purify natural water from organic effluents (Ismail et al., 2019). An aqueous extract of *Cordia myxa* L leaves was used as a reduction, stabilization and capping agent, and a simple and low-cost approach to CuNP preparation was reported (Thamer et al., 2018). A novel green route for the synthesis of CuNPs was reported using an aqueous extract from a *Curcuma longa* solution. CuNPs capped with *C. longa* extract have shown promising antibacterial activity against Gram-positive and Gram-negative bacteria. Compared to Gram-negative bacteria, the inhibition zone of CuNPs for Gram-positive bacteria is greater (Jayarambabu et al., 2020). *Eryngium campestre* leaf extract was used to investigate the synthesis of CuNPs. The synthesis of CuNPs was completed with pH 6, temperature 35°C and time 80 min (Khodaie & Ghasemi, 2018). A hydroalcoholic extract of *Moringa oleifera* leaves was reported to biosynthesize CuNPs. The synthesized CuNPs have antioxidant, antibacterial, and antifungal properties (Das et al., 2020). *Seedless date* extract contains a high content of phenolics, and flavonoids were used to biosynthesize $Cu/Cu_2O$ NPs (Mohamed, 2020). Green synthesis of CuNPs using aqueous extracts of *bilberry* waste residues from the production of fruit juices and nonedible false bilberry fruits was tested. CuNPs can be considered a cost-effective sanitization method with a broad range of action from green synthesis (Benassai et al., 2021).

A simple, beneficial and green method for CuNP biosynthesis using *Eucalyptus* and *Mint* plant extracts was reported. The CuNPs were tested and proven to be highly successful for their efficacy against the fungus *C. capsici* (Iliger et al., 2021). An aqueous extract of *E. caucasicum Trautv* leaves was used as a reducing agent for green and environmentally friendly CuNP biosynthesis. Antibacterial characterization of CuNPs revealed remarkable antibacterial properties against *E. coli, S. typhimurium, B. cereus, and S. aureus* (Hasheminya & Dehghannya, 2020). Using ginger lily leaf extract as a natural reducing and stabilizing agent, a simple, renewable, eco-caring technique was reported to phytogenically synthesize luminescent CuO NPs (Nagore et al., 2020). A green method for AgNP and CuNP preparation was reported using water and ethanol extracts of dried *hawthorn* fruit extract as a stabilizer and reducant. They have antibacterial activity against selected pathogenic microorganisms (*A niger, E. coli and S. cerevisiae*) (Długosz et al., 2020). An eco-friendly approach for CuNP synthesis was carried out by hydrothermal processes using carboxymethyl gum konagogu as a capping agent from the bark of *Cochlospermum gossypium*. CuNPs had excellent antibacterial and antifungal activity against pathogenic strains (Seku et al., 2018).

A simple green chemistry method for CuNP biosynthesis using *J. curcas* leaf extract was reported. CuNPs have potential photocatalytic activity against methylene blue (MB) (Ghosh et al., 2020). Biosynthesis of CuO NPs using *Magnolia champaca* floral extract as a reducing agent and stabilizing agent was reported (Santhosh Kumar & Shanmugam, 2020). *Mitragyna parvifolia* plant aqueous bark extract was used as a reducing agent for the green synthesis of CuNPs. The antibacterial activity of CuNPs was determined (Kotval et al., 2018). A cost-effective and environmentally friendly technique for the synthesis of bentonite-supported copper nanoparticles (bentonite/CuNPs) using *Thymus vulgaris* leaf extract as a mild, sustainable and nontoxic reducing agent and an efficient stabilizer without adding any surfactants was reported. They had excellent catalytic activity in the presence of $NaBH_4$ in water at room temperature for Congo Red and Methylene Blue degradation. (Issaabadi et al., 2017). Copper oxide nanoparticles were synthesized using *Phoenix dactylifera L* leaves from the El Oued area (Southeast of Algeria) (Berra et al., 2018). The use of *Plectranthus amboinicus* extracts as reducing agents for CuNP biosynthesis was explored (Parthasarathy et al., 2020). A green method was used to prepare CuO NPs using *Rheum palmatum L.* root extract as a reducing and stabilizing agent. Very mild reaction conditions, cleanliness, no need for organic solvents, poisonous and unsafe reagents, and simplicity are some of the many benefits of this process. The CuO NPs demonstrated strong activity as a heterogeneous catalyst in the reduction of 4-nitrophenol, methylene blue, and rhodamine B at ambient temperature. (Bordbar et al., 2017).

Green synthesis of CuNPs without any harmful materials using *Citrus limon* fruit extract was reported. CuNPs demonstrated excellent antibacterial activity against pathogenic *E. coli* and *S. aureus* bacteria (Amer & Awwad, 2020). A cost-effective, eco-friendly, green approach for CuO NP synthesis from *S. alternifolium* stem bark extract was reported, which had antimicrobial and anticancer action (Yugandhar et al., 2017). Using *Syzygium aromaticum* bud extract as a reducing agent, CuNPs were synthesized at room temperature through a simple, fast and eco-friendly pathway. The CuNPs showed excellent bactericidal action against *Bacillus* spp. and marked fungicidal activity against *Penicillium* spp. of microorganisms (Rajesh et al., 2018).

The biosynthesized CuNPs capped with *Tinospora cardifolia* were formulated and integrated into the fabric. The antimicrobial behavior of formulated CuNPs along with fabric was evaluated, which showed that the percent efficacy of copper nanoparticle-coated fabric for Gram-positive and Gram-negative bacteria was 101% and 74% at the maximum concentration, respectively (Sharma et al., 2019). The CuNPs were biosynthesized using *Azadirachta indica* leaf broth as a reducing agent in which synthesized CuNPs were strongly crystalline and cubical (Nagar & Devra, 2018). The *E. prostrata* leaf extract was used as a reducing agent for biosynthesizing CuNPs. The biosynthesized CuNPs showed significant antioxidant potential, whereas cytotoxicity against tested HepG2 cells has been demonstrated in in vitro anticancer studies (Chung et al., 2017).

The biological extract of *Nerium oleander* leaves was used as a reducing agent for the biosynthesis of CuNPs. They were used as nanoadsorbents for methylene blue and crystal violet (Sebeia et al., 2019). Green synthesis of CuNPs by in situ $CuCl_2$ reduction using *Plantago asiatica* leaf extract as a reductant and stabilizer was explored. The catalytic activity of the CuNPs was evaluated (Nasrollahzadeh et al., 2017). A green chemistry approach for Cu-NP biosynthesis using *Ziziphus spina-christi* fruit extract was explored. They act as a green and effective nanoadsorbent for crystal violet and also had antibacterial activity (Khani et al., 2018).

## Ferrous Nanoparticles

Metallic iron oxide nanoparticles have a variety of scientific and technological applications, such as biosensors, antimicrobial activity, food preservation, magnetic storage media, ferrofluids, magnetic refrigeration, magnetic resonance imaging, hyperthermic cancer treatments, cell sorting and targeted drug delivery (Behera et al., 2012; Kumar, Singh, & Pandey, 2015; Kumar, Suresh, Nagabhushana, et al., 2015; Wei et al., 2016; Nagajyothi et al., 2017; Asghar et al., 2018; Devi et al., 2019). A summary of the green phytochemical methods used for FeNP biosynthesis is presented in Table 14.3, which summarizes the different precursors, plant parts used for biosynthesis, extracts, and their concentrations as well as the application of FeNPs.

Iron nanoparticles synthesized by the green method were found to be toxic to bacteria such as *Escherichia coli* and *Staphylococcus aureus* (Pandey et al., 2012). Green synthesis of iron oxide nanoparticles using *Carica papaya* leaf extract was reported for photocatalytic degradation of remazol yellow RR dye and antibacterial activity (Bhuiyan et al., 2020). Iron nanoparticles were applied for the remediation of heavy metals from soil and disinfection of water (Mahdavi et al., 2013). Membrane-supported nanoparticles were used for the degradation of the common and highly important pollutant trichloroethylene (TCE) (Kharissova et al., 2013).

Using *Agrewia optiva* and *Prunus persica* leaf extracts, biogenic synthesis of FeNPs was carried out using a simple, nontoxic, sustainable, low-cost, quick and environmentally friendly process (Mirza et al., 2018). The FeNPs were

**Table 14.3** Different precursors, plants, plant parts, and ratios of extract and metal salts used for the biosynthesis of FeNPs (Pawar, 2021)

| Sr. no | Precursor | Precursor concentration (mM) | Plant used | Plant part used | Ratio of extract to metal salt (ml) | Nanoparticle size (nm) | Applications | Reference |
|---|---|---|---|---|---|---|---|---|
| 1 | $FeCl_2 \cdot 4H_2O$ | 100 | *Agrewia optiva* and *Prunus persica* | Leaves | 30 | 15–60<br>13–70 | Antioxidant activity, antibacterial activity | Mirza et al. (2018) |
| 2 | $Fecl_3$ | 10 | *Cinnamomum Verum* | Bark | 1:1 | 20–50 | Antioxidant, anti-inflammatory and anti-diabetic | Sivakami et al. (2020) |
| 3 | $Fe_3O_4$ | --- | *Padina pavonica*<br>*Sargassum acinarium* | Seaweeds | 1:1 | 10–19.5<br>21.6–27.4 | Biomedical uses | El-Kassas et al. (2016) |
| 4 | $FeSO_4 \cdot 7H_2O$ | 100 | *Mangifera Indica*<br>*Murraya Koenigii*<br>*Azadiracta indica*<br>*Magnolia champaca* | Leaves | 1:2 | 100–150 | For domestic wastewater treatment | Devatha et al. (2016) |
| 5 | $FeSO_4$ | 20 | *Mimosa pudica* | Root | 3 | 60–80 | Antibacterial activity targeted drug delivery | Niraimathee et al. (2016) |
| 6 | $FeSO_4$ | 1 | *Ruellia tuberosa* | Leaf | 50 | 20–80 | Antibacterial activity | Vasantharaj et al. (2019) |
| 7 | $FeCl_3 \cdot 6H_2O$ | 100 | *Falcaria vulgaris* | Leaf | 9 | 25 | Antioxidant, noncytotoxic, skin wound healing, antifungal, and antibacterial | Zangeneh et al. (2019) |
| 8 | $FeCl_3$ | 10 | *Eucalyptus globulus* | Leaf | 20 | 100 | | Balamurugan et al. (2014) |
| 9 | $FeCl_2$ | 1 | *Punica granatum* | Seed | 1:2 | 25–55 | Photocatalytic activity, antioxidant and antimicrobial activities | Bibi et al. (2019) |

(continued)

**Table 14.3** (continued)

| Sr. no | Precursor | Precursor concentration (mM) | Plant used | Plant part used | Ratio of extract to metal salt (ml) | Nanoparticle size (nm) | Applications | Reference |
|---|---|---|---|---|---|---|---|---|
| 10 | $FeSO_4$ | 10 | *Lawsonia inermis* and *Gardenia jasminoides* | Leaves | 50 | 21 and 32 | Antibacterial activity | Naseem and Farrukh (2015) |
| 11 | $FeCl_2$ | 500 | *Terminalia bellirica* (TB); *Moringa oleifera fruit* (MOF) and *Moringa oleifera* | Fruit and Leaves | | 21.32 and 45 | Antimicrobial and anti-oxidant activity | Jegadeesan et al. (2019) |
| 12 | $FeCl_2 \cdot 4H_2O$ | 100 | *Nettle* and *Thyme* | Leaves | Ratio of 2:3 | 20–80 | Antibiotic activity | Leili et al. (2018) |
| 13 | $FeCl_3 \cdot 6H_2O$ | 100 | *Couroupita guianensis Aubl.* | Fruit | 1:1 | $17 \pm 10$ | Antimicrobial activity | Sathishkumar et al. (2018) |
| 14 | Ag- $Fe_2O_3$ | 100 | *Psidium guajava* | Leaf | 1:1 | 50–90 | Not reported | Biswal et al. (2020) |
| 15 | $FeCl_3$ | 100 | *Moringa oleifera* | Leaf and Seed | – | | Antibacterial activity | Katata-Seru et al. (2018) |
| 16 | $FeCl_2 \cdot 4H_2O$ | 100 | *Rosa damascene 33* (RD), *Thymus vulgaris* (TV), and *Urtica dioica* (UD | Leaves | – | 100 | Antioxidant activity | Fazlzadeh et al. (2017) |
| 17 | — | – | *Chromolaena odorata* | Root | – | 5.6–16.8 | Magnetic resonance imaging, cancel cell amelioration, biosensors | Nnadozie and Ajibade (2020) |
| 18 | $FeCl_3 \cdot 6H_2O$ | 200 | *U. dioica* | Leaves | 9 | 21–71 | Anti-inflammatory, anti-hyperglycaemic, antioxi-dant, diuretic, natriuretic, antiulcer, and analgesic | Ebrahiminezhad et al. (2017) |

| | | | | | | | | |
|---|---|---|---|---|---|---|---|---|
| 19 | $FeCl_3$ | 10 | *Trigonella foenum-graecum* | Seed | 5 | 7–14 | Antibacterial activity | Radini et al. (2018) |
| 20 | $FeCl_3.\ 6H_2O$ | 100 | *Laurus nobilis L.* | Leaves | 30 | 8.03 ± 8.99 | Antimicrobial activity | Jamzad and Kamari (2020) |
| 21 | $FeCl_3$ | 400 | *Phoenix dactylifera* | Leaf | 50 | 2–30 | Antioxidant activity | Abdullah et al. (2020) |
| 22 | $FeCl_3{\cdot}6H_2O$ | 100 | *Daphne mezereum* | Leaf | 9 | 6.5–14.9 | Not reported | Beheshtkhoo et al. (2018) |
| 23 | FeCl3 | 1 | *Murrayakoenigii* | Leaves | 10 | 12 | Not reported | Prasad et al. (2017) |
| 24 | $FeCl_3$ | 10 | *Green tea* | Leaves | 1:1 | 116 | Antioxidant | Gottimukkala et al. (2017) |
| 25 | $Fe(NO_3)_3{\cdot}9H_2O$ | – | *Platanus orientalis* | Leaf | 10 | 38 | Antifungal activity, antimicrobial | Devi et al. (2019) |
| 26 | $FeSO_4$ | – | *green tea* | green tea extract | 1:2 | 40–50 | Not reported | Huang et al. (2014) |
| 27 | $FeCl_3$ | – | *Eichhornia crassipes* | Leaf | 1:1 | 20–80 | Antioxidant activity. | Wei et al. (2017) |
| 28 | $FeSO_4{\cdot}7H_2O$ | – | *Eriobotrya japonica* | Leaves | 1:1 | 171.2 | Not reported | Önal et al. (2017) |
| 29 | $FeCl_3$ | 100 | *Nephrolepis auriculata* | Plant | 2:1 | 40–70 | Not reported | Yi et al. (2019) |
| 30 | $FeCl_3$ | 100 | *Cassia fistula Linn* | Leaf | 50 | 20–50 | Anti-inflammatory and antibacterial activity | Ranjith et al. (2018) |
| 31 | $FeSO_4$ | 100 | *eucalyptus* | Leaf | 2:1 | 20–80 | Antioxidant activity | Wang et al. (2014) |
| 32 | $FeCl_3\ {\cdot}6H_2\ O$ | 10 | *Lagenaria siceraria* | Leaves | 1:1 | 30–100 | Antimicrobial Activity | Kanagasubbulakshmi and Kadirvelu (2017) |
| 33 | $FeCl_3$ | 500 | *Amaranthus dubius* | Leaves | 40 | 43–220 | Antioxidant activity | Harshiny et al. (2015) |
| 34 | $FeCl_3$ | 1 | *Azadirachta indica* | Leaf | 1:1 | 50–100 | Antifungal, antidiabetic, antibacterial, | Pattanayak and Nayak (2013) |
| 35 | $FeCl_3{\cdot}6H_2O$ | – | *Aerva lanta* | Flower | 50 | 38 | Anti-inflammatory | Clarina et al. (2018) |

biosynthesized using *Cinnamomum verum* bark extract as a reducing agent reported for biomedical applications. The antioxidant, anti-inflammatory and antidiabetic assays demonstrated the productive actions of low concentration Fe NPs (Sivakami et al., 2020). The $Fe_3O_4$-NPs were synthesized using a completely green biosynthetic method by reduction of ferric chloride solution using *brown seaweed* water extracts. The biosynthesized $Fe_3O_4$-NPs showed bioremoval capacity for Pb (91%) (El-Kassas et al., 2016). A quick, fast and eco-friendly method was used to synthesize FeNPs using different leaf extracts, such as *Mangifera indica, Murraya Koenigii, Azadiracta indica* and *Magnolia champaca,* which were used for domestic wastewater treatment. *Azadiracta indica* showed 98.08% phosphate, 84.32% ammonia nitrogen and 82.35% chemical oxygen demand removal among the various plant-mediated synthesized iron nanoparticles (Devatha et al., 2016). Superparamagnetic FeO NPs have been synthesized successfully from the root extract of *M. Pudica* (Niraimathee et al., 2016).

FeONPs were synthesized using *Ruellia tuberosa* leaf aqueous extract as a reducing and stabilizing agent. Higher antibacterial activity against *Escherichia coli and Klebsiella pneumoniae* and less antibacterial activity against *Staphylococcus aureus* were shown by FeONPs. The synthesized FeONP photocatalytic potential was demonstrated by up to 80% of the degrading crystal violet dye under solar irradiation (Vasantharaj et al., 2019). The potential of *F. vulgaris* aqueous extract for the biosynthesis of FeNPs was explored. It has useful antioxidant, noncytotoxic, skin wound healing, antifungal, and antibacterial impacts (Zangeneh et al., 2019). The β-iron oxide NPs were synthesized by adding the leaf extract of *Eucalyptus globulus* to an aqueous solution of ferric chloride. This method was capable of producing crystalline β-$Fe_2O_3$ (Balamurugan et al., 2014). $Fe_2O_3$ NPs were developed using pomegranate (*Punica granatum*) seed extracts via the green path. The synthesized $Fe_2O_3$ showed excellent photocatalytic activity with a maximum degradation of 95.08% (Bibi et al., 2019). Green biosynthesis of FeNPs using *Lawsonia inermis* and *Gardenia jasminoides* leaf extracts showed excellent antibacterial activity against the human pathogens *Escherichia coli* and *Staphylococcus aureus* (Naseem & Farrukh, 2015). A simple biogenic synthesis of FeNPs from three plant sources using aqueous extracts, *Terminalia bellirica* (TB), *Moringa oleifera* fruit (MOF), and *Moringa oleifera* leaves (MOL), was carried out. Biosynthesized FeNPs have excellent thermal, antimicrobial and antioxidant properties (Jegadeesan et al., 2019). For the synthesis of novel nano zero-valent FeNPs for the removal of cephalexin adsorption, *nettle and thyme* leaf extracts have been used. These materials could be considered promising adsorbents for the removal of antibiotics and probably other organic contaminants from aqueous solutions (Leili et al., 2018).

To produce magnetic $Fe_3O_4$ nanoparticles using edible *C. guianensis* (CGFE) aqueous fruit extract, a sustainable green chemistry approach was been developed. The synthesized $CGFe_3O_4$ NPs induce a substantially enhanced cytotoxic effect against $HepG_2$ cells at a minimum dose of 44.5 μg/ml (Sathishkumar et al., 2018).

For the preparation of α-$Fe_2O_3$-Ag nanocomposite, guava (*Psidium guajava*) leaf extract was used, which can be used as an adsorbent for the decontamination of medium-sized aqueous chromium (VI) ions (Biswal et al., 2020). The green synthesis of FeNPs using *Moringa oleifera* leaf and seed extracts was reported by mixing different ratios of plant extracts with iron chloride solution. Furthermore, the use of biosynthesized FeNPs for the removal of nitrate ions ($NO_3$) from surface and groundwater was explored (Katata-Seru et al., 2018). Nano zero-valent iron (NZVI) particles were synthesized from plant extracts such as *Rosa damascene*, *Thymus vulgaris*, and *Urtica dioica*, which have the potential to effectively adsorb hexavalent chromium even at high concentration levels (Fazlzadeh et al., 2017). FeNPs were prepared using the root extract of *Chromolaena odorata* (Nnadozie & Ajibade, 2020). By using *U. Dioica* leaf extract as a capping and shielding agent, zero-valent FeNPs have been synthesized. (Ebrahiminezhad et al., 2017). *Trigonella foenum*-graecum was found in the seed extract and was used for biosynthesis of FeNPs, which was very efficient for methyl orange dye degradation. The minimum inhibitory concentrations (MICs) of Fe NPs for *E. coli* and *S. aureus* were 32 μg/ml and 64 *μg/ml*, respectively (Radini et al., 2018). The FeONPs were synthesized in the phase of hematite ($\alpha$-$Fe_2O_3$) by the aqueous extract of *Laurus nobilis L.* leaves in a simple and eco-friendly method. The nanoparticles were moderately selective on *Listeria monocytogenes*, a Gram-positive bacterium, and *Aspergillus flavus* and *Penicillium spinulosumum* fungi (Jamzad & Bidkorpeh, 2020). The biosynthesis of stable FeNPs was carried out using *Phoenix dactylifera L.* as a reducing agent. The antioxidant and antiradical activities of the bioreduced iron nanoparticles were measured (Abdullah et al., 2020). FeNPs were synthesized using *Daphne mezereum* aqueous leaf extract as a reducing and stabilizing agent. The synthesized FeNPs were more effective in degrading methyl orange dye catalysts (Beheshtkhoo et al., 2018). *Murraya konenigii* leaf extract was used to form an RGO/$Fe_3O_4$ nanocomposite that was capable of removing Pb(II) from aqueous solution (Prasad et al., 2017). *Camellia sinensis* leaf extract was used to synthesize FeNPs in an environmentally friendly and long-term manner (Gottimukkala et al., 2017). An aqueous phase green synthesis of FeNPs using *Platanus orientalis* leaf extract as a reduction and capping agent was performed. The NPs were stable, crystalline, and impurity-free and had effective antimicrobial activity. The antifungal action of the FeNPs against *M. Piriformis* was 1.6-fold higher than that against *A. niger* (Devi et al., 2019).

The green synthesis of FeNPs using tea extracts was performed, and the dye degradation capacity was evaluated. The high degradation of malachite green (81.56%) was obtained, which may be due to the high polyphenol/caffeine content in green tea extracts (Huang et al., 2014). The biosynthesis of FeNPs using aqueous extracts of *Eichhornia crassipes* was reported, which showed 89.9% Cr (VI) extraction (Wei et al., 2017). The green synthesis of FeNP aqueous extract from *Eriobotrya japonica* leaves was used to biosynthesize FeNPs could be an excellent catalyst for the decolorization and removal of toxic dyes from aqueous solutions (Önal et al., 2017). Using *Nephrolepis auriculata* extracts, the efficient synthesis of FeNPs was reported, which showed 90.93% Cr(VI) removal (Yi et al., 2019). FeNPs were successfully synthesized using an aqueous solution of *Cassia fistula* Linn leaf extract by reducing the ferric chloride solution. The procedure was nonlethal, straightforward, environmentally friendly, and relatively inexpensive. The synthesized FeNPs had antibacterial activity against pathogenic bacteria such as

*Staphylococcus* sp, *Pseudomonas* sp, *Bacillus* sp and *Klebsiella* sp. (Ranjith et al., 2018). The spheroidal FeNPs were biosynthesized via a one-step process using *eucalyptus* leaf extracts at room temperature. FeNPs showed that 71.7% of total N and 84.5% of COD were extracted and used for in situ eutrophic wastewater remediation (Wang et al., 2014).

The *Lagenaria siceraria* leaf extract was capable of green synthesis of $Fe_3O_4$-NPs. The antimicrobial properties of synthesized $Fe_3O_4$-NPs were assessed against Gram-negative *Escherichia coli* and Gram-positive *Staphylococcus aureus* (Kanagasubbulakshmi & Kadirvelu, 2017). The biosynthesis of FeNPs was reported using aqueous extracts of *Amaranthus dubius* leaves reduced from ferric chloride. Synthesis of *A. dubius* leaf extract was a safer alternative to toxic reduction, including $NaBH_4$ (Harshiny et al., 2015). Nanoscaled zero-valent FeNPs were synthesized under atmospheric conditions from *Neem* plant leaf extract (Pattanayak & Nayak, 2013). The eco-friendly chemical-free biosynthesis of FeNPs using *polpala* flower extract as a reducing agent was simple and green and used as a recyclable magnetic nanocatalyst for organic reactions (Clarina et al., 2018).

## *Gold Nanoparticles (AuNPs)*

Gold nanoparticles (AuNPs) are widely studied for their biological and diagnostic properties. Gold nanoparticles are considered promising new metal NPs because of their efficient use in catalysis, gene expression profiling, optoelectronics, optics, and disease diagnostics and treatment. A number of medicinal plants have been used for the biosynthesis of AuNPs. AuNPs are made using either phytochemicals or other extract constituents that are not stable for a long time. Table 14.4 provides a detailed outlook of AuNP biosynthesis using various Indian medicinal plants.

## *Zinc Nanoparticles (ZnNPs)*

Zinc (Zn) is an inorganic metal oxide that forms diverse types of nanomaterials, such as nanotubes, nanowires, nanoclusters and zinc nanoparticles (ZnNPs). ZnNPs have attracted considerable attention because of their large surface area to volume ratio and low cost. The biological activities of ZnNPs include UV-filtering, antifungal, antioxidant, antibacterial and catalytic properties. Plant-based synthesis of ZnNPs has been widely studied (Lingaraju et al., 2016; Choudhary et al., 2017).

Phytoconstituents present in plant extracts act as reducing agents (reduction of metallic salt to zero-valent metal nanoparticles) and capping agents/stabilizing agents (provides nanoparticle surface capping) (Mittal et al., 2013; Baker et al., 2013). Various organic reducing agents, such as amino acids, vitamins, polysaccharides, polyphenols, organic acids, citric acid, dehydrogenase, flavonoids, functional groups (alcohol, aldehyde, amines, carboxylic acids, ketones and sulfahydryl), heterocyclic compounds hydrogenase, intracellular $CO_2$, membrane proteins,

**Table 14.4** Different precursors, plants, plant parts, and the ratio of extract and metal salts used for the biosynthesis of AuNPs

| Sr. no. | Precursor metal | Precursor concentration (mM) | Plant used | Plant part used | Size (nm) | Application | Reference |
|---|---|---|---|---|---|---|---|
| 1. | $AuCl_3$ | 500 | *Aegle marmelos* | Leaves | 4–10 | Antioxidant activity | Noruzi et al. (2011) |
| 2. | $AuCl_3$ | 100 | Aloe vera | Plant extract | 50/350 | Antifungal, antidiabetic, antibacterial | Chandran et al. (2006) |
| 3. | $AuCl_2$ | 1 | *Canna indica, Quisqualis indica* | Leaves and flower | 30–130 | Antifungal, antidiabetic, antibacterial | Shukla and Vankar (2012) |
| 4. | $AuCl_3$ | 100 | Cinnamon | Bark | 35 | Antioxidant activity | El Mitwalli et al. (2020) |
| 5. | $AuCl_2$ | 20 | *Croton Caudatus Geisel* | Leaves | 20 | Antioxidant activity | Giljohann et al. (2010) |
| 6. | $AuCl_2$ | 400 | *Cuminum cyminum* | Seeds | 1–10 | Antifungal activity, antimicrobial | Sneha et al. (2011) |
| 7. | $AuCl_2$ | 10 | *Curcuma pseudomontana* | Rhizome | 20 | Antioxidant activity | Kumari and Arti (2014) |
| 8. | $AuCl_3$ | | *Lawsonia inermis* | Leaves | 20 | Antifungal activity, antimicrobial | El Mitwalli et al. (2020) |
| 9. | $AuCl_3$ | 100 | *Mangifera indica* | Leaves | 17–20 | Antibacterial activity | Philip (2010) |
| 10. | $AuCl_3$ | 10 | *Mentha, Ocimum,* Eucalyptus | Leaves | 3–16 | Anti-inflammatory | Jha and Prasad (2012) |
| 11. | $AuCl_2$ | 100 | *Momordica charantia* | Fruit | 30–40 | Antioxidant activity | Pandey et al. (2012) |
| 12. | $AuCl_2$ | 10 | *Murraya koenigii* | Leaves | 20 | Antibacterial activity | Philip et al. (2011) |
| 13. | $AuCl_2$ | 200 | *Nyctanthes arbortristis* | Flower extract | 19.8 | Antifungal activity, antimicrobial | Pandey et al. (2012) |
| 14. | $AuCl_2$ | 100 | *Parkia biglobosa* | Leaves | 1–35 | Antibacterial activity | Giljohann et al. (2010) |
| 15. | $AuCl_3$ | 100 | *Phyllanthus amarus* | Leaves | 65–99 | Antifungal activity, antimicrobial | Annamalai et al. (2011) |

(continued)

**Table 14.4** (continued)

| Sr. no. | Precursor metal | Precursor concentration (mM) | Plant used | Plant part used | Size (nm) | Application | Reference |
|---|---|---|---|---|---|---|---|
| 16. | $AuCl_3$ | 100 | *Rosa hybrid* | Rose petals | 10 | Antifungal activity, antimicrobial | Kumar et al. (2012) |
| 17. | $AuCl_2$ | 100 | *Sorbus aucuparia* | Leaf extract | 16–18 | Antibacterial activity | Dubey et al. (2010) |
| 18. | $AuCl_2$ | - | *Stevia rebaudiana* | Leaves | 8–20 | Antioxidant activity | Mishra et al. (2010) |
| 19. | $AuCl_2$ | 1 | *Tamarind* | Leaves | 20–40 | Antifungal activity, antimicrobial | Giljohann et al. (2010) |
| 20. | $AuCl_3$ | 100 | *Tanacetum vulgare* | Fruit | 11 | Antioxidant activity | Giljohann et al. (2010) |
| 21. | $AuCl_2$ | 100 | *Terminalia chebula* | Plant extract | 6–60 | Antioxidant activity | Kumar et al. (2012) |
| 22. | $AuCl_3$ | 10 | *Trigonella foneum-graecum* | Leaves | 15–25 | Antioxidant activity | Pandey et al. (2012)] |

NADP reductase, peptide, phenolics, polyphenol, polyols, proteins, quercetin, reducing sugars, saponins, secondary metabolites, sesquiterpenes, tannic acid, tartaric acid and terpenoids, are reducing and stabilizing agents in the reaction system (Akhtar et al., 2013; Baker et al., 2013). These metabolites are synthesized by various plant tissues, such as roots, stems, leaves, fruits and seeds. These phytochemicals lower the metal's valence to zero and then calcinate it to add oxide. Zinc ions interact with the polyphenols, teprnoids, and flavones in the plant extract to form a complex. After that, zinc hydroxide ($Zn(OH)_2$) is formed via hydrolysis, and then ZnO nanoparticles are synthesized (Choudhary et al., 2017). Several members of the Fabaceae, Rutaceae, Apocynaceae, Solanaceae and Lamiaceae families are most commonly employed for the production of ZnNPs (Table 14.5).

A number of plants from the family Apocynaceae, Lamiaceae, have been used to produce ZnO nanoparticles of different sizes and shapes and biological activities. All experiments displayed nanoparticles in the same size range with spherical and hexagonal disc shapes, which were characterized by XRD and TEM analysis. One of the important members from the Apocynaceae family, *Calotropis procera*, has been potentially used to synthesize ZnNPs with spherical shapes in the range of 5–40 nm in size (Singh et al., 2011). Some reports on the plant-assisted synthesis of zinc nanoparticles are listed below in Table 14.5.

## *Titanium Nanoparticles (TiNPs)*

Titanium and dioxide nanoparticles (TiNPs and $TiO_2$NPs) are becoming popular due to their electrical band structure, quantum efficacy, high surface area and stability, and chemical innertness. Therefore, TiNPs are considered the most versatile and prominent oxides, with an annual production of over 20,000 tons in 2020. TiNPs have wide insecticidal and pharmaceutical applications. For example, TiNPs are used to lower the toxicity of reactive dyes (Pawar et al., 2011), as antimicrobial and pharmaceutical dosage forms (Ajmal et al., 2019), in wastewater treatment and cosmetics (Ziental et al., 2020). TiNP synthesis is largely carried out using biological methods and has attracted scientific attention recently because of its low cost, environmental friendliness and reproducibility. Microbes such as bacteria, fungi, algae and plants are widely used for the biosynthesis of TiNPs. A number of non-Apocynaceae plants have also been reported for the biosynthesis of TiNPs. The extract of *Eclipta prostrata* is used to produce TiNPs with spherical shapes and sizes ranging from 36 nm to 68 nm (Nakayama et al., 2020). Subhashini and Nachiyar (2017) studied TiNPs using the leaf extract of *Albizia saman* as well as by using the extract of *Curcuma longa*. TiNPs with various shapes and sizes have been synthesized by a number of green synthesis methods. TiNPs have been characterized using a number of spectroscopic and microscopic techniques, including XRD, FTIR, SEM and EDX, which reveal the structure and particle size of the TiNPs. Table 14.6 provides a detailed outlook of the plant-assisted synthesis of TiNPs.

**Table 14.5** Different precursors, plants, plant parts, and the ratio of extract and metal salts used for the biosynthesis of ZnO nanoparticles

| Sr. no. | Precursor metal | Precursor concentration (mM) | Plant used | Plant part used | Size (nm) | Application | Reference |
|---|---|---|---|---|---|---|---|
| 1. | ZnO | -- | *Acalypha Indica* | Leaves | 100–200 | Antibacterial and antifungal | Jamdagni et al. (2018) |
| 2. | ZnO | 100 | *Aloe vera* | Leaves | 22.18 | Environmental applications | Varghese and George (2015) |
| 3. | ZnO | 100 | *Artemisia pallens* | Leaves along with stem | 50–100 | Studying glucosemetabolism | Gomathi and Suhana (2021) |
| 4. | ZnO | 10 | *Artocarpus gomezianus* | Fruit | 50 | Studying metabolism of intracellular substances | Oudhia et al. (2015) |
| 5. | ZnO | 100 | *Azadirachta indica* | Leaves | 50 | Antioxidant activity. | Noorjahan et al. (2015) |
| 6. | ZnO | 20 | *Azadirachta indica* | Leaves | 25 | antifungal, antidiabetic, antibacterial, | Oudhia et al. (2015) |
| 7. | ZnO | 20 | *Beta vulgaris* | Root | 52–76 | Studying glucose metabolism | Kumar, Singh, and Pandey (2015), Kumar, Suresh, Nagabhushana, et al. (2015) |
| 8. | ZnO | 100 | *Brassica oleraceae* | Leaves | 1–100 | Studying glucose metabolism | Oudhia et al. (2015) |
| 9. | ZnO | 100 | *Calotropis gigantea* | Leaves | 1.5–8.5 | Studying glucosemetabolism | Choudhary et al. (2017) |
| 10. | ZnO | 10 | *Catharanthus roseus* | Leaves | 23–57 | Studying glucosemetabolism | Bhumi and Savithramma (2014) |
| 11. | ZnO | 20 | *Cayratia pedata* | Leaves | 52.24 | Studying glucose metabolism | Jayachandran et al. (2021) |
| 12. | ZnO | 10 | *Coptidis rhizome* | Rhizome | 2.9–25.2 | Antibacterial and antifungal | Oudhia et al. (2015) |

| | | | | | | | |
|---|---|---|---|---|---|---|---|
| 13. | ZnO | 1 | *Curcuma longa* | Root | 20–80 | Antioxidant activity | Raut et al. (2015) |
| 14. | ZnO | 100 | *Eucalyptus globules* | Leaves | 52–70 | Antifungal, antidiabetic antibacterial | Ahmad et al. (2020) |
| 15. | ZnO | 1 | Euphorbia *hirta* | Leaves | 20–50 | Antioxidant activity | Ahmad Kalra (2020) |
| 16. | ZnO | 100 | Ginger | Rhizome | 23–26 | Antihelminths | Awwad et al. (2014) |
| 17. | ZnO | 100 | *Hibiscus rosa-sinensis* | Leaves | 30–35 | Studying metabolism of intracellular substances | Devi and Gayathri (2014) |
| 18. | ZnO | 100 | *Nephelium lappaceum* | Peel | 20 | Antifungal, antidiabetic, antibacterial | Azizi et al. (2014) |
| 19. | ZnO | 1 | *Nyctanthes arbor-tristis* | Flowers | 12–32 | Studying metabolism of intracellular substances | Jamdagni et al. (2018) |
| 20. | ZnO | 100 | *Ocimum tenuiflorum* | Leaves | 11–25 | Environmental applications and global warming | Raut et al. (2015) |
| 21. | ZnO | 100 | *Olea europea* | Leaves | 18–30 | Antioxidant activity | Awwad et al. (2014) |
| 22. | ZnO | 200 | *Plectranthus amboinicus* | Leaves | 20–50 | Antihelminths | Devi and Gayathri (2014) |
| 23. | ZnO | 100 | *Ruta graveolens* | Stem | 28 | Studying microbial denitrification and global warming | Lingaraju et al. (2016) |
| 24. | ZnO | – | *Sargassum muticum* | Leaves | 30–57 | To create strategies for reducing the effects of ZnO NPs on microbial denitrification | Azizi et al. (2014) |
| 25. | ZnO | 1 | *Senna auriculata* | Leaves | 2 | Studying metabolism of intracellular substances | Jamdagni et al. (2018) |
| 26. | ZnO | 1 | *Solanum nigrum* | Leaves | 20–30 | Studying glucose metabolism | Ramesh et al. (2015) |

(continued)

**Table 14.5** (continued)

| Sr. no. | Precursor metal | Precursor concentration (mM) | Plant used | Plant part used | Size (nm) | Application | Reference |
|---|---|---|---|---|---|---|---|
| 27. | ZnO | 10 | *Tecoma castanifolia* | Leaves | 70–75 | Antifungal, antidiabetic, antibacterial | Sharmila et al. (2019) |
| 28. | ZnO | 1 | *Zingiber officinale* | Root | 30–50 | Antioxidant activity | Raj and Jayalakshmy (2015) |

**Table 14.6** Different precursors, plants, plant parts, and the ratio of extract and metal salts used for the biosynthesis $TiO_2$ nanoparticles

| Sr. no. | Precursor metal | Precursor concentration (mM) | Plant used | Plant part used | Size (nm) | Application | Reference |
|---|---|---|---|---|---|---|---|
| 1. | $TiO_2$ | 1.5 mM | *Annona squamosal* | Leaves | 43–45 | To kill breast cancer cells | Prasad et al. (2017) |
| 2. | $TiO_2$ | 1.5 mM | *Azadirachta indica* | Leaves | 15–50 | Plastics, anticancer | Wei et al. (2017) |
| 3. | $TiO_2$ | 0.80 mM | *Azadirachta indica* | Leaves | 124 | Antimicrobial | Harshiny et al. (2015) |
| 4. | $TiO_2$ | 1 mM | *Calotropis gigantea* | Flower | 10–52 | Antifungal | Ranjith et al. (2018) |
| 5. | $TiO_2$ | 0.5 mM | *Cassia auriculata* | Leaves | 38 | Antibacterial | Yi et al. (2019) |
| 6. | $TiO_2$ | 0.5 mM | *Citrus sinensis* | Fruit peel | 19 | Antifungal | Wang et al. (2014) |
| 7. | $TiO_2$ | 1.5 mM | *Euphorbia prostrata* | Leaves | 81–84 | Antimicrobial | Ranjith et al. (2018) |
| 8. | $TiO_2$ | 2.5 mM | *Jatropha curcas* | Latex | 25–50 | Textile | Yi et al. (2019) |
| 9. | $TiO_2$ | 2.5 mM | *Jatropha curcas, citrus aurantium* | Leaves | 25–50 | Plastics, anticancer | Gottimukkala et al. (2017) |
| 10. | $TiO_2$ | 1 mM | *Ledebouria revoluta* | Bulb | 47 | Toothpaste, coatings, | Prasad et al. (2017) |
| 11. | $TiO_2$ | 0.80 mM | *Mentha arvensis* | Leaves | 20–70 | Cosmetics | Huang et al. (2014) |
| 12. | $TiO_2$ | 2.5 mM | *Nyctanthes arbor-tristis* | Leaves | 100–150, 100–200 | Antimicrobial | Yi et al. (2019) |
| 13. | $TiO_2$ | 1.8 mM | *Ocimum basilicum* | Leaves | 50 | Antimicrobial | Ranjith et al. (2018) |
| 14. | $TiO_2$ | 0.5 mM | *Pouteria campechiana* | Leaves | 73–140 | Pharmaceuticals papers inks | Gottimukkala et al. (2017) |
| 15. | $TiO_2$ | 2.0 mM | *Psidium guajava* | Leaves | 32.58 | Textile | Önal et al. (2017) |
| 16. | $TiO_2$ | 0.5 mM | *Salvia officinalis* | Leaves | 15–20 | Antibacterial | Wang et al. (2014) |
| 17. | $TiO_2$ | 0.55 mM | *Solanum trilobatum* | Leaves | 70 | Antimicrobial | Kanagasubbulakshmi and Kadirvelu (2017) |
| 18. | $TiO_2$ | 0.55 mM | *Syzygium cumini* | Leaves | 22 | Food products | Devi et al. (2019) |

## Practical Approaches for the Biosynthesis of Metal Nanoparticles Using Apocynaceae Plants

Generally, the aqueous extract of part of the selected plant is used for the biosynthesis of metal nanoparticles. The plant parts include leaves, roots, stems, rhizomes, tubers, etc., of the selected plant were chopped into thin slices and dried at room temperature in the dark. The dried material was ground to a fine powder using a grinder and sieved to obtain uniformly sized particles. A specific amount of powdered material was boiled in water, and the extract was filtered. The rich plant metabolites can be used for further experimentation as well biosynthesis studies. The extract obtained was also subjected to phytochemical analysis. The extracts were analysed for the presence of various phytochemicals, such as carbohydrates, amino acids, steroids, flavonoids, and alkaloids (Pawar, 2021).

The process of phytosynthesis of metal nanoparticles is summarized in Fig. 14.3. The plant extract is used as a reducing agent for the biosynthesis of metal nanoparticles. In brief, the aqueous extract of selected plants (usually 2–10% w/v) is added to the solution of selected metal salt (for example, 1 mM $AgNO_3$ can be diluted to 0.1 mM, 1 mM, 2 mM, 5 mM, 10 mM and 20 mM) in an Erlenmeyer flask (Pawar & Patil, 2020). The reaction mixture was kept at room temperature for 5 h under static conditions. The bioreduction of silver nitrate can be preliminarily evaluated through visual observation. The color of the reaction mixture changes from colorless to brown. The reaction can also be monitored by scanning UV absorbance (200–800 nm) at 1 h intervals using a UV–visible spectrophotometer. After incubation for 5–6 h, the reaction mixture is centrifuged to settle the biosynthesized nanoparticles at the bottom of the centrifugation tube. The pellet is redispersed in double-distilled water, and recentrifugation is carried out for better separation of nanoparticles. The collected pellets of biosynthesized AgNPs were dried and powdered samples of AgNPs were used for further experimentation (Pawar & Patil, 2020).

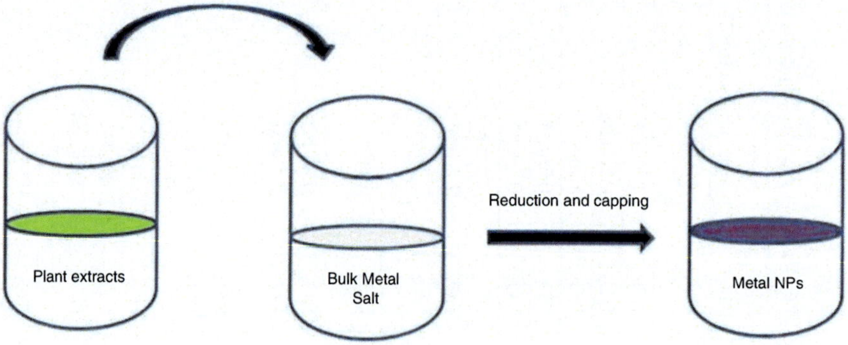


**Fig. 14.3** General procedure of biosynthesis of metal nanoparticles using plant extracts

Plants or plant parts can be effectively used as reducing and capping agents in the biosynthesis of metal nanoparticles. The phytochemicals present in the tuber extract played a significant role in the bioreduction and stabilization of the synthesized MNPs. A number of plants of Apocynaceae have been explored for the biosynthesis of metal and metal oxide nanoparticles. Some important members of this family, viz. *C. roseus, C. procera, C. giganta, A. scholaris, R. serpentina, H. antidysentriaca, N. indicum, C. carandas, Cassia auriculata, Wrightia tinctoria, Cynanchum sarcomedium* and T. peruviana have been widely used for the biosynthesis of metal and metal oxide nanoparticles. Table 14.7 highlights Apocynaceae members used for the biosynthesis of various metal and metal oxide nanoparticles.

## Characterization of Biosynthesized MNPs

The characterization is important to ensure that the prepared particles are at the nanoscale. In material science, "characterization" includes the structure and properties of the material to be explored through general and broad processes. The characterization of biosynthesized NPs can be carried out using various spectroscopic and microscopic techniques. The spectroscopic method involved UV–Visible and FT-IR spectroscopy, microscopy-based nanomaterial characterization using SEM, and EDX. X-ray-related characterization techniques such as X-ray diffraction analysis (XRD), dynamic light scattering (DLS), zeta potential, etc., were also carried out to determine the shape and size of the NPs (Titus & Samuel, 2019). The details of characterization are mentioned in the following paragraphs. The characterization techniques used for biosynthesized metal nanoparticles are highlighted in Fig. 14.4.

### *UV–Visible Spectrophotometer*

A UV–vis visible spectrophotometer with a wavelength range of 200–1100 nm consisting of a deuterium and tungsten lamp as a light source with 1 nm resolution is used for measuring the absorption. The matching pair of quartz cuvettes with a path length of 1 cm can be used for analysis, and the data are processed using the software. The reaction mixture formed in the biosynthesis of metal nanoparticles is the combination of metal salts, and plant extracts can be directly scanned in the UV–vis visible range of 200–800 nm. The UV–vis visible absorbance spectra of various biosynthesized MNPs (AgNPs, CuNPs and FeNPs) using plant extracts were recorded and compared with control samples of the respective metal salts, and the test extracts were scanned in the same range. The reaction mixtures are monitored periodically.

**Table 14.7** Apocynaceae members used for biosynthesis of nanoparticles

| Sr. no. | Precursor metal | Precursor concentration (mM) | Plant used | Plant part used | Size (nm) | Application | Reference |
|---|---|---|---|---|---|---|---|
| 1. | $AgNO_3$ | 1.5 mM | Calotropis gigantea. | Leaves | 43–45 | Antibacterial activity against Gram-positive bacteria. | Mathew et al. (2020) |
| 2. | Chitosan | 1.5 mM | Catharanthus roseus | Leaves | 15–50 | Protection against salinity stress. | Saxena et al. (2021) |
| 3. | $TiO_2$ | 1 mM | Calotropis gigantea | Flower | 10–52 | Antifungal | Ranjith et al. (2018) |
| 4. | $TiO_2$ | 0.5 mM | Cassia auriculata | Leaves | 38 | Antibacterial | Yi et al. (2019) |
| 5. | Pd | 0.5 mM | Allium fistulosum | Leaves | 19 | Antioxidant, antifungal and antibacterial activity. | Vinodhini et al. (2022) |
| 6. | Pd | 1.5 mM | Basella alba | Leaves | 81–84 | Antioxidant, antifungal and antibacterial activity. | Vinodhini et al. (2022) |
| 7. | Pd | 2.5 mM | Tabernaemontana divaricate | Leaves | 25–50 | Antioxidant, antifungal and antibacterial activity. | Vinodhini et al. (2022) |
| 8. | $ZrO_2$ | 2.5 mM | Wrightia tinctoria | Leaves | 25–50 | Gram-positive and Gram-negative bacteria. | Naeimi et al. (2021) |
| 9. | Ni | 1 mM | Calotropis procera | wood | 47 | Role in catalytic performance, oxidation | Naeimi et al. (2021) |
| 10. | NiO | 0.80 mM | Calotropis procera | wood | 20–70 | Role in catalytic performance, oxidation | Naeimi et al. (2021) |
| 11. | ZnO | 2.5 mM | Carissa spinarum | Leaves | 100–150, 100–200 | Antimicrobial against Bacillus and Staphylococcus. | Saka et al. (2022) |
| 12. | $AgNO_3$ | 1 mM | Cynanchum viminale. | Leaves | 50 | Biomedical applications | Bhagyanathan and Thoppil (2018) |
| 13. | $AgNO_3$ | 1 mM | Cynanchum sarcomedium | Leaves | 73–140 | Biomedical applications | Bhagyanathan and Thoppil (2018) |

| | | | | | | | |
|---|---|---|---|---|---|---|---|
| 14. | $TiO_2$ | 2.0 mM | Catharanthus roseus | Leaves | 32.58 | Biomedical applications | Bhagyanathan and Thoppil (2018) |
| 15. | $TiO_2$ | 0.5 mM | Salvia officinalis | Leaves | 15–20 | Antibacterial | Wang et al. (2014) |
| 16. | $TiO_2$ | 0.55 mM | Catharanthus roseus | Leaves | 70 | Antimicrobial | Kanagasubbulakshmi and Kadirvelu (2017) |
| 17. | $TiO_2$ | 0.55 mM | Syzygium cumini | Leaves | 22 | Food products | Devi et al. (2019) |

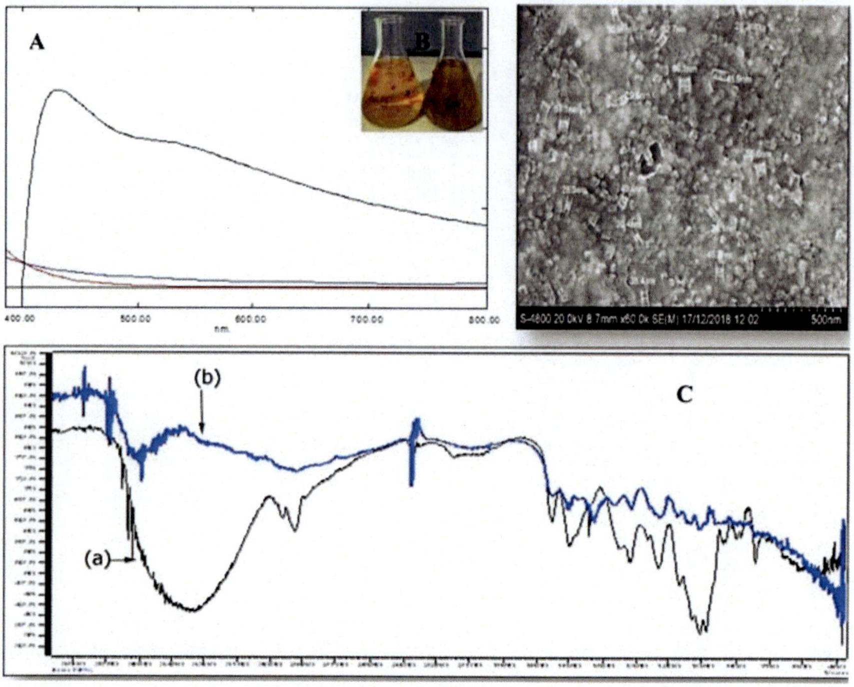


**Fig. 14.4** Characterization techniques used for biosynthesized metal nanoparticles

## *Fourier Transform Infrared Spectroscopy (FTIR)*

FT-IR spectroscopy was performed to determine the presence or absence of functional groups using an FT-IR spectrometer. The diffuse reflectance scan (DRS) sampling method can be used in the 4000–400 $cm^{-1}$ range. To analyse the sample by the DRS method, a solid dispersion of the sample can be prepared with KBr (IR spectroscopy grade) (Pawar & Patil, 2020).

The biosynthesized MNPs obtained can be centrifuged at 10,000 rpm for 20 min at room temperature, and the pellet of each MNP is collected. The process of centrifugation and redispersion of the collected pellet in water is repeated three times to ensure better separation of nanoparticles from other phytochemical adsorbed material. The pellet of each MNP was dried, and a powder sample of each MNP was used for further experimentation (Ghosh et al., 2012).

The dried powder sample of each MNP was mixed with KBr (IR spectroscopy grade) at a proportion of 1:100 using a small mortar and pestle to obtain a small particle size distribution. The solid dispersion of each MNP in KBr was used as the sample for FT-IR spectroscopy. The mixture is then placed in a sample cup 3–4 mm deep and approximately 10–15 mm in diameter to record the FT-IR spectra of each sample. For FT-IR spectroscopy of the plant extract, 1 ml of tuber extract was dried to obtain a powder. The plant tuber extract powder was mixed with KBr at a proportion of 1:100. The sample was ground and mixed uniformly using a small mortar and pestle to obtain a small particle size distribution (Pawar, 2021).

## *Scanning Electron Microscopy (SEM)*

A scanning electron microscope with a resolution of 3.5 nm and an accelerating voltage in the range from 0.5 kV to 30 kV is commonly used to study the surface morphology and size of biosynthesized MNPs. The JSM-5600 is a conventional SEM with a tungsten filament electron source (Pawar & Patil, 2020).

Conductive tape (double-sided carbon tape) is affixed on a polished aluminum surface (Pawar & Patil, 2020). Samples of biosynthesized MNPs are placed at the center of the carbon tape individually while performing the analysis. The pressure is applied, and excess free particles are removed by blowing. The samples were kept in a vacuum chamber for one h and then loaded on the SEM platform (Pawar & Patil, 2020). It is equipped with a secondary electron detector for topographic contrast imaging and an Oxford Inca EDX system for compositional analysis (Pawar, 2021).

## *Energy Dispersive X-ray Analysis (EDX)*

EDX is used in conjunction with SEM. This technique gives an overall mapping of the sample by analysing near-surface elements and estimates the elemental proportion at different positions (Titus & Samuel, 2019).

An electron beam with an energy of 10–20 keV strikes the conducting sample's surface, causing X-rays to emit from the material, and the energy of the emitted X-rays depends on the material under examination (Titus & Samuel, 2019).

## *X-ray Diffraction (XRD)*

The degree of crystallinity, element identification and composition, and structural state can be derived from XRD data analysis (Titus & Samuel, 2019).

The MNP sample was prepared by spreading the sample on a microscopic glass slide and exposed to diffraction at an operating voltage of 40 kV, a current of 30 mA and a scan rate of 10°/min with Cu-K$\alpha$ wavelength radiation in a 2$\theta$ configuration from 20–80°. The software modules for easy X-ray powder diffraction data acquisition and evaluation can be used for interpretation of the data.

### *Dynamic Light Scattering (DLS)*

Light scattered from a laser, which travels through the colloidal solution, is measured by DLS. Modulation of the intensity of the scattered light is analysed as a function of time, and the particle size can be derived based on this information (Titus & Samuel, 2019).

### *Zeta Potential*

Zeta potential is a measure of the effective electric charge on the nanoparticle surface and quantifies the charge stability of colloidal particles (Titus & Samuel, 2019). Biosynthesized MNPs using plant extracts are dispersed in double-distilled water. The instrument software then automatically determines the appropriate electric field strength, adjusts the reference beam intensity to ensure the optimal signal-to-noise ratio, and collects and analyses the data.

## Conclusion

Plants of the Apocynaceae family can be effectively used as reducing and capping agents in the biosynthesis of metal nanoparticles. The exploration of plant capability to reduce metal salt to zero-valent metal depends on many experimental parameters. The biosynthesis of metal nanoparticles using these plants is simple, cost-effective and eco-friendly. Biosynthesis of AgNPs, CuNPs, ZnONPs, AuNPs and FeNPs using a number of plants of this family is well studied. The phytochemicals present in the plant extract played a significant role in the bioreduction and stabilization of the synthesized MNPs. The synthesized MNPs were characterized by UV–VIS, FT-IR, SEM, EDX and X-ray diffraction spectroscopy. Biosynthesized metal nanoparticles have a wide range of biological activities; therefore, Apocynaceae plants can be considered valuable resources for future studies in nanobiotechnology.

## References

Abdel-Fattah, W. I., & Ali, G. W. (2018). On the anti-cancer activities of silver nanoparticles. *Journal of Applied Biotechnology & Bioengineering, 5*(2), 1–4.

Abdullah, J. A., Eddine, L. S., Abderrhmane, B., Alonso-González, M., Guerrero, A., & Romero, A. (2020). Green synthesis and characterization of iron oxide nanoparticles by pheonix dactylifera leaf extract and evaluation of their antioxidant activity. *Sustainable Chemistry and Pharmacy, 17*, 100280.

Aher, Y. B., Jain, G. H., Patil, G. E., Savale, A. R., Ghotekar, S. K., Pore, D. M., Pansambal, S. S., & Deshmukh, K. K. (2017). Biosynthesis of copper oxide nanoparticles using leaves extract of Leucaena leucocephala L. and their promising upshot against diverse pathogens. *International Journal of Molecular and Clinical Microbiology, 7*(1), 776–786.

Ahmad, H., Venugopal, K., Rajagopal, K., De Britto, S., Nandini, B., Pushpalatha, H. G., Konappa, N., Udayashankar, A. C., Geetha, N., & Jogaiah, S. (2020). Green synthesis and characterization of zinc oxide nanoparticles using Eucalyptus globules and their fungicidal ability against pathogenic fungi of apple orchards. *Biomolecules, 10*(3), 425.

Ahmad, N., Sharma, S., Alam, M. K., Singh, V. N., Shamsi, S. F., Mehta, B. R., & Fatma, A. (2010). Rapid synthesis of silver nanoparticles using dried medicinal plant of basil. *Colloids and Surfaces B: Biointerfaces, 81*(1), 81–86.

Ahmad, W., & Kalra, D. (2020). Green synthesis, characterization and anti microbial activities of ZnO nanoparticles using Euphorbia hirta leaf extract. *Journal of King Saud University-Science, 32*(4), 2358–2364.

Ahmed, S., Ahmad, M., Swami, B. L., & Ikram, S. (2016). A review on plants extract mediated synthesis of silver nanoparticles for antimicrobial applications: A green expertise. *Journal of Advanced Research, 7*(1), 17–28.

Ajmal, N., Saraswat, K., Bakht, M. A., Riadi, Y., Ahsan, M. J., & Noushad, M. (2019). Cost-effective and eco-friendly synthesis of titanium dioxide ($TiO_2$) nanoparticles using fruit's peel agro-waste extracts: characterization, in vitro antibacterial, antioxidant activities. *Green Chemistry Letters and Reviews, 12*(3), 244–254.

Akhtar, M. S., Panwar, J., & Yun, Y. S. (2013). Biogenic synthesis of metallic nanoparticles by plant extracts. *ACS Sustainable Chemistry & Engineering, 1*(6), 591–602.

Alam, M. N., Roy, N., Mandal, D., & Begum, N. A. (2013). Green chemistry for nanochemistry: exploring medicinal plants for the biogenic synthesis of metal NPs with fine-tuned properties. *RSC Advances, 3*(30), 11935–11956.

Alsammarraie, F. K., Wang, W., Zhou, P., Mustapha, A., & Lin, M. (2018). Green synthesis of silver nanoparticles using turmeric extracts and investigation of their antibacterial activities. *Colloids and Surfaces B: Biointerfaces, 171*, 398–405.

Amer, M., & Awwad, A. (2020). Green synthesis of copper nanoparticles by Citrus limon fruits extract, characterization and antibacterial activity. *Chemistry International, 7*(1), 1–8.

Amin, M., Anwar, F., Janjua, M. R., Iqbal, M. A., & Rashid, U. (2012). Green synthesis of silver nanoparticles through reduction with Solanum xanthocarpum L. berry extract: characterization, antimicrobial and urease inhibitory activities against Helicobacter pylori. *International Journal of Molecular Sciences, 13*(8), 9923–9941.

Angajala, G., Pavan, P., & Subashini, R. (2014). One-step biofabrication of copper nanoparticles from Aegle marmelos correa aqueous leaf extract and evaluation of its anti-inflammatory and mosquito larvicidal efficacy. *RSC Advances, 4*(93), 51459–51470.

Ankanna, S. T., Prasad, T. N., Elumalai, E. K., & Savithramma, N. (2010). Production of biogenic silver nanoparticles using Boswellia ovalifoliolata stem bark. *Digest Journal of Nanomaterials and Biostructures, 5*(2), 369–372.

Annamalai, A., Babu, S. T., Jose, N. A., Sudha, D., & Lyza, C. V. (2011). Biosynthesis and characterization of silver and gold nanoparticles using aqueous leaf extraction of Phyllanthus amarus Schum & Thonn. *World Applied Sciences Journal, 13*(8), 1833–1840.

Asghar, M. A., Zahir, E., Shahid, S. M., Khan, M. N., Asghar, M. A., Iqbal, J., & Walker, G. (2018). Iron, copper and silver nanoparticles: Green synthesis using green and black tea leaves extracts and evaluation of antibacterial, antifungal and aflatoxin B1 adsorption activity. *Lwt, 90*, 98–107.

Aslan, K., Zhang, J., Lakowicz, J. R., & Geddes, C. D. (2004). Saccharide sensing using gold and silver nanoparticles: A review. *Journal of Fluorescence, 14*, 391–400.

Awwad, M., & A, Albiss B, L Ahmad A. (2014). Green synthesis, characterization and optical properties of zinc oxide nanosheets using Olea europea leaf extract. Advanced. *Materials Letters, 5*(9), 520–524.

Azizi, S., Ahmad, M. B., Namvar, F., & Mohamad, R. (2014). Green biosynthesis and characterization of zinc oxide nanoparticles using brown marine macroalga Sargassum muticum aqueous extract. *Materials Letters, 116*, 275–277.

Baker, S., Rakshith, D., Kavitha, K. S., Santosh, P., Kavitha, H. U., Rao, Y., & Satish, S. (2013). Plants: Emerging as nanofactories towards facile route in synthesis of nanoparticles. *BioImpacts: BI, 3*(3), 111.

Balamurugan, M., Saravanan, S., & Soga, T. (2014). Synthesis of iron oxide nanoparticles by using Eucalyptus globulus plant extract. *e-Journal of Surface Science and Nanotechnology, 12*, 363–367.

Bar, H., Bhui, D. K., Sahoo, G. P., Sarkar, P., De, S. P., & Misra, A. (2009). Green synthesis of silver nanoparticles using latex of Jatropha curcas. *Colloids and Surfaces A: Physicochemical and Engineering Aspects, 339*(1–3), 134–139.

Behera, S. S., Patra, J. K., Pramanik, K., Panda, N., & Thatoi, H. (2012). Characterization and evaluation of antibacterial activities of chemically synthesized iron oxide nanoparticles. *World Journal of Nano Science and Engineering, 2*(4), 196–200. https://doi.org/10.4236/wjnse.2012.24026

Beheshtkhoo, N., Kouhbanani, M. A., Savardashtaki, A., Amani, A. M., & Taghizadeh, S. (2018). Green synthesis of iron oxide nanoparticles by aqueous leaf extract of Daphne mezereum as a novel dye removing material. *Applied Physics A, 124*, 1–7.

Benassai, E., Del Bubba, M., Ancillotti, C., Colzi, I., Gonnelli, C., Calisi, N., Salvatici, M. C., Casalone, E., & Ristori, S. (2021). Green and cost-effective synthesis of copper nanoparticles by extracts of nonedible and waste plant materials from Vaccinium species: Characterization and antimicrobial activity. *Materials Science and Engineering: C, 119*, 111453.

Berra, D., Laouini, S. E., Benhaoua, B., Ouahrani, M. R., Berrani, D., & Rahal, A. (2018). Green synthesis of copper oxide nanoparticles by Pheonix dactylifera L leaves extract. *Digest Journal of Nanomaterials and Biostructures, 13*(4), 1231–1238.

Bhagyanathan, N. K., & Thoppil, J. E. (2018). Plant-mediated synthesis of Silver nanoparticles by two species of Cynanchum L.(Apocynaceae): A comparative approach on its physical characteristics. *International Journal of Nano Dimension, 9*(2), 104–111.

Bhakya, S., Muthukrishnan, S., Sukumaran, M., Muthukumar, M., Kumar, S. T., & Rao, M. V. (2015). Catalytic degradation of organic dyes using synthesized silver nanoparticles: a green approach. *Journal of Bioremediation & Biodegredation, 6*(5), 1.

Bhuiyan, M. S., Miah, M. Y., Paul, S. C., Aka, T. D., Saha, O., Rahaman, M. M., Sharif, M. J., Habiba, O., & Ashaduzzaman, M. (2020). Green synthesis of iron oxide nanoparticle using Carica papaya leaf extract: application for photocatalytic degradation of remazol yellow RR dye and antibacterial activity. *Heliyon, 6*(8), e04603.

Bhumi, G., & Savithramma, N. (2014). Biological synthesis of zinc oxide nanoparticles from Catharanthus roseus (L.) G. Don. Leaf extract and validation for antibacterial activity. *International Journal of Drug Development and Research, 6*(1), 208–214.

Bhuyan, T., Mishra, K., Khanuja, M., Prasad, R., & Varma, A. (2015). Biosynthesis of zinc oxide nanoparticles from Azadirachta indica for antibacterial and photocatalytic applications. *Materials Science in Semiconductor Processing, 32*, 55–61.

Bibi, I., Nazar, N., Ata, S., Sultan, M., Ali, A., Abbas, A., Jilani, K., Kamal, S., Sarim, F. M., Khan, M. I., & Jalal, F. (2019). Green synthesis of iron oxide nanoparticles using pomegranate seeds extract and photocatalytic activity evaluation for the degradation of textile dye. *Journal of Materials Research and Technology, 8*(6), 6115–6124.

Biswal, S. K., Panigrahi, G. K., & Sahoo, S. K. (2020). Green synthesis of $Fe_2O_3$-Ag nanocomposite using Psidium guajava leaf extract: An eco-friendly and recyclable adsorbent for remediation of Cr (VI) from aqueous media. *Biophysical Chemistry, 263*, 106392.

Bordbar, M., Sharifi-Zarchi, Z., & Khodadadi, B. (2017). Green synthesis of copper oxide nanoparticles/clinoptilolite using Rheum palmatum L. root extract: high catalytic activity for reduction of 4-nitro phenol, rhodamine B, and methylene blue. *Journal of Sol-Gel Science and Technology, 81*, 724–733.

Bunghez, I. R., Barbinta Patrascu, M. E., Badea, N. M., Doncea, S. M., Popescu, A., & Ion, R. M. (2012). Antioxidant silver nanoparticles green synthesized using ornamental plants. *Journal of Optoelectronics and Advanced Materials, 14*(11), 1016.

Chandran, S. P., Chaudhary, M., Pasricha, R., Ahmad, A., & Sastry, M. (2006). Synthesis of gold nanotriangles and silver nanoparticles using Aloevera plant extract. *Biotechnology Progress, 22*(2), 577–583.

Choudhary, R. C., Kumaraswamy, R. V., Kumari, S., Pal, A., Raliya, R., Biswas, P., & Saharan, V. (2017). Synthesis, characterization, and application of chitosan nanomaterials loaded with zinc and copper for plant growth and protection. In *Nanotechnology: An agricultural paradigm* (pp. 227–247).

Chung, I. M., Abdul Rahuman, A., Marimuthu, S., Vishnu Kirthi, A., Anbarasan, K., Padmini, P., & Rajakumar, G. (2017). Green synthesis of copper nanoparticles using Eclipta prostrata leaves extract and their antioxidant and cytotoxic activities. *Experimental and Therapeutic Medicine, 14*(1), 18–24.

Clarina, T., Flomina, P. J., Thangeswari, P., & Rama, V. (2018). Polpala flower extract mediated one step green synthesis and characterization of magnetite ($Fe_3O_4$) nanoparticles. *Asian Journal of Research in Chemistry, 11*(2), 459–462.

Dai, J., & Bruening, M. L. (2002). Catalytic nanoparticles formed by reduction of metal ions in multilayered polyelectrolyte films. *Nano Letters, 2*(5), 497–501.

Dakal, T. C., Kumar, A., Majumdar, R. S., & Yadav, V. (2016). Mechanistic basis of antimicrobial actions of silver nanoparticles. *Frontiers in Microbiology, 7*, 1831.

Das, D., Nath, B. C., Phukon, P., & Dolui, S. K. (2013). Synthesis and evaluation of antioxidant and antibacterial behavior of CuO nanoparticles. *Colloids and Surfaces B: Biointerfaces, 101*, 430–433.

Das, P. E., Abu-Yousef, I. A., Majdalawieh, A. F., Narasimhan, S., & Poltronieri, P. (2020). Green synthesis of encapsulated copper nanoparticles using a hydroalcoholic extract of Moringa oleifera leaves and assessment of their antioxidant and antimicrobial activities. *Molecules, 25*(3), 555.

Devatha, C. P., Thalla, A. K., & Katte, S. Y. (2016). Green synthesis of iron nanoparticles using different leaf extracts for treatment of domestic waste water. *Journal of Cleaner Production, 139*, 1425–1435.

Devi, H. S., Boda, M. A., Shah, M. A., Parveen, S., & Wani, A. H. (2019). Green synthesis of iron oxide nanoparticles using Platanus orientalis leaf extract for antifungal activity. *Green Processing and Synthesis, 8*(1), 38–45.

Devi, R. S., & Gayathri, R. (2014). Green synthesis of zinc oxide nanoparticles by using Hibiscus rosa-sinensis. *International Journal of Current Engineering and Technology, 4*(4), 2444–2446.

Din, M. I., & Rehan, R. (2017). Synthesis, characterization, and applications of copper nanoparticles. *Analytical Letters, 50*(1), 50–62.

Długosz, O., Chwastowski, J., & Banach, M. (2020). Hawthorn berries extract for the green synthesis of copper and silver nanoparticles. *Chemical Papers, 74*, 239–252.

Doria, G., Conde, J., Veigas, B., Giestas, L., Almeida, C., Assunção, M., Rosa, J., & Baptista, P. V. (2012). Noble metal nanoparticles for biosensing applications. *Sensors, 12*(2), 1657–1687.

Dubey, S. P., Lahtinen, M., & Sillanpää, M. (2010). Green synthesis and characterizations of silver and gold nanoparticles using leaf extract of Rosa rugosa. *Colloids and Surfaces A: Physicochemical and Engineering Aspects, 364*(1-3), 34–41.

Ebrahiminezhad, A., Zare-Hoseinabadi, A., Berenjian, A., & Ghasemi, Y. (2017). Green synthesis and characterization of zero-valent iron nanoparticles using stinging nettle (Urtica dioica) leaf extract. *Green Processing and Synthesis, 6*(5), 469–475.

El Mitwalli, O. S., Barakat, O. A., Daoud, R. M., Akhtar, S., & Henari, F. Z. (2020). Green synthesis of gold nanoparticles using cinnamon bark extract, characterization, and fluorescence activity in Au/eosin Y assemblies. *Journal of Nanoparticle Research, 22*, 1–9.

Elghanian, R., Storhoff, J. J., Mucic, R. C., Letsinger, R. L., & Mirkin, C. A. (1997). Selective colorimetric detection of polynucleotides based on the distance-dependent optical properties of gold nanoparticles. *Science, 277*(5329), 1078–1081.

El-Kassas, H. Y., Aly Eldeen, M. A., & Gharib, S. M. (2016). Green synthesis of iron oxide (Fe 3° 4) nanoparticles using two selected brown seaweeds: Characterization and application for lead bioremediation. *Acta Oceanologica Sinica, 35*, 89–98.

El-Refai, A. A., Ghoniem, G. A., El-Khateeb, A. Y., & Hassaan, M. M. (2018). Eco-friendly synthesis of metal nanoparticles using ginger and garlic extracts as biocompatible novel antioxidant and antimicrobial agents. *Journal of Nanostructure in Chemistry, 8*, 71–81.

Fahmy, H. M., Mohamed, F. M., Marzouq, M. H., Mustafa, A. B., Alsoudi, A. M., Ali, O. A., Mohamed, M. A., & Mahmoud, F. A. (2018). Review of green methods of iron nanoparticles synthesis and applications. *BioNanoScience, 8*, 491–503.

Fayaz, A. M., Balaji, K., Girilal, M., Yadav, R., Kalaichelvan, P. T., & Venketesan, R. (2010). Biogenic synthesis of silver nanoparticles and their synergistic effect with antibiotics: A study against Gram-positive and Gram-negative bacteria. *Nanomedicine: Nanotechnology, Biology and Medicine, 6*(1), 103–109.

Fazlzadeh, M., Rahmani, K., Zarei, A., Abdoallahzadeh, H., Nasiri, F., & Khosravi, R. (2017). A novel green synthesis of zero valent iron nanoparticles (NZVI) using three plant extracts and their efficient application for removal of Cr (VI) from aqueous solutions. *Advanced Powder Technology, 28*(1), 122–130.

Fendler, J. H., & Tian, Y. (1998). Nanoparticles and nanostructured films: current accomplishments and future prospects. In *Nanoparticles and nanostructured films: Preparation, characterization and applications* (pp. 429–461).

Feng, Q. L., Wu, J., Chen, G. Q., Cui, F. Z., Kim, T. N., & Kim, J. O. (2000). A mechanistic study of the antibacterial effect of silver ions on Escherichia coli and Staphylococcus aureus. *Journal of Biomedical Materials Research, 52*(4), 662–668.

Feynman, R. P. (1959). Plenty of room at the bottom. In *APS annual meeting*

Gahlawat, G., & Choudhury, A. R. (2019). A review on the biosynthesis of metal and metal salt nanoparticles by microbes. *RSC Advances, 9*(23), 12944–12967.

Ganapuram, B. R., Alle, M., Dadigala, R., Dasari, A., Maragoni, V., & Guttena, V. (2015). Catalytic reduction of methylene blue and Congo red dyes using green synthesized gold nanoparticles capped by salmalia malabarica gum. *International Nano Letters, 5*, 215–222.

Ghorbani, H. R. (2014). A review of methods for synthesis of Al nanoparticles. *Oriental Journal of Chemistry, 30*(4), 1941–1949.

Ghosh, M. K., Sahu, S., Gupta, I., & Ghorai, T. K. (2020). Green synthesis of copper nanoparticles from an extract of Jatropha curcas leaves: Characterization, optical properties, CT-DNA binding and photocatalytic activity. *RSC Advances, 10*(37), 22027–22035.

Ghosh, S., Gurav, S. P., Harke, A. N., Chacko, M. J., Joshi, K. A., Dhepe, A., Charolkar, C., Shinde, V. S., Kitture, R., Parihar, V. S., & Banerjee, K. (2016). Dioscorea oppositifolia mediated synthesis of gold and silver nanoparticles with catalytic activity. *Journal of Nanomedicine & Nanotechnology, 7*(05), 2.

Ghosh, S., More, P., Nitnavare, R., Jagtap, S., Chippalkatti, R., Derle, A., Kitture, R., Asok, A., Kale, S., Singh, S., & Shaikh, M. L. (2015). Antidiabetic and antioxidant properties of copper nanoparticles synthesized by medicinal plant Dioscorea bulbifera. *Journal of Nanomedicine & Nanotechnology, S6*, 1.

Ghosh, S., Patil, S., Ahire, M., Kitture, R., Jabgunde, A., Kale, S., Pardesi, K., Bellare, J., Dhavale, D. D., & Chopade, B. A. (2011). Synthesis of gold nanoanisotrops using Dioscorea bulbifera tuber extract. *Journal of Nanomaterials, 2011*, 1–8.

Ghosh, S., Patil, S., Ahire, M., Kitture, R., Kale, S., Pardesi, K., Cameotra, S. S., Bellare, J., Dhavale, D. D., Jabgunde, A., & Chopade, B. A. (2012). Synthesis of silver nanoparticles using Dioscorea bulbifera tuber extract and evaluation of its synergistic potential in combination with antimicrobial agents. *International Journal of Nanomedicine, 7*, 483–496.

Giljohann, D. A., Seferos, D. S., Daniel, W. L., Massich, M. D., Patel, P. C., & Mirkin, C. A. (2010). Gold nanoparticles for biology and medicine. *Angewandte Chemie International Edition, 49*(19), 3280–3294.

Gomathi, R., & Suhana, H. (2021). Green synthesis, characterization and antimicrobial activity of zinc oxide nanoparticles using Artemisia pallens plant extract. *Inorganic and Nano-Metal Chemistry, 51*(12), 1663–1672.

Gondwal, M., & Joshi nee Pant, G. (2018). Synthesis and catalytic and biological activities of silver and copper nanoparticles using Cassia occidentalis. *International Journal of Biomaterials, 2018*, 6735426.

Gottimukkala, K. S., Harika, R. P., & Zamare, D. (2017). Green synthesis of iron nanoparticles using green tea leaves extract. *Journal of Nanomedicine and Biotherapeutic Discovery, 7*, 151.

Gupta, D., & Chauhan, P. (2017). Green synthesis of silver nanoparticles involving extract of plants of different taxonomic groups. *Journal of Nanomedicine Research, 5*(2), 00110.

Harikumar, P. S., & Aravind, A. (2016). Antibacterial activity of copper nanoparticles and copper nanocomposites against Escherichia coli bacteria. *International Journal of Science, 2*, 83–90.

Harshiny, M., Iswarya, C. N., & Matheswaran, M. (2015). Biogenic synthesis of iron nanoparticles using Amaranthus dubius leaf extract as a reducing agent. *Powder Technology, 286*, 744–749.

Hasheminya, S. M., & Dehghannya, J. (2020). Green synthesis and characterization of copper nanoparticles using Eryngium caucasicum Trautv aqueous extracts and its antioxidant and antimicrobial properties. *Particulate Science and Technology, 38*(8), 1019–1026.

Hatchett, D. W., & White, H. S. (1996). Electrochemistry of sulfur adlayers on the low-index faces of silver. *The Journal of Physical Chemistry, 100*(23), 9854–9859.

He, C., Sasaki, T., Shimizu, Y., & Koshizaki, N. (2008). Synthesis of ZnO nanoparticles using nanosecond pulsed laser ablation in aqueous media and their self-assembly towards spindle-like ZnO aggregates. *Applied Surface Science, 254*(7), 2196–2202.

Hemalatha, K., Shantakani, S., & Botcha, S. (2019). Green synthesis of silver nanoparticles using aqueous fruit and tuber extracts of Momordica cymbalaria. *Journal of Plant Biochemistry and Biotechnology, 30*, 1–9.

Huang, J., Li, Q., Sun, D., Lu, Y., Su, Y., Yang, X., Wang, H., Wang, Y., Shao, W., He, N., & Hong, J. (2007). Biosynthesis of silver and gold nanoparticles by novel sundried *Cinnamomum camphora* leaf. *Nanotechnology, 18*(10), 105.

Huang, J., Zhan, G., Zheng, B., Sun, D., Lu, F., Lin, Y., Chen, H., Zheng, Z., Zheng, Y., & Li, Q. (2011). Biogenic silver nanoparticles by Cacumen platycladi extract: synthesis, formation mechanism, and antibacterial activity. *Industrial & Engineering Chemistry Research, 50*(15), 9095–9106.

Huang, L., Weng, X., Chen, Z., Megharaj, M., & Naidu, R. (2014). Green synthesis of iron nanoparticles by various tea extracts: comparative study of the reactivity. *Spectrochimica Acta Part A: Molecular and Biomolecular Spectroscopy, 130*, 295–301.

Husen, A., & Siddiqi, K. S. (2014). Phytosynthesis of nanoparticles: concept, controversy and application. *Nanoscale Research Letters, 9*, 1–24.

Ibrahim, H. M. (2015). Green synthesis and characterization of silver nanoparticles using banana peel extract and their antimicrobial activity against representative microorganisms. *Journal of Radiation Research and Applied Science, 8*(3), 265–275.

Ijaz, F., Shahid, S., Khan, S. A., Ahmad, W., & Zaman, S. (2017). Green synthesis of copper oxide nanoparticles using Abutilon indicum leaf extract: Antimicrobial, antioxidant and photocatalytic dye degradation activitie. *Tropical Journal of Pharmaceutical Research, 16*(4), 743–753.

Iliger, K. S., Sofi, T. A., Bhat, N. A., Ahanger, F. A., Sekhar, J. C., Elhendi, A. Z., Al-Huqail, A. A., & Khan, F. (2021). Copper nanoparticles: Green synthesis and managing fruit rot disease of chilli caused by Colletotrichum capsici. *Saudi Journal of Biological Sciences, 28*(2), 1477–1486.

Imran Din, M., & Rani, A. (2016). Recent advances in the synthesis and stabilization of nickel and nickel oxide nanoparticles: A green adeptness. *International Journal of Analytical Chemistry, 2016*, 3512145.

Indana, M. K., Gangapuram, B. R., Dadigala, R., Bandi, R., & Guttena, V. (2016). A novel green synthesis and characterization of silver nanoparticles using gum tragacanth and evaluation of their potential catalytic reduction activities with methylene blue and Congo red dyes. *Journal of Analytical Science and Technology, 7*, 1–9.

Iravani, S. (2011). Green synthesis of metal nanoparticles using plants. *Green Chemistry, 13*(10), 2638–2650.

Iravani, S., Korbekandi, H., Mirmohammadi, S. V., & Zolfaghari, B. (2014). Synthesis of silver nanoparticles: Chemical, physical and biological methods. *Research in Pharmaceutical Sciences, 9*(6), 385.

Ismail, M., Gul, S., Khan, M. I., Khan, M. A., Asiri, A. M., & Khan, S. B. (2019). Green synthesis of zerovalent copper nanoparticles for efficient reduction of toxic azo dyes congo red and methyl orange. *Green Processing and Synthesis, 8*(1), 135–143.

Issaabadi, Z., Nasrollahzadeh, M., & Sajadi, S. M. (2017). Green synthesis of the copper nanoparticles supported on bentonite and investigation of its catalytic activity. *Journal of Cleaner Production, 142*, 3584–3591.

Jahan, I., Erci, F., & Isildak, I. (2021). Facile microwave-mediated green synthesis of nontoxic copper nanoparticles using Citrus sinensis aqueous fruit extract and their antibacterial potentials. *Journal of Drug Delivery Science and Technology, 61*, 102172.

Jain, D., Daima, H. K., Kachhwaha, S., & Kothari, S. L. (2009). Synthesis of plant-mediated silver nanoparticles using papaya fruit extract and evaluation of their anti microbial activities. *Digest Journal of Nanomaterials and Biostructures, 4*(3), 557–563.

Jamdagni, P., Khatri, P., & Rana, J. S. (2018). Green synthesis of zinc oxide nanoparticles using flower extract of Nyctanthes arbor-tristis and their antifungal activity. *Journal of King Saud University-Science, 30*(2), 168–175.

Jamzad, M., & Bidkorpeh, M. K. (2020). Green synthesis of iron oxide nanoparticles by the aqueous extract of Laurus nobilis L. leaves and evaluation of the antimicrobial activity. *Journal of Nanostructure in Chemistry, 10*, 193–201.

Jay Chithra, M., Sathya, M., & Pushpanathan, K. J. (2015). Effect of pH on crystal size and photoluminescence property of ZnO nanoparticles prepared by chemical precipitation method. *Acta Metallurgica Sinica (English Letters), 28*, 394–404.

Jayachandran, A., Aswathy, T. R., & Nair, A. S. (2021). Green synthesis and characterization of zinc oxide nanoparticles using Cayratia pedata leaf extract. *Biochemistry and Biophysics Reports, 26*, 100995.

Jayarambabu, N., Akshaykranth, A., Rao, T. V., Rao, K. V., & Kumar, R. R. (2020). Green synthesis of Cu nanoparticles using Curcuma longa extract and their application in antimicrobial activity. *Materials Letters, 259*, 126813.

Jegadeesan, G. B., Srimathi, K., Srinivas, N. S., Manishkanna, S., & Vignesh, D. (2019). Green synthesis of iron oxide nanoparticles using Terminalia bellirica and Moringa oleifera fruit and leaf extracts: Antioxidant, antibacterial and thermoacoustic properties. *Biocatalysis and Agricultural Biotechnology, 21*, 101354.

Jha, A. K., & Prasad, K. (2012). Biosynthesis of gold nanoparticles using common aromatic plants. *International Journal of Green Nanotechnology, 4*(3), 219–224.

Kanagasubbulakshmi, S., & Kadirvelu, K. (2017). Green synthesis of iron oxide nanoparticles using Lagenaria siceraria and evaluation of its antimicrobial activity. *Defence Life Science Journal, 2*(4), 422–427.

Kannan, R. R., Arumugam, R., Ramya, D., Manivannan, K., & Anantharaman, P. (2013). Green synthesis of silver nanoparticles using marine macroalga Chaetomorpha linum. *Applied Nanoscience, 3*, 229–233.

Karazhanov, S. Z., & Ravindran, P. (2010). Ab initio study of double oxides ZnX2O4 (X= Al, Ga, In) having spinel structure. *Journal of the American Ceramic Society, 93*(10), 3335–3341.

Katata-Seru, L., Moremedi, T., Aremu, O. S., & Bahadur, I. (2018). Green synthesis of iron nanoparticles using Moringa oleifera extracts and their applications: removal of nitrate from water and antibacterial activity against Escherichia coli. *Journal of Molecular Liquids, 256*, 296–304.

Khani, R., Roostaei, B., Bagherzade, G., & Moudi, M. (2018). Green synthesis of copper nanoparticles by fruit extract of Ziziphus spina-christi (L.) Willd.: Application for adsorption of triphenylmethane dye and antibacterial assay. *Journal of Molecular Liquids, 255*, 541–549.

Kharissova, O. V., Dias, H. R., Kharisov, B. I., Pérez, B. O., & Pérez, V. M. (2013). The greener synthesis of nanoparticles. *Trends in Biotechnology, 31*(4), 240–248.

Khodaie, M., & Ghasemi, N. (2018). Green synthesis and characterization of copper nanoparticles using Eryngium campestre leaf extract. *Bulgarian Chemical Communications, 50*, 244–250.

Kim, J. S., Kuk, E., Yu, K. N., Kim, J. H., Park, S. J., Lee, H. J., Kim, S. H., Park, Y. K., Park, Y. H., Hwang, C. Y., & Kim, Y. K. (2007). Antimicrobial effects of silver nanoparticles. *Nanomedicine: Nanotechnology, Biology and Medicine, 3*(1), 95–101.

Klabunde, K. J., & Richards, R. M. (Eds.). (2009). *Nanoscale materials in chemistry*. John Wiley & Sons.

Kołodziejczak-Radzimska, A., & Jesionowski, T. (2014). Zinc oxide—from synthesis to application: a review. *Materials, 7*(4), 2833–2881.

Kolya, H., Maiti, P., Pandey, A., & Tripathy, T. (2015). Green synthesis of silver nanoparticles with antimicrobial and azo dye (Congo red) degradation properties using Amaranthus gangeticus Linn leaf extract. *Journal of Analytical Science and Technology, 6*, 1–7.

Kotval, S. C., John, T., & Parmar, K. A. (2018). Green synthesis of copper nanoparticles using Mitragyna parvifolia plant bark extract and its antimicrobial study. *Journal of Nanoscience and Technology*, 456–460.

Kreibig, U., & Genzel, L. (1985). Optical absorption of small metallic particles. *Surface Science, 156*, 678–700.

Kumari, B., & Atri, N. S. (2014). Nutritional and nutraceutical potential of wild edible macrolepiotoid mushrooms of north India. *International Journal of Pharmacy and Pharmaceutical Sciences, 6*, 200–204.

Kumar, K. M., Mandal, B. K., Sinha, M., & Krishnakumar, V. (2012). Terminalia chebula mediated green and rapid synthesis of gold nanoparticles. *Spectrochimica Acta Part A: Molecular and Biomolecular Spectroscopy, 86*, 490–494.

Kumar, M. P., Suresh, D., Nagabhushana, H., & Sharma, S. C. (2015). Beta vulgaris aided green synthesis of ZnO nanoparticles and their luminescence, photocatalytic and antioxidant properties. *The European Physical Journal Plus, 130*(6), 109.

Kumar, R., Singh, N., & Pandey, S. N. (2015). Potential of green synthesized zero-valent iron nanoparticles for remediation of lead-contaminated water. *International Journal of Environmental Science and Technology, 12*, 3943–3950.

Kumar, S., Daimary, R. M., Swargiary, M., Brahma, A., Kumar, S., & Singh, M. (2013). Biosynthesis of silver nanoparticles using Premna herbacea leaf extract and evaluation of its antimicrobial activity against bacteria causing dysentery. *International Journal of Pharma and Bio Sciences, 4*(4), 378–384.

Kuppusamy, P., Ichwan, S. J., Al-Zikri, P. N., Suriyah, W. H., Soundharrajan, I., Govindan, N., Maniam, G. P., & Yusoff, M. M. (2016). Vitro anticancer activity of Au, Ag nanoparticles synthesized using Commelina nudiflora L. aqueous extract against HCT-116 colon cancer cells. *Biological Trace Element Research, 173*, 297–305.

Lassoued, A., Dkhil, B., Gadri, A., & Ammar, S. (2017). Control of the shape and size of iron oxide ($\alpha$-$Fe_2O_3$) nanoparticles synthesized through the chemical precipitation method. *The Results in Physics, 7*, 3007–3015.

Leili, M., Fazlzadeh, M., & Bhatnagar, A. (2018). Green synthesis of nanozero-valent iron from Nettle and Thyme leaf extracts and their application for the removal of cephalexin antibiotic from aqueous solutions. *Environmental Technology, 39*(9), 1158–1172.

Lingaraju, K., Raja Naika, H., Manjunath, K., Basavaraj, R. B., Nagabhushana, H., Nagaraju, G., & Suresh, D. (2016). Biogenic synthesis of zinc oxide nanoparticles using Ruta graveolens (L.) and their antibacterial and antioxidant activities. *Applied Nanoscience, 6*, 703–710.

Liu, X., Pan, L., Zhao, Q., Lv, T., Zhu, G., Chen, T., Lu, T., Sun, Z., & Sun, C. (2012). UV-assisted photocatalytic synthesis of ZnO–reduced graphene oxide composites with enhanced photocatalytic activity in reduction of Cr (VI). *Chemical Engineering Journal, 183*, 238–243.

Long, Q., Cai, M., Rogers, J. D., Rong, H., Li, J., & Jiang, L. (2007). Synthesis of nanostructured Pt/$CeO_2$–$ZrO_2$–$Al_2O_3$ catalysts by a two-step sol–gel method. *Nanotechnology, 18*(35), 355601.

Luque, R., & Varma, R. S. (Eds.). (2012). *Sustainable preparation of metal nanoparticles: Methods and applications*. Royal Society of Chemistry.

Mahdavi, M., Namvar, F., Ahmad, M. B., & Mohamad, R. (2013). Green biosynthesis and characterization of magnetic iron oxide ($Fe_3O_4$) nanoparticles using seaweed (Sargassum muticum) aqueous extract. *Molecules, 18*(5), 5954–5964.

Makarov, V. V., Love, A. J., Sinitsyna, O. V., Makarova, S. S., Yaminsky, I. V., Taliansky, M. E., & Kalinina, N. O. (2014). "Green" nanotechnologies: synthesis of metal nanoparticles using plants. *Acta Naturae (англоязычная версия), 6*(1(20)), 35–44.

Malik, P., Shankar, R., Malik, V., Sharma, N., & Mukherjee, T. K. (2014). Green chemistry based benign routes for nanoparticle synthesis. *Journal of Nanoparticles, 2014*.

Maliszewska, I., & Sadowski, Z. (2009). Synthesis and antibacterial activity of silver nanoparticles. In *Journal of physics: Conference series 2009* (Vol. 146, No. 1, p. 012024). IOP Publishing.

Mallikarjuna, K., Narasimha, G., Dillip, G. R., Praveen, B., Shreedhar, B., Lakshmi, C. S., Reddy, B. V., & Raju, B. D. (2011). Green synthesis of silver nanoparticles using Ocimum leaf extract and their characterization. *Digest Journal of Nanomaterials and Biostructures, 6*(1), 181–186.

Marathe, K., Naik, J., & Maheshwari, V. (2021). Biogenic synthesis of silver nanoparticles using Streptomyces spp. and their antifungal activity against *Fusarium verticillioides*. *Journal of Cluster Science, 32*, 1299–1309.

Mathew, S., Victorio, C. P., Sidhi, J., & BH BT. (2020). Biosynthesis of silver nanoparticle using flowers of Calotropis gigantea (L.) WT Aiton and activity against pathogenic bacteria. *Arabian Journal of Chemistry, 13*(12), 9139–9144.

Matsumura, Y., Yoshikata, K., Kunisaki, S. I., & Tsuchido, T. (2003). Mode of bactericidal action of silver zeolite and its comparison with that of silver nitrate. *Applied and Environmental Microbiology, 69*(7), 4278–4281.

Mirza, A. U., Kareem, A., Nami, S. A., Khan, M. S., Rehman, S., Bhat, S. A., Mohammad, A., & Nishat, N. (2018). Biogenic synthesis of iron oxide nanoparticles using Agrewia optiva and Prunus persica phyto species: Characterization, antibacterial and antioxidant activity. *Journal of Photochemistry and Photobiology, B: Biology, 185*, 262–274.

Mishra, A. N., Bhadauria, S., Gaur, M. S., Pasricha, R., & Kushwah, B. S. (2010). Synthesis of gold nanoparticles by leaves of zero-calorie sweetener herb (Stevia rebaudiana) and their nanoscopic characterization by spectroscopy and microscopy. *International Journal of Green Nanotechnology: Physics and Chemistry, 1*(2), P118–P124.

Mittal, A. K., Chisti, Y., & Banerjee, U. C. (2013). Synthesis of metallic nanoparticles using plant extracts. *Biotechnology Advances, 31*(2), 346–356.

Mohamed, E. A. (2020). Green synthesis of copper & copper oxide nanoparticles using the extract of seedless dates. *Heliyon, 6*(1), e03123.

Moravec, P., Keskinen, H., Jyrki, M. M., Bakardjieva, S., & Levdansky, V. V. (2011). NiO x nanoparticle synthesis by chemical vapour deposition from nickel acetylacetonate. *Materials Sciences and Applications, 2*(04), 258.

Morones, J. R., Elechiguerra, J. L., Camacho, A., Holt, K., Kouri, J. B., Ramírez, J. T., & Yacaman, M. J. (2005). The bactericidal effect of silver nanoparticles. *Nanotechnology, 16*(10), 2346.

Mubayi, A., Chatterji, S., Rai, K., & P, Watal G. (2012). Evidence based green synthesis of nanoparticles. Advanced. *Materials Letters, 3*(6), 519–525.

Mukherjee, P., Ahmad, A., Mandal, D., Senapati, S., Sainkar, S. R., Khan, M. I., Parishcha, R., Ajaykumar, P. V., Alam, M., Kumar, R., & Sastry, M. (2001). Fungus-mediated synthesis of silver nanoparticles and their immobilization in the mycelial matrix: A novel biological approach to nanoparticle synthesis. *Nano Letters, 1*(10), 515–519.

Murray, C. B., Sun, S., Doyle, H., & Betley, T. (2001). Monodisperse 3d transition-metal (Co, Ni, Fe) nanoparticles and their assembly intonanoparticle superlattices. *MRS Bulletin, 26*(12), 985–991.

Muthuraman, P., Ramkumar, K., & Kim, D. H. (2014). Analysis of dose-dependent effect of zinc oxide nanoparticles on the oxidative stress and antioxidant enzyme activity in adipocytes. *Applied Biochemistry and Biotechnology, 174*, 2851–2863.

Naeimi, A., Ekrami-Kakhki, M. S., & Donyagard, F. (2021). Enhanced electrocatalytic performance of Pt nanoparticles immobilized on novel electrospun PVA@ Ni/NiO/Cu complex bionanofiber/chitosan based on Calotropis procera plant for methanol electro-oxidation. *International Journal of Hydrogen Energy, 46*(36), 18949–18963.

Nagajyothi, P. C., Pandurangan, M., Kim, D. H., Sreekanth, T. V., & Shim, J. (2017). Green synthesis of iron oxide nanoparticles and their catalytic and in vitro anticancer activities. *Journal of Cluster Science, 28*, 245–257.

Nagar, N., & Devra, V. (2018). Green synthesis and characterization of copper nanoparticles using Azadirachta indica leaves. *Materials Chemistry and Physics, 213*, 44–51.

Nagore, P. B., Ghoti, A. J., Salve, A. P., & Mane, K. G. (2020). Retracted article: Green synthesis of luminescent copper oxide nanoparticles using ginger lily leaves extract. *Journal of Inorganic and Organometallic Polymers and Materials, 30*(10), 4229.

Nakayama, M., Smith, C. L., Feltis, B. N., Piva, T. J., Tabatabaie, F., Harty, P. D., et al. (2020). Samarium doped titanium dioxide nanoparticles as theranostic agents in radiation therapy. *Physica Medica, 75*, 69–76.

Nakhjiri, F., Mirhosseini, M., & Mozaheb, M. A. (2017). Green synthesis of silver nanoparticle using echinops extract and its antibacterial activity. *Nanomedicine Journal, 4*(2), 98–106.

Nangare, S. N., & Patil, P. O. (2020). Affinity-based nanoarchitectured biotransducer for sensitivity enhancement of surface plasmon resonance sensors for in vitro diagnosis: A review. *ACS Biomaterials Science & Engineering, 7*(1), 2–30.

Narayanan, K. B., & Sakthivel, N. (2010). Biological synthesis of metal nanoparticles by microbes. *Advances in Colloid and Interface Science, 156*(1-2), 1–3.

Narayanan, R., & El-Sayed, M. A. (2004). Shape-dependent catalytic activity of platinum nanoparticles in colloidal solution. *Nano Letters, 4*(7), 1343–1348.

Naseem, T., & Farrukh, M. A. (2015). Antibacterial activity of green synthesis of iron nanoparticles using Lawsonia inermis and Gardenia jasminoides leaves extract. *Journal of Chemistry, 2015*.

Nasrollahi, A., Pourshamsian, K.H. and Mansourkiaee, P., 2011. Antifungal activity of silver nanoparticles on some of fungi.

Nasrollahzadeh, M., Momeni, S. S., & Sajadi, S. M. (2017). Green synthesis of copper nanoparticles using Plantago asiatica leaf extract and their application for the cyanation of aldehydes using K4Fe (CN) 6. *Journal of Colloid and Interface Science, 506*, 471–477.

Nath, N., & Chilkoti, A. (2002). A colorimetric gold nanoparticle sensor to interrogate biomolecular interactions in real time on a surface. *Analytical Chemistry, 74*, 504–509.

Nath, D., & Banerjee, P. (2013). Green nanotechnology–a new hope for medical biology. *Environmental Toxicology and Pharmacology, 36*(3), 997–1014.

Nguyen, P. A., Nguyen, A. V., Dang-Bao, T., Phan, H. P., Nguyen, T. T., Tran, B. A., Huynh, T. L., Hoang, T. C., Huynh, V. T., & Nguyen, T. (2020). Green synthesis of copper nanoparticles using Cocoa pod extract and its catalytic activity in deep oxidation of aromatic hydrocarbons. *SN Applied Sciences, 2*, 1–3.

Niraimathee, V. A., Subha, V., Ravindran, R. E., & Renganathan, S. (2016). Green synthesis of iron oxide nanoparticles from Mimosa pudica root extract. *International Journal of Environment and Sustainable Development, 15*(3), 227–240.

Nnadozie, E. C., & Ajibade, P. A. (2020). Green synthesis and characterization of magnetite (Fe3O4) nanoparticles using Chromolaena odorata root extract for smart nanocomposite. *Materials Letters, 263*, 127145.

Noorjahan, C. M., Shahina, S. J., Deepika, T., & Rafiq, S. (2015). Green synthesis and characterization of zinc oxide nanoparticles from Neem (Azadirachta indicia). *International Journal of Scientific Engineering and Technology Research, 4*(30), 5751–5753.

Noruzi, M., Zare, D., Khoshnevisan, K., & Davoodi, D. (2011). Rapid green synthesis of gold nanoparticles using Rosa hybrida petal extract at room temperature. *Spectrochimica Acta Part A: Molecular and Biomolecular Spectroscopy, 79*(5), 1461–1465.

Olajire, A. A., Ifediora, N. F., Bello, M. D., & Benson, N. U. (2018). Green synthesis of copper nanoparticles using Alchornea laxiflora leaf extract and their catalytic application for oxidative desulfurization of model oil. *Iranian Journal of Science and Technology, Transactions A: Science, 42*, 1935–1946.

Önal, E. S., Yatkin, T., Ergüt, M., & Özer, A. (2017). Green synthesis of iron nanoparticles by aqueous extract of Eriobotrya japonica leaves as a heterogeneous fenton-like catalyst: degradation of basic red 46. *International Journal of Chemical Engineering and Applications, 8*(5), 327–333.

Oudhia, A., Kulkarni, P., & Sharma, S. (2015). Green synthesis of ZnO nanotubes for bioapplications. *International Journal of Current Engineering and Technology, 1*, 280–281.

Panáček, A., Kvitek, L., Prucek, R., Kolář, M., Večeřová, R., Pizúrová, N., Sharma, V. K., Nevěčná, T. J., & Zbořil, R. (2006). Silver colloid nanoparticles: synthesis, characterization, and their antibacterial activity. *The Journal of Physical Chemistry B, 110*(33), 16248–16253.

Pandey, S., Oza, G., Mewada, A., & Sharon, M. (2012). Green synthesis of highly stable gold nanoparticles using Momordica charantia as nano fabricator. *Archives of Applied Science Research, 4*(2), 1135–1141.

Parthasarathy, S., Jayacumar, S., Chakraborty, S., Soundararajan, P., Joshi, D., Gangwar, K., Bhattacharjee, A., & Venkatesh, M. P. D. (2020). *Fabrication and characterization of copper nanoparticles by green synthesis approach using Plectranthus amboinicus leaves extract.* https://doi.org/10.21203/rs.3.rs-55984/v1

Pattanayak, M., & Nayak, P. L. (2013). Green synthesis and characterization of zero valent iron nanoparticles from the leaf extract of Azadirachta indica (Neem). *World Journal of Nano Science & Technology, 2*(1), 06–09.

Pawar, J. S. (2021). *Phytosynthesis of metal nanoparticles using plant tubers and their biological activities*. PhD Thesis. KBC North Maharashtra University.

Pawar, J. S., & Patil, R. H. (2020). Green synthesis of silver nanoparticles using Eulophia herbacea (Lindl.) tuber extract and evaluation of its biological and catalytic activity. *SN Applied Sciences, 2*, 52.

Pawar, S. G., Patil, S. L., Chougule, M. A., Jundale, D. M., & Patil, V. B. (2011). Synthesis and characterization of nanocrystalline $TiO_2$ thin films. *Journal of Materials Science: Materials in Electronics, 22*, 260–264.

Pethakamsetty, L., Kothapenta, K., Nammi, H. R., Ruddaraju, L. K., Kollu, P., Yoon, S. G., & Pammi, S. V. (2017). Green synthesis, characterization and antimicrobial activity of silver nanoparticles using methanolic root extracts of Diospyros sylvatica. *Journal of Environmental Sciences, 55*, 157–163.

Philip, D. (2010). Rapid green synthesis of spherical gold nanoparticles using Mangifera indica leaf. *Spectrochimica Acta Part A: Molecular and Biomolecular Spectroscopy, 77*(4), 807–810.

Philip, D., Unni, C., Aromal, S. A., & Vidhu, V. K. (2011). Murraya koenigii leaf-assisted rapid green synthesis of silver and gold nanoparticles. *Spectrochimica Acta Part A: Molecular and Biomolecular Spectroscopy, 78*(2), 899–904.

Prabhu, S., & Poulose, E. K. (2012). Silver nanoparticles: mechanism of antimicrobial action, synthesis, medical applications, and toxicity effects. *International Nano Letters, 2*, 1–10.

Prasad, C., Murthy, P. K., Krishna, R. H., Rao, R. S., Suneetha, V., & Venkateswarlu, P. (2017). Bioinspired green synthesis of RGO/Fe3O4 magnetic nanoparticles using Murrayakoenigii leaves extract and its application for removal of Pb (II) from aqueous solution. *Journal of Environmental Chemical Engineering, 5*(5), 4374–4380.

Premkumar, J., Sudhakar, T., Dhakal, A., Shrestha, J. B., Krishnakumar, S., & Balashanmugam, P. (2018). Synthesis of silver nanoparticles (AgNPs) from cinnamon against bacterial pathogens. *Biocatalysis and Agricultural Biotechnology, 15*, 311–316.

Radini, I. A., Hasan, N., Malik, M. A., & Khan, Z. (2018). Biosynthesis of iron nanoparticles using Trigonella foenum-graecum seed extract for photocatalytic methyl orange dye degradation and antibacterial applications. *Journal of Photochemistry and Photobiology, B: Biology, 183*, 154–163.

Raghunandan, D., Ravishankar, B., Sharanbasava, G., Mahesh, D. B., Harsoor, V., Yalagatti, M. S., Bhagawanraju, M., & Venkataraman, A. (2011). Anticancer studies of noble metal nanoparticles synthesized using different plant extracts. *Cancer Nanotechnology, 2*, 57–65.

Rai, M., Yadav, A., & Gade, A. (2008). CRC 675 current trends in phytosynthesis of metal nanoparticles. *Critical Reviews in Biotechnology, 28*(4), 277–284. 104.

Rai, M., Yadav, A., & Gade, A. (2009). Silver nanoparticles as a new generation of antimicrobials. *Biotechnology Advances, 27*(1), 76–83.

Raj, L. F., & Jayalakshmy, E. (2015). Biosynthesis and characterization of zinc oxide nanoparticles using root extract of Zingiber officinale. *Oriental Journal of Chemistry, 31*(1), 51–56.

Rajesh, K. M., Ajitha, B., Reddy, Y. A., Suneetha, Y., & Reddy, P. S. (2018). Assisted green synthesis of copper nanoparticles using Syzygium aromaticum bud extract: Physical, optical and antimicrobial properties. *Optik, 154*, 593–600.

Rajeshkumar, S., Menon, S., Kumar, S. V., Tambuwala, M. M., Bakshi, H. A., Mehta, M., Satija, S., Gupta, G., Chellappan, D. K., Thangavelu, L., & Dua, K. (2019). Antibacterial and antioxidant potential of biosynthesized copper nanoparticles mediated through Cissus arnotiana plant extract. *Journal of Photochemistry and Photobiology B: Biology, 197*, 111531.

Rajput, N. (2015). Methods of preparation of nanoparticles: A review. *International Journal of Advances in Engineering & Technology, 7*(6), 1806.

Ramesh, M., Anbuvannan, M., & Viruthagiri, G. J. (2015). Green synthesis of ZnO nanoparticles using Solanum nigrum leaf extract and their antibacterial activity. *Spectrochimica Acta Part A: Molecular and Biomolecular Spectroscopy, 136*, 864–870.

Ranjit, K., & Baquee, A. A. (2013). Nanoparticle: An overview of preparation, characterization and application. *International Research Journal of Pharmacy, 4*(4), 47–57.

Ranjith, S., Arungandhi, K., Sunmathi, D., Priyanga, S., & Monisha, S. (2018). Green synthesis of iron nanoparticles from cassia fistula linn leaf extract and its antibacterial activity. *EJBPS, 5*, 490–495.

Raut, R. W., Mendhulkar, V. D., & Kashid, S. B. (2015). Photosensitized synthesis of silver nanoparticles using Withania somnifera leaf powder and silver nitrate. *Journal of Photochemistry and Photobiology B: Biology, 132*, 45–55.

Rautela A, Rani J. Green synthesis of silver nanoparticles from Tectona grandis seeds extract: Characterization and mechanism of antimicrobial action on different microorganisms. Journal of Analytical Science and Technology 2019;10(1):1-0.

Reynolds, R. A., Mirkin, C. A., & Letsinger, R. L. (2000). Homogeneous, nanoparticle-based quantitative colorimetric detection of oligonucleotides. *Journal of the American Chemical Society, 122*(15), 3795–3796.

Saikia, J. P., Paul, S., Konwar, B. K., & Samdarshi, S. K. (2010). Ultrasonication: Enhances the antioxidant activity of metal oxide nanoparticles. *Colloids and Surfaces. B, Biointerfaces, 79*(2), 521–523.

Saka, A., Tesfaye, J. L., Gudata, L., Shanmugam, R., Dwarampudi, L. P., Nagaprasad, N., Krishnaraj, R., & Rajeshkumar, S. (2022). Synthesis, characterization, and antibacterial activity of ZnO nanoparticles from fresh leaf extracts of Apocynaceae, Carissa spinarum L.(Hagamsa). *Journal of Nanomaterials, 2022*, 1–6.

Santhosh Kumar, J., & Shanmugam, V. (2020). Green synthesis of copper oxide nanoparticles from magnolia champaca floral extract and its antioxidant & toxicity assay using Danio Rerio. *International Journal of Recent Technology and Engineering, 8*, 5444–5449.

Saratale, R. G., Benelli, G., Kumar, G., Kim, D. S., & Saratale, G. D. (2018). Biofabrication of silver nanoparticles using the leaf extract of an ancient herbal medicine, dandelion (Taraxacum officinale), evaluation of their antioxidant, anticancer potential, and antimicrobial activity against phytopathogens. *Environmental Science and Pollution Research, 25*, 10392–10406.

Sasaki, T., Shimizu, Y., & Koshizaki, N. (2006). Preparation of metal oxide-based nanomaterials using nanosecond pulsed laser ablation in liquids. *Journal of Photochemistry and Photobiology A: Chemistry, 182*(3), 335–341.

Sastry, M., Ahmad, A., Khan, M. I., & Kumar, R. (2003). Biosynthesis of metal nanoparticles using fungi and actinomycete. *Current Science*, 162–170.

Sathishkumar, G., Logeshwaran, V., Sarathbabu, S., Jha, P. K., Jeyaraj, M., Rajkuberan, C., Senthilkumar, N., & Sivaramakrishnan, S. (2018). Green synthesis of magnetic $Fe_3O_4$ nanoparticles using Couroupita guianensis Aubl. fruit extract for their antibacterial and cytotoxicity activities. *Artificial Cells, Nanomedicine, and Biotechnology, 46*(3), 589–598.

Sathishkumar, M., Sneha, K., & Yun, Y. S. (2009). Palladium nanocrystal synthesis using Curcuma longa tuber extract. *International Journal of Materials Sciences, 4*(1), 11–17.

Saxena, V., Hasan, A., & Pandey, L. M. (2021). Antibacterial nano-biocomposite scaffolds of Chitosan, Carboxymethyl Cellulose and Zn & Fe integrated Hydroxyapatite (Chitosan-CMC-FZO@ HAp) for bone tissue engineering. *Cellulose, 28*, 9207–9226.

Sebeia, N., Jabli, M., & Ghith, A. (2019). Biological synthesis of copper nanoparticles, using Nerium oleander leaves extract: characterization and study of their interaction with organic dyes. *Inorganic Chemistry Communications, 105*, 36–46.

Seku, K., Ganapuram, B. R., Pejjai, B., Kotu, G. M., & Narasimha, G. (2018). Hydrothermal synthesis of Copper nanoparticles, characterization and their biological applications. *International Journal of Nano Dimension, 9*(1), 7–14.

Seralathan, J., Stevenson, P., Subramaniam, S., Raghavan, R., Pemaiah, B., Sivasubramanian, A., & Veerappan, A. (2014). Spectroscopy investigation on chemo-catalytic, free radical scavenging and bactericidal properties of biogenic silver nanoparticles synthesized using Salicornia brachiata aqueous extract. *Spectrochimica Acta Part A: Molecular and Biomolecular Spectroscopy, 118*, 349–355.

Shameli, K., Ahmad, M. B., Zamanian, A., Sangpour, P., Shabanzadeh, P., Abdollahi, Y., & Zargar, M. (2012). Green biosynthesis of silver nanoparticles using Curcuma longa tuber powder. *International Journal of Nanomedicine*, 5603–5610.

Sharma, P., Bhatt, D., Zaidi, M. G., Saradhi, P. P., Khanna, P. K., & Arora, S. (2012). Silver nanoparticle-mediated enhancement in growth and antioxidant status of Brassica juncea. *Applied Biochemistry and Biotechnology, 167*, 2225–2233.

Sharma, P., Pant, S., Dave, V., Tak, K., Sadhu, V., & Reddy, K. R. (2019). Green synthesis and characterization of copper nanoparticles by Tinospora cardifolia to produce nature-friendly copper nanocoated fabric and their antimicrobial evaluation. *Journal of Microbiological Methods, 160*, 107–116.

Sharmila, G., Thirumarimurugan, M., & Muthukumaran, C. (2019). Green synthesis of ZnO nanoparticles using Tecoma castanifolia leaf extract: characterization and evaluation of its antioxidant, bactericidal and anticancer activities. *Microchemical Journal, 145*, 578–587.

Shende, S., Ingle, A. P., Gade, A., & Rai, M. (2015). Green synthesis of copper nanoparticles by Citrus medica Linn.(Idilimbu) juice and its antimicrobial activity. *World Journal of Microbiology and Biotechnology, 31*, 865–873.

Shi, J., Zhu, Y., Zhang, X., Baeyens, W. R., & García-Campaña, A. M. (2004). Recent developments in nanomaterial optical sensors. *TrAC Trends in Analytical Chemistry, 23*, 351–360.

Shukla, D., & Vankar, P. S. (2012). Synthesis of plant parts mediated gold nanoparticles. *International Journal of Green Nanotechnology, 4*(3), 277–288.

Singaravelan, R., & Bangaru Sudarsan Alwar, S. (2015). Electrochemical synthesis, characterization and phytogenic properties of silver nanoparticles. *Applied Nanoscience, 5*, 983–991.

Singh, K., Panghal, M., Kadyan, S., & Yadav, J. P. (2014). Evaluation of antimicrobial activity of synthesized silver nanoparticles using Phyllanthus amarus and Tinospora cordifolia medicinal plants. *Journal of Nanomedicine & Nanotechnology, 5*(6), 1.

Singh, R. P., Magesh, S., & Rakkiyappan, C. (2011). Ginger (Zingiber officinale) root extract: A source of silver nanoparticles and their application. *International Journal of Bio-Engineering, Sciences and Technology, 2*(3), 75–80.

Singh, V., & Chauhan, P. (2009). Structural and optical characterization of CdS nanoparticles prepared by chemical precipitation method. *Journal of Physics and Chemistry of Solids, 70*(7), 1074–1079.

Sivakami, M., Renuka, R., & Thilagavathi, T. (2020). Green synthesis of magnetic nanoparticles via Cinnamomum verum bark extract for biological application. *Journal of Environmental Chemical Engineering, 8*(5), 104420.

Sivakumar, T., Gajalakshmi, D., Subramanian, V. K., & Palanisamy, K. (2015). Tuber extract mediated biosynthesis of silver nanoparticles and its antioxidant, antibacterial activity. *Journal of Biological Sciences, 15*(2), 68.

Sneha, K., Sathishkumar, M., Lee, S. Y., Bae, M. A., & Yun, Y. S. (2011). Biosynthesis of Au nanoparticles using cumin seed powder extract. *Journal of Nanoscience and Nanotechnology, 11*(2), 1811–1814.

Sondi, I., & Salopek-Sondi, B. (2004). Silver nanoparticles as antimicrobial agent: a case study on E. coli as a model for Gram-negative bacteria. *Journal of Colloid and Interface Science, 275*(1), 177–182.

Storhoff, J. J., Lazarides, A. A., Mucic, R. C., Mirkin, C. A., Letsinger, R. L., & Schatz, G. C. (2000). What controls the optical properties of DNA-linked gold nanoparticle assemblies? *Journal of the American Chemical Society, 122*(19), 4640–4650.

Subhankari, I., & Nayak, P. L. (2013). Synthesis of copper nanoparticles using Syzygium aromaticum (Cloves) aqueous extract by using green chemistry. *World Journal of Nanoscience and Technology, 2*(1), 14–17.

Subhashini, D., & Valli Nachiyar, C. (2017). Albizia saman: a Green route for the reduction of bulk TiO2. *International Journal of ChemTech Research, 6*(12), 5137–5141.

Sukumar, S., Rudrasenan, A., & Padmanabhan, N. D. (2020). Green-synthesized rice-shaped copper oxide nanoparticles using Caesalpinia bonducella seed extract and their applications. *ACS Omega, 5*(2), 1040–1051.

Tahvilian, R., Zangeneh, M. M., Falahi, H., Sadrjavadi, K., Jalalvand, A. R., & Zangeneh, A. (2019). Green synthesis and chemical characterization of copper nanoparticles using Allium saralicum leaves and assessment of their cytotoxicity, antioxidant, antimicrobial, and cutaneous wound healing properties. *Applied Organometallic Chemistry, 33*(12), e5234.

Talebi, S., Ramezani, F., & Ramezani, M. (2010). Biosynthesis of metal nanoparticles by micro-organisms. *Nanocon Olomouc, Czech Republic, EU, 10*, 12–18.

Thakkar, K. N., Mhatre, S. S., & Parikh, R. Y. (2010). Biological synthesis of metallic nanoparticles. *Nanomedicine: Nanotechnology, Biology and Medicine, 6*(2), 257–262.

Thakur, S., Sharma, S., Thakur, S., & Rai, R. (2018). Green synthesis of copper nano-particles using Asparagus adscendens roxb. Root and leaf extract and their antimicrobial activities. *International Journal of Current Microbiology and Applied Sciences, 7*(4), 683–694.

Thamer, N. A., Muftin, N. Q., & Al-Rubae, S. (2018). Optimization properties and characterization of green synthesis of copper oxide nanoparticles using aqueous extract of Cordia myxa L. leaves. *Asian Journal of Chemistry, 30*(7), 1559–1563.

Thanh, N. T., & Rosenzweig, Z. (2002). Development of an aggregation-based immunoassay for anti-protein A using gold nanoparticles. *Analytical Chemistry, 74*(7), 1624–1628.

Thatikayala, D., Jayarambabu, N., Banothu, V., Ballipalli, C. B., Park, J., & Rao, K. V. (2019). Biogenic synthesis of silver nanoparticles mediated by Theobroma cacao extract: Enhanced antibacterial and photocatalytic activities. *Journal of Materials Science: Materials in Electronics, 30*, 17303–17313.

Thiruvengadam, M., Chung, I. M., Gomathi, T., Ansari, M. A., Gopiesh Khanna, V., Babu, V., & Rajakumar, G. (2019). Synthesis, characterization and pharmacological potential of green synthesized copper nanoparticles. *Bioprocess and Biosystems Engineering, 42*, 1769–1777.

Titus, D., & Samuel, E. J. (2019). Photocatalytic degradation of azo dye using biogenic $SnO_2$ nanoparticles with antifungal property: RSM optimization and kinetic study. *Journal of Cluster Science, 30*, 1335–1345.

Valgimigli, L., Baschieri, A., & Amorati, R. (2018). Antioxidant activity of nanomaterials. *Journal of Materials Chemistry B, 6*(14), 2036–2051.

Varghese, E., & George, M. (2015). Green synthesis of zinc oxide nanoparticles. *International Journal of Advance Research in Science and Engineering, 4*(01), 307–314.

Vasantharaj, S., Sathiyavimal, S., Senthilkumar, P., LewisOscar, F., & Pugazhendhi, A. (2019). Biosynthesis of iron oxide nanoparticles using leaf extract of Ruellia tuberosa: antimicrobial properties and their applications in photocatalytic degradation. *Journal of Photochemistry and Photobiology B: Biology, 192*, 74–82.

Veeraapandian, S., Sawant, S. N., & Doble, M. (2012). Antibacterial and antioxidant activity of protein capped silver and gold nanoparticles synthesized with Escherichia coli. *Journal of Biomedical Nanotechnology, 8*(1), 140–148.

Venugopalan, R., Pitchai, S., Devarayan, K., & Swaminathan, V. C. (2020). Biogenic synthesis of copper nanoparticles using Borreria hispida (Linn.) extract and its antioxidant activity. *Materials Today: Proceedings, 33*, 4023–4025.

Vijayakumar, S., Vaseeharan, B., Malaikozhundan, B., Gopi, N., Ekambaram, P., Pachaiappan, R., Velusamy, P., Murugan, K., Benelli, G., Kumar, R. S., & Suriyanarayanamoorthy, M. (2017). Therapeutic effects of gold nanoparticles synthesized using Musa paradisiaca peel extract against multiple antibiotic resistant Enterococcus faecalis biofilms and human lung cancer cells (A549). *Microbial Pathogenesis, 102*, 173–183.

Vijayan, R., Joseph, S., & Mathew, B. (2019). Green synthesis of silver nanoparticles using Nervalia zeylanica leaf extract and evaluation of their antioxidant, catalytic, and antimicrobial potentials. *Particulate Science and Technology, 37*(7), 809–819.

Vinodhini, S., Vithiya, B. S., & Prasad, T. A. (2022). Green synthesis of palladium nanoparticles using aqueous plant extracts and its biomedical applications. *Journal of King Saud University-Science, 34*(4), 102017.

Wang, T., Jin, X., Chen, Z., Megharaj, M., & Naidu, R. (2014). Green synthesis of Fe nanoparticles using eucalyptus leaf extracts for treatment of eutrophic wastewater. *Science of the Total Environment, 466*, 210–213.

Wang, X., Wang, Y., Chen, Z. G., & Shin, D. M. (2009). Advances of cancer therapy by nanotechnology. *Cancer Research and Treatment, 41*(1), 1–1.

Wang, X., Yang, L., Chen, Z., & Shin, D. M. (2008). Application of nanotechnology in cancer therapy and imaging. *CA: a Cancer Journal for Clinicians, 58*(2), 97–110.

Wei, Y., Fang, Z., Zheng, L., Tan, L., & Tsang, E. P. (2016). Green synthesis of Fe nanoparticles using Citrus maxima peels aqueous extracts. *Materials Letters, 185*, 384–386.

Wei, Y., Fang, Z., Zheng, L., & Tsang, E. P. (2017). Biosynthesized iron nanoparticles in aqueous extracts of Eichhornia crassipes and its mechanism in the hexavalent chromium removal. *Applied Surface Science, 399*, 322–329.

Yallappa, S., Manjanna, J., Sindhe, M. A., Satyanarayan, N. D., Pramod, S. N., & Nagaraja, K. (2013). Microwave assisted rapid synthesis and biological evaluation of stable copper nanoparticles using T. arjuna bark extract. *Spectrochimica Acta Part A: Molecular and Biomolecular Spectroscopy, 110*, 108–115.

Yang, W. H., Schatz, G. C., & Van duyne, R. P. (1995). Discrete dipole approximation for calculating extinction and Raman intensities for small particles with arbitrary shapes. *The Journal of Chemical Physics, 103*, 869–875.

Yi, Y., Tu, G., Tsang, P. E., Xiao, S., & Fang, Z. (2019). Green synthesis of iron-based nanoparticles from extracts of Nephrolepis auriculata and applications for Cr (VI) removal. *Materials Letters, 234*, 388–391.

Yugandhar, P., Vasavi, T., Uma Maheswari Devi, P., & Savithramma, N. (2017). Bioinspired green synthesis of copper oxide nanoparticles from Syzygium alternifolium (Wt.) Walp: Characterization and evaluation of its synergistic antimicrobial and anticancer activity. *Applied Nanoscience, 7*, 417–427.

Zangeneh, M. M., Zangeneh, A., Pirabbasi, E., Moradi, R., & Almasi, M. (2019). Falcaria vulgaris leaf aqueous extract mediated synthesis of iron nanoparticles and their therapeutic potentials under in vitro and in vivo condition. *Applied Organometallic Chemistry, 33*(12), e5246.

Zhang, L., Wei, Y., Wang, H., Wu, F., Zhao, Y., Liu, X., Wu, H., Wang, L., & Su, H. (2020). Green synthesis of silver nanoparticles using mushroom Flammulina velutipes extract and their antibacterial activity against aquatic pathogens. *Food and Bioprocess Technology, 13*, 1908–1917.

Ziental, D., Czarczynska-Goslinska, B., Mlynarczyk, D. T., Glowacka-Sobotta, A., Stanisz, B., Goslinski, T., & Sobotta, L. (2020). Titanium dioxide nanoparticles: prospects and applications in medicine. *Nanomaterials, 10*(2), 387.